Fundamentals of Biomechanics

Fundamentals of Biomechanics

Duane Knudson

California State University at Chico
Chico, California

Kluwer Academic / Plenum Publishers
New York, Boston, Dordrecht, London, Moscow

Library of Congress Cataloging-in-Publication Data

Knudson, Duane V., 1961–
    Fundamentals of biomechanics/Duane Knudson.
        p.   cm.
    Includes bibliographic references and index.
    ISBN 0-306-47474-3
    1. Biomechanics.   I. Title.

QP303 .K588 2003
612.7′6—dc21

2002040785

ISBN 0-306-47474-3

© 2003 Kluwer Academic/Plenum Publishers, New York
233 Spring Street, New York, New York 10013

http://www.wkap.com

10 9 8 7 6 5 4 3 2 1

A C.I.P. record for this book is available from the Library of Congress

Permissions for books published in Europe: *permissions@wkap.nl*
Permissions for books published in the United States of America: *permissions@wkap.com*

Printed in the United States of America

Contents

CHAPTER 7
ANGULAR KINETICS

CHAPTER 8
FLUID MECHANICS

PART IV
APPLICATIONS OF BIOMECHANICS IN QUALITATIVE ANALYSIS

CHAPTER 9
APPLYING BIOMECHANICS IN PHYSICAL EDUCATION

CHAPTER 10
APPLYING BIOMECHANICS IN COACHING

CHAPTER 11
APPLYING BIOMECHANICS IN STRENGTH AND CONDITIONING

CHAPTER 12
APPLYING BIOMECHANICS IN SPORTS MEDICINE AND REHABILITATION

Preface

This book is written for students taking the introductory biomechanics course in Kinesiology/HPERD. The book is designed for majors preparing for all kinds of human movement professions and therefore uses a wide variety of movement examples to illustrate the application of biomechanics. While this approach to the application of biomechanics is critical, it is also important that students be introduced to the scientific support or lack of support for these qualitative judgments. Throughout the text extensive citations are provided to support the principles developed and give students references for further study. Algebraic level mathematics is used to teach mechanical concepts. The focus of the mathematical examples is to understand the mechanical variables and to highlight the relationship between various biomechanical variables, rather than to solve quantitative biomechanical word problems. It is obvious from research in physics instruction that solving quantitative word problems does not increase the conceptual understanding of important mechanical laws (Elby, 2001; Lawson & McDermott, 1987; Kim & Pak, 2002).

So why another textbook on the biomechanics of human motion? There are plenty of books that are really anatomy books with superficial mechanics, that teach mechanics with sport examples, or are sport books that use some mechanics to illustrate technique points. Unfortunately, there are not many books that truly integrate the biological and mechanical foundations of human movement and show students how to apply and integrate biomechanical knowledge in improving human movement. This book was

written to address these limitations in previous biomechanics texts. The text presents a clear conceptual understanding of biomechanics and builds nine principles for the application of biomechanics. These nine principles form the *applied biomechanics tools* kinesiology professionals need. The application of these biomechanical principles is illustrated in qualitative analysis of a variety of human movements in several contexts for the kinesiology professional: physical education, coaching, strength and conditioning, and sports medicine. This qualitative analysis approach meets the NASPE Guidelines and Standards (Kinesiology Academy, 1992) for an introductory biomechanics course, and clearly shows students how biomechanical knowledge must be applied when kinesiology professionals improve human movement.

The text is subdivided into four parts: Introduction, Biological/Structural Bases, Mechanical Bases, and Applications of Biomechanics in Qualitative Analysis. Each part opener provides a concise summary of the importance and content of that section of text. The text builds from familiar anatomical knowledge, to new biomechanical principles and their application.

This book has several features that are designed to help students link personal experience to biomechanical concepts and that illustrate the application of biomechanics principles. First, nine general principles of biomechanics are proposed and developed throughout the text. These principles are the application link for the biomechanical concepts used to improve movement or reduce injury risk. Some texts have application chapters at the end of the book, but an application approach and examples are built in throughout *Fundamentals of Biomechanics*. Second, there are *activity boxes* that provide opportunities for students to see and feel the biomechanical variables discussed. Third, there are *practical application boxes* that highlight the applications of biomechanics in improving movement and in treating and preventing injury. Fourth, the *interdisciplinary issues boxes* show how biomechanics is integrated with other sport sciences in addressing human movement problems. Fifth, all chapters have associated *lab activities* (located at the end of the book, after the index) that use simple movements and measurements to explore concepts and principles. These lab activities do not require expensive lab equipment, large blocks of time, or dedicated lab space. Finally, Part IV (chapters 9 through 12) provides real-life case studies of how the biomechanical principles can be qualitatively applied to improve human movement in a variety of professions. No other text provides as many or as thorough guided examples of applying biomechanical principles in actual human movement situations. These application chapters also provide *discussion questions* so that students

and instructors can extend the discussion and debate on professional practice using specific examples.

There are also features that make it easy for students to follow the material and study for examinations. Extensive use of graphs, photographs, and illustrations are incorporated throughout. Aside from visual appeal, these figures illustrate important points and relationships between biomechanical variables and performance. The book provides an extensive glossary of key terms and biomechanics research terminology so that students can read original biomechanical research. Each chapter provides a *summary*, extensive citations of important biomechanical research, and *suggested readings*. The chapters in Parts I, II, and III conclude with *review questions* for student study and review. The lists of *web links* offer students the internet addresses of significant websites and professional organizations.

I hope that you master the fundamentals of biomechanics, integrate biomechanics into your professional practice, and are challenged to continuously update your biomechanical toolbox. Some of you will find advanced study and a career in biomechanics exciting opportunities.

Acknowledgments

The author would like to thank the many people who have contributed to this book. I am indebted to many biomechanics colleagues who have shared their expertise with me, given permission to share their work, and contributed so much to students and our profession. I would like to thank Tim Oliver for his expert editing, formatting, and design of the book, Katherine Hanley-Knutson for many fine illustrations, and Aaron Johnson at Kluwer for his vision to make this book happen.

To the ones I truly love—Lois, Josh, and Mandy—thanks for being such great people and for sharing the computer. Finally, I would like to thank God for knitting all of us so "fearfully and wonderfully made."

Fundamentals of Biomechanics

INTRODUCTION

Kinesiology is the scholarly study of human movement, and biomechanics is one of the many academic subdisciplines of kinesiology. Biomechanics in kinesiology involves the precise description of human movement and the study of the causes of human movement. The study of biomechanics is relevant to professional practice in many kinesiology professions. The physical educator or coach who is teaching movement technique and the athletic trainer or physical therapist treating an injury use biomechanics to qualitatively analyze movement. The chapters in part I demonstrate the importance of biomechanics in kinesiology and introduce you to key biomechanical terms and principles that will be developed throughout the text. The lab activities associated with part I relate to finding biomechanical knowledge and identifying biomechanical principles in action.

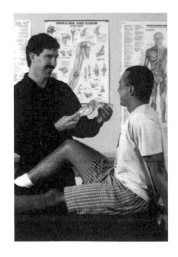

CHAPTER 1

Introduction to Biomechanics of Human Movement

Most people are extremely skilled in many everyday movements like standing, walking, or climbing stairs. By the time children are two, they are skilled walkers with little instruction from parents aside from emotional encouragement. Unfortunately, modern living does not require enough movement to prevent several chronic diseases associated with low physical activity (USD-HHS, 1996). Fortunately, many human movement professions help people to participate in beneficial physical activities. Physical Educators, coaches, athletic trainers, strength & conditioning coaches, personal trainers, and physical therapists all help people reap the benefits of physical activity. These human movement professions rely on undergraduate training in kinesiology, and typically require coursework in biomechanics. **Kinesiology** is the term referring to the whole scholarly area of human movement study, while **biomechanics** is the study of motion and its causes in living things. Biomechanics provides key information on the most effective and safest movement patterns, equipment, and relevant exercises to improve human movement. In a sense, kinesiology professionals solve human movement problems every day, and one of their most important tools is biomechanics. This chapter outlines the field of biomechanics, why biomechanics is such an important area to the kinesiology professional, and where biomechanics information can be found.

WHAT IS BIOMECHANICS?

Biomechanics has been defined as *the study of the movement of living things using the science of mechanics* (Hatze, 1974). Mechanics is a branch of physics that is concerned with the description of motion and how forces create motion. Forces acting on living things can create motion, be a healthy stimulus for growth and development, or overload tissues, causing injury. Biomechanics provides conceptual and mathematical tools that are necessary for understanding how living things move and how kinesiology professionals might improve movement or make movement safer.

Most readers of this book will be majors in departments of Kinesiology, Human Performance, or HPERD (Health, Physical Education, Recreation, and Dance). Kinesiology comes from two Greek verbs that translated literally means "the study of movement." Most American higher education programs in HPERD now use "kinesiology" in the title of their department because this term has come to be known as the academic area for the study of human movement (Corbin & Eckert, 1990). This change in terminology can be confusing because "kinesiology" is also the title of a foundational course on applied anatomy that was commonly required for a physical education degree in the first half of the twentieth century. This older meaning of kinesiology persists even today, possibly because biomechanics has only recently (since 1970s) become a recognized specialization of scientific study (Atwater, 1980; Wilkerson, 1997).

This book will use the term kinesiology in the modern sense of the whole academic area of the study of human movement. Since kinesiology majors are pursuing careers focused on improving human movement, you and almost all kinesiology students are required to take at least one course on the biomechanics of human movement. It is a good thing that you are studying biomechanics. Once your friends and family know you are a kinesiology major, you will invariably be asked questions like: should I get one of those new rackets, why does my elbow hurt, or how can I stop my drive from slicing? Does it sometimes seem as if your friends and family have regressed to that preschool age when every other word out of their mouth is "why"? What is truly important about this common experience is that it is a metaphor for the life of a human movement professional. Professions require formal study of theoretical and specialized knowledge that allows for the reliable solution to problems. This is the traditional meaning of the word "professional," and it is different than its common use today. Today people refer to professional athletes or painters because people earn a living with these jobs, but I believe that kinesiology careers should strive to be more like true professions such as medicine or law.

People need help in improving human movement and this help requires knowledge of "why" and "how" the human body moves. Since biomechanics gives the kinesiology professional much of the knowledge and many of the skills necessary to answer these "what works?" and "why?" questions, biomechanics is an important science for solving human movement problems. However, biomechanics is but one of many sport and human movement science tools in a kinesiology professional's toolbox. This text is also based on the philosophy that your biomechanical tools must be combined with tools from other kinesiology sciences to most effectively deal with human movement problems. Figure 1.1a illustrates the typical scientific subdisciplines of kinesiology. These typically are the core sciences all kinesiology majors take in their undergraduate preparations. This overview should not be interpreted to diminish the other academic subdisciplines common in kinesiology departments like sport history, sport philosophy, dance, and sport administration/management, just to name a few.

The important point is that information from all the subdisciplines must be integrated in professional practice since problems in human movement are multifaceted, with many interrelated factors. For the most part, the human movement problems you face as a kinesiology professional will be like those "trick" questions professors ask on exams: they are complicated by many factors and tend to defy simple, dualistic (black/white) answers. While the application examples discussed in this text will emphasize biomechanical principles, readers should bear in mind that this biomechanical information should be integrated with professional experience and the other subdisciplines of kinesiology. It is this **interdisciplinary** approach (Figure 1.1b) that is essential to finding the best interventions to help people more effectively and safely. Dotson (1980) suggests that true kinesiology professionals can integrate the many factors that interact to affect movement, while the layman typically looks at things one factor at time. Unfortunately, this interdisciplinary approach to kinesiology instruction in higher education has been elusive (Harris, 1993). Let's look at some examples of human movement problems where it is particularly important to

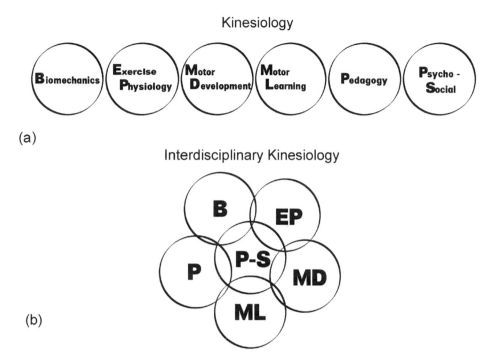

Figure 1.1. (a) The major academic subdisciplines or sciences of kinesiology. (b) Schematic of the integration of all the sciences in an interdisciplinary approach to solving human movement problems in kinesiology.

integrate biomechanical knowledge into the qualitative analysis.

WHY STUDY BIOMECHANICS?

Scientists from many different areas (e.g., kinesiology, engineering, physics, biology, zoology) are interested in biomechanics. Why are scholars from so many different academic backgrounds interested in animal movement? Biomechanics is interesting because many people marvel at the ability and beauty in animal movement. Some scholars have purely theoretical or academic interests in discovering the laws and principles that govern animal movement. Within kinesiology, many biomechanists have been interested in the application of biomechanics to sport and exercise. The applications of biomechanics to human

movement can be classified into two main areas: the improvement of performance and the reduction or treatment of injury (Figure 1.2).

Improving Performance

Human movement performance can be enhanced many ways. Effective movement involves anatomical factors, neuromuscular skills, physiological capacities, and psychological/cognitive abilities. Most kinesiology professionals prescribe technique changes and give instructions that allow a person to improve performance. Biomechanics is most useful in improving performance in sports or activities where technique is the dominant factor rather than physical structure or physiological capacity. Since biomechanics is essentially the

Applications of Biomechanics

Improved Performance

Preventing or Treating Injury

Figure 1.2. The two major applications of biomechanics are to improve human movement and the treatment or prevention of injury.

body arch are performed poorly. The coach's experience tells him that this athlete is strong enough to perform this skill, but they must decide if the gymnast should concentrate on her takeoff angle or more back hyperextension in the block. The coach uses his knowledge of biomechanics to help in the qualitative analysis of this situation. Since the coach knows that a better arch affects the force the gymnast creates against the mat and affects the angle of takeoff of the gymnast, he decides to help the gymnast work on her "arch" following the round off.

Biomechanics research on sports techniques sometimes tends to lag behind the changes that are naturally occurring in sports. Athletes and coaches experiment with new techniques all the time. Students of biomechanics may be surprised to find that there are often limited biomechanical

science of movement technique, biomechanics is the main contributor to one of the most important skills of kinesiology professionals: the qualitative analysis of human movement (Knudson & Morrison, 2002).

Imagine a coach is working with a gymnast who is having problems with her back handspring (Figure 1.3). The coach observes several attempts and judges that the angle of takeoff from the round off and

Figure 1.3. Biomechanics principles must be integrated with other kinesiology sciences to solve human movement problems, like in the qualitative analysis a round off and back handspring.

studies on many techniques in many popular sports. The vast number of techniques, their variations, and their high rates of change and innovation tend to outdistance biomechanics research resources. Sport biomechanics research also lags behind the coaches and athletes because scientific research takes considerable time to conduct and report, and there is a lack of funding for this important research. There is less funding for biomechanical studies aimed at improving performance compared to studies focused on preventing and treating injuries. Students looking for biomechanical research on improving sports technique often will have less research than students researching the biomechanics of injury.

While technique is always relevant in human movement, in some activities the psychological, anatomical, or physiological factors are more strongly related to success. Running is a good example of this kind of movement. There is a considerable amount of research on the biomechanics of running so coaches can fine tune a runner's technique to match the profile of elite runners (Cavanagh, Andrew, Kram, Rogers, Sanderson, & Hennig, 1985; Buckalew, Barlow, Fischer, & Richards, 1985; Williams, Cavanagh, & Ziff, 1987). While these technique adjustments make small improvements in performance, most of running performance is related to physiological abilities and their training. Studies that provide technique changes in running based on biomechanical measurements have found minimal effects on running economy (Cavanagh, 1990; Lake & Cavanagh, 1996; Messier & Cirillo, 1989). This suggests that track coaches can use biomechanics to refine running technique, but they should only expect small changes in performance from these modifications.

Human performance can also be enhanced by improvements in the design of equipment. Many of these improvements are related to new materials and engineer-ing designs. When these changes are integrated with information about the human performer, we can say the improvements in equipment were based on biomechanics. Engineers interested in sports equipment often belong to the International Sports Engineering Association (http://www.sports-engineering.co.uk/) and publish research in ISEA proceedings (Subic & Haake, 2000) or the *Sports Engineering* journal. Research on all kinds of equipment is conducted in biomechanics labs at most major sporting goods manufacturers. Unfortunately, much of the results of these studies are closely guarded trade secrets, and it is difficult for the layperson to determine if marketing claims for "improvements" in equipment design are real biomechanical innovations or just creative marketing.

There are many examples of how applying biomechanics in changing equipment designs has improved sports performance. When improved javelin designs in the early 1980s resulted in longer throws that endangered other athletes and spectators, redesigns in the weight distribution of the "new rules" javelin again shortened throws to safer distances (Hubbard & Alaways, 1987). Biomechanics researchers (Elliott, 1981; Ward & Groppel, 1980) were some of the first to call for smaller tennis rackets that more closely matched the muscular strength of young players (Figure 1.4). Chapter 8 will discuss how changes in sports equipment are used to change fluid forces and improve performance.

While breaking world records using new equipment is exciting, not all changes in equipment are welcomed with open arms by sport governing bodies. Some equipment changes are so drastic they change the very nature of the game and are quickly outlawed by the rules committee of the sport. One biomechanist developed a way to measure the stiffness of basketball goals, hoping to improve the consistency of

Figure 1.4. The design of sports equipment must be appropriate for an athlete, so rackets for children are shorter and lighter than adult rackets. Photo used with permission from Getty Images.

of the engineers than athletes (Bjerklie, 1993).

Another way biomechanics research improves performance is advances in exercise and conditioning programs. Biomechanical studies of exercise movements and training devices serve to determine the most effective training to improve performance (Figure 1.5). Biomechanical research on exercises is often compared to research on the sport or activity that is the focus of training. Strength and conditioning professionals can better apply the principle of specificity when biomechanical research is used in the development of exercise programs. Computer-controlled exercise and testing machines are another example of how biomechanics contributes to strength and conditioning (Ariel, 1983). In the next section the application of biomechanics in the medical areas of **orthotics** and **prosthetics** will be mentioned in relation to preventing injury, but many prosthetics are now being designed to improve the performance of disabled athletes.

their response but found considerable resistance from basketball folks who liked their unique home court advantages. Another biomechanist recently developed a new "klap" speed skate that increased the time and range of motion of each push off the ice, dramatically improving times and breaking world records (de Koning, Houdijk, de Groot, & Bobbert, 2000). This gave quite an advantage to the country where these skates were developed, and there was controversy over the amount of time other skaters were able to practice with the new skates before competition. These dramatic equipment improvements in many sports have some people worried that winning Olympic medals may be more in the hands

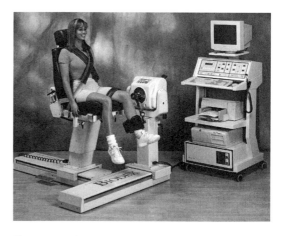

Figure 1.5. A computerized testing and exercise dynamometer by Biodex. The speed, muscle actions (isometric, concentric, eccentric), and pattern of loading (isokinetic, isotonic) can be selected. Image courtesy of Biodex Medical Systems.

Preventing and Treating Injury

Movement safety, or injury prevention/treatment, is another primary area where biomechanics can be applied. Sports medicine professionals have traditionally studied injury data to try to determine the potential causes of disease or injury (epidemiology). Biomechanical research is a powerful ally in the sports medicine quest to prevent and treat injury. Biomechanical studies help prevent injuries by providing information on the mechanical properties of tissues, mechanical loadings during movement, and preventative or rehabilitative therapies. Biomechanical studies provide important data to confirm potential injury mechanisms hypothesized by sports medicine physicians and epidemiological studies. The increased participation of girls and women in sports has made it clear that females are at a higher risk for anterior cruciate ligament (ACL) injuries than males due to several biomechanical factors (Boden, Griffin, & Garrett, 2000). Continued biomechanical and sports medicine studies may help unravel the mystery of this high risk and suggest prevention strategies.

Engineers and occupational therapists use biomechanics to design work tasks and assistive equipment to prevent overuse injuries related to specific jobs. Combining biomechanics with other sport sciences has aided in the design of shoes for specific sports (Segesser & Pforringer, 1989), especially running shoes (Frederick, 1986; Nigg, 1986). Since the 1980s the design and engineering of most sports shoes has included research in company biomechanics labs. The biomechanical study of auto accidents has resulted in measures of the severity of head injuries, which has been applied in biomechanical testing, and in design of many kinds of helmets to prevent head injury (Calvano & Berger, 1979; Norman, 1983; Torg, 1992). When accidents result in amputation, **prosthetics** or artificial limbs can be designed to match the mechanical properties of the missing limb (Klute Kallfelz, & Czerniecki, 2001). Preventing acute injuries is also another area of biomechanics research. Forensic biomechanics involves reconstructing the likely causes of injury from accident measurements and witness testimony.

Biomechanics helps the physical therapist prescribe rehabilitative exercises, assistive devices, or orthotics. **Orthotics** are support objects/braces that correct deformities or joint positioning, while assistive devices are large tools to help patient function like canes or walkers. Qualitative analysis of gait (walking) also helps the therapist decide whether sufficient muscular strength and control have been regained in order to permit safe or cosmetically normal walking (Figure 1.6). An athletic trainer might observe the walking pattern for signs of pain and/or limited range of motion in an athlete undergoing long-term conditioning for future return to the field. An athletic coach might use a similar qualitative analysis of the warm-up activities of

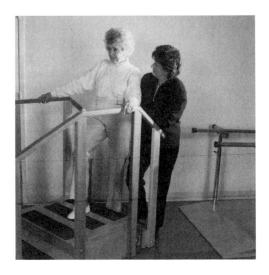

Figure 1.6. Qualitative analysis of gait (walking) is of importance in physical therapy and the treatment of many musculoskeletal conditions.

the same athlete several weeks later to judge their readiness for practice or competition. Many biomechanists work in hospitals providing quantitative assessments of gait function to document the effectiveness of therapy. The North American group interested in these quantitative assessments for medical purposes is the Gait and Clinical Movement Analysis Society (GCMAS) at http://www.gcmas.org/. Good sources for the clinical and biomechanical aspects of gait are Perry (1992) and Whittle (1996), and the clinical gait analysis website: http://aisr1.lib.tju.edu/cga/.

Dramatic increases in computer memory and power have opened up new areas of application for biomechanists. Many of these areas are related to treating and preventing human injury. Biomechanical studies are able to evaluate strategies for preventing falls and fractures in the elderly (Robinovitch, Hsiao, Sandler, Cortez, Liu, & Paiement, 2000). Biomechanical computer models can be used to simulate the effect of various orthopaedic surgeries (Delp, Loan, Hoy, Zajac, & Rosen, 1990) or to educate with computer animation. Some biomechanists have developed software used to adapt human movement kinematic data so that computer game animations have the look of truly human movement, but with

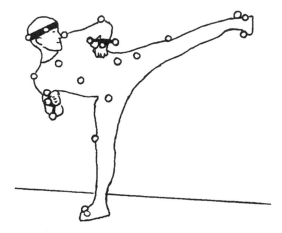

Figure 1.7. Biomechanical measurements and software can be used to make accurate animations of human motion that can be used for technique improvement, cinema special effects, and computer games. Drawing based on image provided by Vicon Motion Systems.

the superhuman speed that makes games exciting (Figure 1.7). Some people use biomechanics to perform forensic examinations. This reconstruction of events from physical measurements at the scene is combined with medical and other evidence to determine the likely cause of many kinds of accidents.

Application

A variety of professions are interested in using biomechanics to modify human movement. A person that fabricates prosthetics (artificial limbs) would use biomechanics to understand the normal functioning of joints, the loadings the prosthetic must withstand, and how the prosthetic can be safely attached to the person. List possible questions biomechanics could answer for a(n):

   Athletic Coach?
   Orthopaedic Surgeon?
   Physical Educator?
   Physical Therapist?
   Athletic Trainer?
   Strength & Conditioning Professional?
   Occupational Fitness Consultant?

You? What question about human movement technique are you curious about?

Qualitative and Quantitative Analysis

Biomechanics provides information for a variety of kinesiology professions to analyze human movement to improve effectiveness or decrease the risk of injury. How the movement is analyzed falls on a continuum between a qualitative analysis and a quantitative analysis. **Quantitative analysis** involves the measurement of biomechanical variables and usually requires a computer to do the voluminous numerical calculations performed. Even short movements will have thousands of samples of data to be collected, scaled, and numerically processed. In contrast, **qualitative analysis** has been defined as the "systematic observation and introspective judgment of the quality of human movement for the purpose of providing the most appropriate intervention to improve performance" (Knudson & Morrison, 2002, p. 4). Analysis in both quantitative and qualitative contexts means identification of the factors that affect human movement performance, which is then interpreted using other higher levels of thinking (synthesis, evaluation) in applying the information to the movement of interest. Solving problems in human movement involves high levels of critical thinking and an interdisciplinary approach, integrating the many kinesiology sciences.

The advantages of numerical measurements of quantitative over those of qualitative analysis are greater accuracy, consistency, and precision. Most quantitative biomechanical analysis is performed in research settings; however, more and more devices are commercially available that inexpensively measure some biomechanical variables (e.g., radar, timing lights, timing mats, quantitative videography systems). Unfortunately, the greater accuracy of quantitative measures comes at the cost of technical skills, calibration, computational

and processing time, as well as dangers of increasing errors with the additional computations involved. Even with very fast modern computers, quantitative biomechanics is a labor-intensive task requiring considerable graduate training and experience. For these reasons and others, qualitative analysis of human movement remains the main approach kinesiology professionals use in solving most human movement problems. Qualitative analysis will be the main focus of the applications of biomechanics presented in this book. Whether your future jobs use qualitative or quantitative biomechanical analysis, you will need to be able to access biomechanical knowledge. The next section will show you many sources of biomechanical knowledge.

Activity: Videotape Replay

Tape a sporting event from a TV broadcast on a VCR. Find a sequence in the video where there is a movement of interest to you and where there is a good close-up shot of the action. You could also video yourself performing a movement using a camcorder. Watch the replay at real-time speed and try to estimate the percentage of time taken up by the major phases of the movement. Most skills can be broken down into three phases—preparation, action, and follow-through—but you can have as many phases as you think apply to the movement of interest. Rewind the tape and use the "pause" and "frame" advance functions to count the number of video frames in the skill and calculate the times and percentages for each phase of the skill. Most VCRs show every other field, giving you a video "clock" with 30 pictures per second. Note, however, that some VCRs show you every field (half of interlaced video) so your clock will be accurate to 1/60th of a second. How could you check what your or the classes' VCR does in frame advance mode? How close was your qualitative judgment to the more accurate quantitative measure of time?

Application

Even though qualitative and quantitative analyses are not mutually exclusive, assume that qualitative-versus-quantitative biomechanical analysis is an either/or proposition in the following exercise. For the sports medicine and athletics career areas, discuss with other students what kind of analysis is most appropriate for the questions listed. Come to a consensus and be prepared to give your reasons (cost, time, accuracy, need, etc.) for believing that one approach might be better than another.

Sport Medicine

1. Is the patient doing the lunge exercise correctly?
2. Is athlete "A" ready to play following rehab for their injured ACL?

Athletics

1. Should pole vaulter "B" change to a longer pole?
2. Is athlete "A" ready to play following rehab for their injured ACL?

WHERE CAN I FIND OUT ABOUT BIOMECHANICS?

This text provides a general introduction to the biomechanics of human movement in kinesiology. Many students take advanced courses in biomechanics and do library research for term projects. This text will provide quite a few references on many topics that will help students find original sources of biomechanical data. The relative youth of the science of biomechanics and the many different academic areas interested in biomechanics (among others, biology, engineering, medicine, kinesiology, physics) makes the search for biomechanical knowledge challenging for many students. This section will give you a brief tour of some of the major fields where biomechanics research is of interest.

Where you find biomechanics research depends on the kind of data you are interested in. Many people are curious about human movement, but there are also many scholars who are interested in the biomechanics of a wide variety of animals. An excellent way to study the theoretical aspects of biomechanics is to study animals that

have made adaptations to be good at certain kinds of movements: like fish, kangaroos, or frogs. Much of this biomechanical research on animals is relevant to the study of human movement.

Professionals from many fields are interested in human movement, so there is considerable interest and research in human biomechanics. As a science biomechanics is quite young (infant), but biomechanics is more like the middle child within the subdisciplines of kinesiology. Biomechanics is not as mature as Exercise Physiology or Motor Learning but is a bit older than Sport Psychology and other subdisciplines. Basic biomechanics research on many popular sport techniques will have been conducted in the early to mid-20th century. Biomechanics research in kinesiology since the 1970s has tended to become more narrowly focused and specialized, and has branched into areas far beyond sport and education. As a result, students with basic sport technique interests now have to integrate biomechanics research over a 50-year period.

Depending on the depth of analysis and the human movement of interest, a student of biomechanics may find himself reading literature in biomechanical, medical, physiological, engineering, or other specialized journals. The smaller and more narrow the area of biomechanical interest (for example, specific fibers, myofibrils, ligaments, tendons), the more likely there will be very recent research on the topic. Research on the effect of computerized retail check-out scanners would likely be found in recent journals related to engineering, human factors, and ergonomics. A student interested in a strength and conditioning career might find biomechanical studies on exercises in medical, physical education, physiology, and specialized strength and conditioning journals. Students with clinical career interests who want to know exactly what muscles do during movement may put together data from studies dealing with a variety of animals. Clues can come from classic research on the muscles of the frog (Hill, 1970), the cat (Gregor & Abelew, 1994) and turkeys (Roberts, Marsh, Weyand, & Taylor, 1997), as well as human muscle (Ito, Kawakami, Ichinose, Fukashiro, & Fukunaga, 1998). While muscle force-measuring devices have been implanted in humans, the majority of the invasive research to determine the actions of muscles in movement is done on animals (Figure 1.8).

Scholarly Societies

There are scholarly organizations exclusively dedicated to biomechanics. Scholarly societies typically sponsor meetings and publications to promote the development of their fields. Students of sport biomechanics should know that the International Society of Biomechanics in Sports (ISBS) is devoted to promotion of sport biomechanics research and to helping coaches apply biomechanical knowledge in instruction, training, and conditioning for sports. The ISBS publishes scholarly papers on sports biomechanics that are accepted from papers presented at their annual meetings and the journal *Sports Biomechanics*. Their website (http://www.uni-stuttgart.de/External/isbs/) provides links to a variety of information on sport biomechanics.

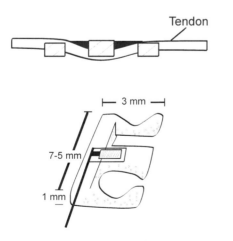
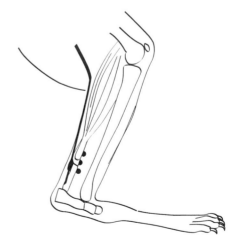

Figure 1.8. Schematic of a buckle transducer for *in vivo* measurement of muscle forces in animal locomotion. Adapted with permission from Biewener and Blickhan (1988).

The International Society of Biomechanics (ISB) is the international society of scholars interested in biomechanics from all kinds of academic fields. The ISB hosts international meetings, sponsors journals, and has a website at http://isb.ri.ccf.org/. Some examples of regional biomechanics societies include the American Society of Biomechanics (ASB), the Canadian Society of Biomechanics, and the European Society of Biomechanics. The ASB website (http://asb-biomech.org/) has several links, including a list of graduate programs and papers accepted for presentation at ABS annual meetings. Another related scholarly society is the International Society for Electrophysiology and Kinesiology (ISEK), which promotes the **electromyographic** (EMG) study of human movement (http://shogun.bu.edu/isek/). Engineers interested in equipment design, sport, and human movement have founded a Sport Engineering Society (http://www.sports-engineering.co.uk/). There are other scholarly organizations that have biomechanics interest groups related to the parent disciplines of medicine, biology, or physics.

Aside from the many specialized biomechanics societies, there are biomechanics interest groups in various scholarly/professional organizations that have an interest in human movement. Two examples are the American Alliance for Health, Physical Education, Recreation, and Dance (AAHPERD) and the American College of Sports Medicine (ACSM). AAHPERD (http://www.aahperd.org/) is the original physical education scholarly/professional organization, founded in 1885. Biomechanists in HPERD can be active in the Biomechanics Academy of the National Association for Sport and Physical Education (NASPE is one of the HPERD associations within the alliance). The American College of Sports Medicine (http://www.acsm.org/) was founded in 1954 by physicians and exercise scientists to be a scholarly society interested in promotion of the study and application of exercise, sports medicine, and sports science. The ACSM substructure interested in biomechanics is the biomechanics interest group (BIG). Other professional organizations in medicine, physical therapy, athletic training, and/or strength and conditioning sponsor biomechanics programs related to their unique interests. Whatever career path you select, it is important that you join and participate in the related scholarly and professional organizations.

Computer Searches

One of the best ways to find information on human biomechanics is to use computerized bibliographies or databases of books, chapters, and articles. Some of the best electronic sources for kinesiology students are *SportDiscus*, *MEDLINE*, and *EMBASE*. *SportDiscus* is the CD-ROM version of the database compiled by the Sport Information Resource Center (SIRC) in Ontario, Canada (http://sirc.ca/sirc/). SIRC has been compiling scholarly sources on sport and exercise science since 1973. Many universities buy access to *SportDiscus* and *Medline* for faculty and student research. *SportDiscus* is quite helpful in locating research papers in the ISBS edited proceedings.

Medical literature has been well cataloged by Index Medicus and the searchable databases *MEDLINE* and *EMBASE*. These databases are quite extensive but do not list all published articles so a search of both is advisable (Minozzi, Pistotti, & Forni, 2000) for literature searches related to sports medicine. Besides access from your university library, the national library of medicine provides free searching of *Medline* at http://www.ncbi.nlm.nih.gov/PubMed/. Very large databases like *SportDiscus*, *Medline*, and *EMBASE* are great research tools if searched intelligently. These databases and others (e.g., *Biological Abstracts, Science Cita-*

tion Index) should be searched by the careful linking of keywords and Boolean (logic: and, or) operators. Remember that much of the power of indexing is the cross-referencing as well as the direct listings for your search items.

Many journals now publish keywords with articles to facilitate the searching for the articles with similar terms. The search request for "biomechanics" in some databases will return all items (probably too many) beginning with these letters in the title, abstract, or keywords including biomechanics or biomechanical. Searching for "kinematic and ankle" will find sources documenting the motion at the ankle joint. Even better would be "kinematic or ankle or subtalar," because any one of the three search terms matching would select a resource. You miss very little with this search, but it is necessary to go through quite a few sources to find the most relevant ones. Be persistent in your search and let your readings refine your search strategy. A student interested in occupational overuse injuries (sports medicine term) will find that the human factors field may refer to this topic as "cumulative trauma disorder," "work-related musculoskeletal disorders," or "occupational overuse syndrome" just to name a few (Grieco, Molteni, DeVito, & Sias, 1998).

There are bibliographies of literature that are in print that list sources relevant to biomechanics. The President's Council on Physical Fitness and Sports publishes *Physical Fitness/Sports Medicine*. The *Physical Education Index* is a bibliographic service for English language publications that is published quarterly by BenOak Publishing. The *PE Index* reviews more than 170 magazines and journals, provides some citations from popular press magazines, and this index can be used to gather "common knowledge." Early sport and exercise biomechanics research has been compiled in several bibliographies published by the University of Iowa (Hay, 1987).

Biomechanics Textbooks

Good sources for knowledge and links (not hyperlinks) to sources commonly missed by students are biomechanics textbooks. Biomechanics students should look up several biomechanics textbooks and review their coverage of a research topic. Scholars often write textbooks with research interests that are blended into their texts, and many authors make an effort to provide extensive reference lists for students. Remember that writing books takes considerable time, so references in a particular text may not be totally up-to-date, but they do give

Interdisciplinary Issue:
Collaborative Biomechanics

Finding biomechanics information is like a scavenger hunt that will lead students all over a library. We have seen that biomechanics research can be found in biology, engineering, medical, and other specialized journals. "Interdisciplinary" means using several different disciplines simultaneously to solve a problem. Do some preliminary research for sources (journals and edited proceedings/books) on a human movement of interest to you. Do the titles and abstracts of the sources you found suggest scholars from different disciplines are working together to solve problems, or are scholars working on a problem primarily from their own area or discipline? What have other students found in their research?

students leads and clues on many good sources. The quality of a biomechanical source will be difficult for many students to judge, so the next section will coach you in evaluating biomechanical sources.

BIOMECHANICAL KNOWLEDGE VERSUS INFORMATION

Biomechanical knowledge is like any other scientific body of knowledge. Biomechanical knowledge is constantly changing, with more and more research being done everyday. For this book biomechanical knowledge can be understood to mean a specific piece of a human movement issue based on a critical review of all the biomechanical evidence. **Knowledge** is the application of our best logic, scientific research, and professional experience applied to data. Students often fail to realize that knowledge is a structure that is constantly being constructed and remodeled in small bits and pieces. Most real-world biomechanical questions have only partial answers based on limited research or knowledge. Although our stack of biomechanical knowledge is not perfect, a critical review of this will be as close as we get to the truth.

The modification of human movement based on biomechanical knowledge is difficult because movement is a multifaceted problem, with many factors related to the performer and activity all interacting to affect the outcome. The next chapter will present nine general principles of biomechanical knowledge that are useful in applying biomechanics in general to improve human movement. There will be a few bits of the knowledge puzzle that are well known and rise to the level of scientific law. While most biomechanical knowledge is not perfect and can only be organized into some general principles, it is much better at

guiding professional practice than merely using information or trail and error.

Living in an information age, it is easy for people to become insensitive to the important distinction between information and knowledge. The most important difference is that information has a much higher chance of being incorrect than knowledge. **Information** is merely access to opinions or data, with no implied degree of accuracy. Information is also much easier to access in the age of the Internet and wireless communications. Do not confuse ease of access with accuracy or value. This distinction is clearer as you look at the hierarchy of the kinds of sources used for scholarly research and a simple strategy for the evaluation of the quality of a source.

Kinds of Sources

When searching for specific biomechanical knowledge it is important to keep in mind the kind of source you are reading. There is a definite hierarchy of the scholarly or academic rigor of published research and writing. Figure 1.9 illustrates typical examples of this hierarchy. Although there are exceptions to most rules, it is generally true that the higher up a source on the hierarchy the better the chance that the information presented is closer to the current state of knowledge and the truth. For this reason professionals and scholars focus their attention on peer-reviewed journals to maintain a knowledge base for practice. Some publishers are now "publishing" electronic versions of their journals on the world wide web (WWW) for subscribers.

Most scholarly journals publish original research that extends the body of knowledge, or review papers that attempt to summarize a body of knowledge. Many journals also publish supplements that contain abstracts (short summaries of a research study) of papers that have been ac-

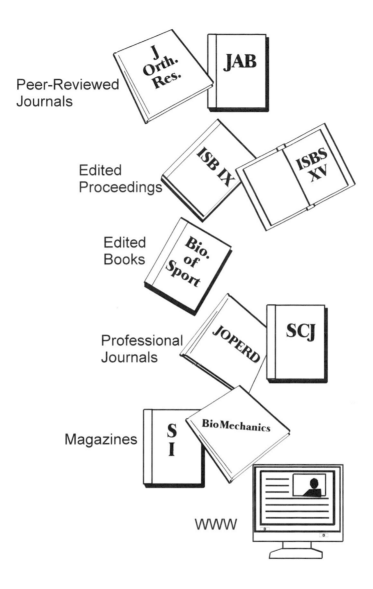

Figure 1.9. The many kinds of biomechanics sources of information and the hierarchy of their academic rigor.

cepted for presentation at a scholarly meeting or were published in another journal. While the review of these abstracts is not as rigorous as a full journal article, abstracts do provide students with clues about what the most recent research is focusing on. Reading biomechanics research will be challenging for most undergraduates. Ap-

pendix A provides a comprehensive glossary of biomechanics terms that will help you when reading the biomechanics literature related to your professional interests.

In the middle of academic rigor are edited proceedings, edited books, and professional journals. These publications have varying degrees of peer review before publication, as well as varying rules on what constitutes acceptable evidence. At the bottom of the credibility chain are popular press publications (magazines/newspapers) and hypertext on the worldwide web. While these sources are appropriate for more subjective observations of laypersons, there are serious threats to the validity of the observations from these sources. The major problems with webpages are their impermanence (unlike archival research literature) and the lack of review (anyone can post a webpage). Another good example is the teaching and coaching tips published by the Physical Education Digest (http://www.pedigest.com). Most of tips and cues are opinions of coaches and teachers in popular press magazines that have not been tested by scientific research. It is possible that some of these opinions are correct and useful, but there is little evidence used to verify the advice, so kinesiology professionals should verify with other primary sources before using the advice. The next section will summarize a quick method for checking the credibility of various sources for biomechanical knowledge.

Evaluating Sources

The previous section clearly suggests that certain sources and kinds of evidence are more likely to be accurate. When evaluating the credibility of sources that fall at similar levels of rigor, the "me" test can be easily applied to judge the chance of the advice being a good and balanced representation of reality. The "m" stands for motivation. What is the motivation for the person or source providing the information? Sources with little financial interest in to making the observations/claims and who are dedicated to advancing a body of knowledge or human potential (scholarly journals) are much more likely to provide accurate information. The motivation of the popular

press (TV, newspapers, magazines) and the internet (WWW) involves profit and self-promotion based on numbers of viewers and, therefore, is more prone to sensationalize and to not weigh all the evidence.

The "e" in the acronym stands for the key element of all science: evidence. Science is based on logical analysis and the balance of many controlled studies. This weighing of all the evidence stands in stark contrast to the more emotional claims of the popular press. The more the emotional and sensational the language, even if it talks about "the latest study," the more likely you are reading only part of the whole picture. Remember that the structure of knowledge is a complicated structure built over time using many small pieces. The "latest" piece of the knowledge puzzle may be in error (see the next section) or will be rejected by most scholars as having flaws that make it less valuable as other research.

One formidable barrier to a kinesiology professional's ability to weigh biomechanical evidence is the technical and specialized terminology employed in most studies. Throughout this text many of these measurement systems and mechanical terms are covered. Appendix A provides an extensive glossary of biomechanical terms and quantitative measurement systems. Two papers that provide good summaries of biomechanical and exercise science terms are available (Knuttgen & Kraemer, 1987; Rogers & Cavanagh, 1984). Students reviewing biomechanical studies should ask their instructor for assistance when the text or these sources do not clear up their understanding.

A Word About Right and Wrong Answers

The increasing amount and complexity of research and technology tends to give many people a false sense of the correctness

Application

On your next trip to a physician or other medical professional's waiting room, evaluate the nature of the articles and advertisements in the magazines and displays you encounter. Do advertisements related to claims in the articles appear near the article? Do the articles talk about several studies, their relative merits, as well as the percentage of subjects with various responses? Does the professional you are visiting sell supplements or products to patients? If so, what does this tell you about motivation and potential conflicts of interest between practice and profits? The biomechanics of most health and human performance problems in human movement are classic examples of complicated problems, with many interrelated factors and variability in the response of individuals to treatment.

of numbers. Few people will question a measurement if the machine printed numbers on a printout, unless they are very familiar with the measurement. Like our knowledge-versus-information discussion, it is very important for kinesiology professionals to understand that the process of reviewing and weighing the evidence is often more important than finding the perfect or "right" answer. Such absolutes in a complicated world are quite rare, usually only occurring when a technique change would run against a law of physics or one of our principles of biomechanics. These principles (and laws) of mechanics are the application tools developed throughout this book.

So the good news is that biomechanics helps kinesiology professionals solve problems, while the bad news is that most of these everyday questions/problems do not have easy, dichotomous (right/wrong) answers. There are many factors that affect most phenomena and there is variation in nearly all phenomena. In fact, all true science is written using statistics to account for this variation. Statistics use estimates of data variation to attach a probability to any yes/no decision about the data. If you read a study that says an observation was significant at the 0.05 level, this only means that the result is not likely a fluke or observation due to chance variation alone. It is possible

that chance alone created this "difference," and $p < 0.05$ means that in the long run there is about a 1-in-20 chance that the observation or decision about the data is wrong. Since most studies use this error standard ($p < 0.05$), this means that, out of twenty studies on a particular topic, one likely reports an incorrect observation from chance variation alone. A common misconception among laypersons is that statistics in a scientific study "proves" things. Statistics only provide tools that allow scientists to place probability values about yes/no decisions on the numbers observed in their research. Proof is a long-term process requiring critical review of the whole body of research on the issue. Remember this when television news broadcasts sensationalize the results of the "latest" study on some health issue or you are tempted to believe that one biomechanical study settles a particular issue.

Biomechanical knowledge is constantly changing and usually cannot be easily classified into always right or wrong answers, so there are two important professional tools you must not forget to use. These tools will work quite well with the biomechanical tools (nine principles) developed in this text. These two tools are the Swiss Army Knives™ or Leathermen™ of your professional toolbox because of they are so flexible and important. One is your *ability to access biomechanical knowledge*, and the other is the *critical thinking* necessary to evaluate and integrate knowledge so it can be applied in solving human movement problems. You are not likely going to remember everything in this book (though you would be wise to), but you should have the knowledge to access, and critical thinking tools that allow you to find, evaluate, and apply biomechanics to human movement. The rest of this text will illustrate and explicate the nine principles of biomechanics, which are tools you would do well to never forget when helping people improve their movement.

SUMMARY

Kinesiology is the scholarly study of human movement. A core science in the academic discipline of kinesiology is biomechanics. Biomechanics in kinesiology is the study of motion and its causes in human movement. The field of biomechanics is relatively new and only has a few principles and laws that can used to inform professional practice. Kinesiology professionals often use biomechanical knowledge in the qualitative analysis of human movement to decide on how to intervene to improve movement and prevent or remediate injury. Applying biomechanics in qualitative analysis is most effective when a professional integrates biomechanical knowledge with professional experience and the other subdisciplines of kinesiology. Biomechanical knowledge is found in a wide variety of journals because there are many academic and professional areas interested in the movement of living things. Students studying human biomechanics might find relevant biomechanical knowledge in books and journals in applied physics, biology, engineering, ergonomics, medicine, physiology, and biomechanics.

REVIEW QUESTIONS

1. What is biomechanics and how is it different from the two common meanings of kinesiology?
2. Biomechanical knowledge is useful for solving what kinds of problems?
3. What are the advantages and disadvantages of a qualitative biomechanical analysis?
4. What are the advantages and disadvantages of a quantitative biomechanical analysis?
5. What kinds of journals publish biomechanics research?
6. What is the difference between knowledge and information?
7. Why should biomechanical knowledge be integrated with other sport and exercise sciences in solving human movement problems?

KEY TERMS

> biomechanics
> electromyography (EMG)
> information
> interdisciplinary
> kinesiology
> knowledge
> orthotics
> prosthetics
> qualitative analysis
> quantitative analysis

SUGGESTED READING

Bartlett, R. M. (1997). Current issues in the mechanics of athletic activities: A position paper. *Journal of Biomechanics*, **30**, 477–486.

Cavanagh, P. R. (1990). Biomechanics: A bridge builder among the sport sciences. *Medicine and Science in Sports and Exercise*. **22**, 546–557.

Chaffin, D., & Andersson, G. (1991). *Occupational biomechanics* (2nd ed.). New York: Wiley.

Elliott, B. (1999). Biomechanics: An integral part of sport science and sport medicine. *Journal of Science and Medicine and Sport*, **2**, 299–310.

Knudson, D. V., & Morrison, C. M. (2002). *Qualitative analysis of human movement* (2nd ed.). Champaign, IL: Human Kinetics.

Kumar, S. (1999). *Biomechanics in ergonomics*. London: Taylor & Francis.

Lees, A. (1999). Biomechanical assessment of individual sports for improved performance. *Sports Medicine*, **28**, 299–305.

Legwold, G. (1984). Can biomechanics produce Olympic medals? *Physician and Sportsmedicine*, **12**(1), 187–189.

LeVeau, B. (1992). *Williams and Lissner's: Biomechanics of human motion* (3rd ed.). Philadelphia: W. B. Sanders.

Segesser, B., & Pforringer, W. (Eds.) (1989). *The shoe in sport*. Chicago: Year Book Medical Publishers.

Yeadon, M. R., & Challis, J. H. (1994). The future of performance-related sports biomechanics research. *Journal of Sports Sciences*, **12**, 3–32.

WEB LINKS

AAHPERD—American Alliance for Health, Physical Education, Recreation, and Dance is the first professional HPERD organization in the United States.
http://www.aahperd.org/

Biomechanics Academy—A biomechanics interest area within AAHPERD and NASPE (National Association for Sport and Physical Education).
http://www.aahperd.org/naspe/template.cfm?template=specialinterests-biomechanics.html

AAKPE—American Academy of Kinesiology and Physical Education is the premier, honorary scholarly society in kinesiology.
http://www.aakpe.org/

ACSM—American College of Sports Medicine is a leader in the clinical and scientific aspects of sports medicine and exercise. ACSM provides the leading professional certifications in sports medicine.
http://acsm.org/

ISB—International Society of Biomechanics was the first biomechanics scholarly society.
http://www.isbweb.org/

ASB—American Society of Biomechanics posts meeting abstracts from a variety of biomechanical scholars.
http://asb-biomech.org/

IFSM—International Federation of Sports Medicine.
http://www.fims.org/

ISBS—International Society of Biomechanics in Sports hosts annual conferences and indexes papers published in their proceedings and journal (*Sports Biomechanics*).
http://www.uni-stuttgart.de/External/isbs/

ISEK—International Society of Electrophysiological Kinesiology is the scholarly society focusing on applied electromyography (EMG) and other electrophysiological phenomena.
http://shogun.bu.edu/isek/

ISI-The Institute for Scientific Information provides a variety of services including rating scholarly journals and authors.
http://www.isinet.com/isi/

Medline—Free searching of this medical database provided by the National Library of Medicine.
http://www.ncbi.nlm.nih.gov/PubMed/

SIRC—The Sport Information Resource Center provides several database services for sport and kinesiology literature like *SportDiscus*. Many college libraries have subscriptions to *SportDiscus*.
http://www.sportquest.com/

CHAPTER 2

FUNDAMENTALS OF BIOMECHANICS AND QUALITATIVE ANALYSIS

In Chapter 1 we found that biomechanics provides tools that are needed to analyze human motion, improve performance, and reduce the risk of injury. In order to facilitate the use of these biomechanical tools, this text will emphasize the qualitative understanding of mechanical concepts. Many chapters, however, will include some quantitative examples using the algebraic definitions of the mechanical variables being discussed. Mathematical formulas are a precise language and are most helpful in showing the importance, interactions, and relationships between biomechanical variables. While more rigorous calculus forms of these equations provide the most accurate answers commonly used by scientists (Beer & Johnson, 1984; Hamill & Knutzen, 1995; Zatsiorsky, 1998, 2002), the majority of kinesiology majors will benefit most from a qualitative understanding of these mechanical concepts. So this chapter begins with key mechanical variables and terminology essential for introducing other biomechanical concepts. This chapter will emphasize the conceptual understanding of these mechanical variables and leave more detailed development and quantitative examples for later in the text. Next, nine general principles of biomechanics are introduced that will be developed throughout the rest of the text. These principles use less technical language and are the tools for applying biomechanical knowledge in the qualitative analysis of human movement. The chapter concludes by summarizing a model of qualitative analysis that is used in the application section of the book.

KEY MECHANICAL CONCEPTS

Mechanics

Before we can begin to understand how humans move, there are several mechanical terms and concepts that must be clarified. **Mechanics** is the branch of physics that studies the motion of objects and the forces that cause that motion. The science of mechanics is divided into many areas, but the three main areas most relevant to biomechanics are: **rigid-body**, **deformable-body**, and **fluids**.

In **rigid-body mechanics**, the object being analyzed is assumed to be rigid and the deformations in its shape so small they can be ignored. While this almost never happens in any material, this assumption is quite reasonable for most biomechanical studies of the major segments of the body. The rigid-body assumption in studies saves considerable mathematical and modeling work without great loss of accuracy. Some biomechanists, however, use deformable-body mechanics to study how biological materials respond to external forces that are applied to them. **Deformable-body** mechanics studies how forces are distributed within a material, and can be focused at many levels (cellular to tissues/organs/system) to examine how forces stimulate growth or cause damage. **Fluid mechanics** is concerned with the forces in fluids (liquids and gasses). A *biomechanist* would use fluid mechanics to study heart valves, swimming, or adapting sports equipment to minimize air resistance.

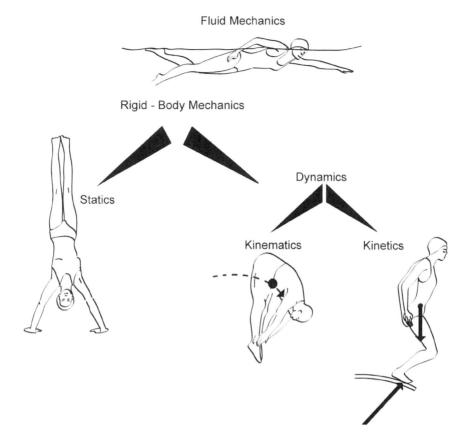

Figure 2.1. The major branches of mechanics used in most biomechanical studies.

Most sports biomechanics studies are based on rigid-body models of the skeletal system. Rigid-body mechanics is divided into **statics** and **dynamics** (Figure 2.1). **Statics** is the study of objects at rest or in uniform (constant) motion. **Dynamics** is the study of objects being accelerated by the actions of forces. Most importantly, dynamics is divided into two branches: **kinematics** and **kinetics**. **Kinematics** is motion description. In kinematics the motions of objects are usually measured in linear (meters, feet, etc.) or angular (radians, degrees, etc.) terms. Examples of the kinematics of running could be the speed of the athlete, the length of the stride, or the angular velocity of hip extension. Most angular mechanical variables have the adjective "angular" before them. **Kinetics** is concerned with determining the causes of motion. Examples of kinetic variables in running are the forces between the feet and the ground or the forces of air resistance. Understanding these variables gives the track coach knowledge of the causes of running performance. Kinetic information is often more powerful in improving human motion because the causes of poor performance have been identified. For example, knowing that the timing and size of hip extensor action is weak in the takeoff phase for a long jumper may be more useful in improving performance than knowing that the jump was shorter than expected.

Basic Units

The language of science is mathematics. Biomechanics often uses some of the most complex kinds of mathematical calculations, especially in deformable-body mechanics. Fortunately, most of the concepts and laws in classical (Newtonian) rigid-body mechanics can be understood in qualitative terms. A conceptual understanding of biomechanics is the focus of this book, but algebraic definitions of mechanical variables will be presented and will make your understanding of mechanical variables and their relationships deeper and more powerful.

First, let's look at how even concepts seemingly as simple as numbers can differ in their complexity. **Scalars** are variables that can be completely represented by a number and the units of measurement. The number and units of measurement (10 kg, 100 m) must be reported to completely identify a scalar quantity. It makes no sense for a track athlete call home and say, "Hey mom, I did 16 and 0"; they need to say, "I made 16 feet with 0 fouls." The number given a scalar quantity represents the magnitude or size of that variable.

Vectors are more complicated quantities, where size, units, and *direction* must be specified. Figure 2.2 shows several scalars and the associated vectors common in biomechanics. For example, **mass** is the scalar quantity that represents the quantity of

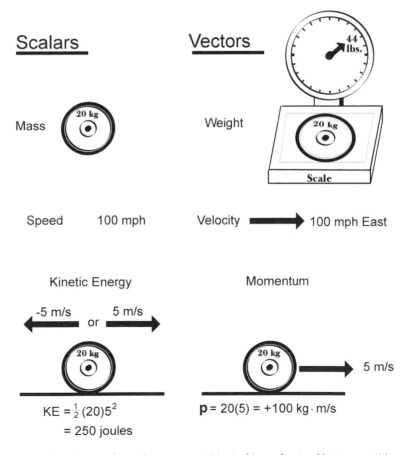

Figure 2.2. Comparison of various scalar and vector quantities in biomechanics. Vector quantities must specify magnitude and direction.

matter for an object. That same object's **weight** is the gravitational force of attraction between the earth and the object. The difference between mass and weight is dramatically illustrated with pictures of astronauts in orbit about the earth. Their masses are essentially unchanged, but their weights are virtually zero because of the microgravity when far from earth.

Biomechanics commonly uses directions at right angles (horizontal/vertical, longitudinal/transverse) to mathematically handle vectors. Calculations of velocity vectors in a two-dimensional (2D) analysis of a long jump are usually done in one direction (e.g., horizontal) and then the other (vertical). The directions chosen depend on the needs of the analysis. Symbols representing vector quantities like velocity (**v**) in this text will be identified with **bold letters**. Physics and mechanics books also use underlining or an arrow over the symbol to identify vector quantities. These and other rules for vector calculations will be summarized in chapter 6. These rules are important because when adding vectors, one plus one is often not two because the directions of the vectors were different. When adding scalars with the same units, one plus one is always equal to two. Another important point related to vectors is that the sign (+ or −) corresponds to directions. A −10 lb force is not less than a +10 lb force; they are the same size but in opposite directions. The addition of vectors to determine their net effect is called the **resultant** and requires right-angle trigonometry. In chapter 6 we will also subtract or break apart a vector into right-angle **components**, to take advantage of these trigonometry relationships to solve problems and to "see" other important pushes/pulls of a force.

There are two important vector quantities at the root of kinetics: **force** and **torque**. A **force** is a straight-line push or pull, usually expressed in pounds (lbs) or Newtons

(N). The symbol for force is **F**. Remember that this push or pull is an interactional effect between two bodies. Sometimes this "push" appears obvious as in a ball hitting a bat, while other times the objects are quite distant as with the "pull" of magnetic or gravitational forces. Forces are vectors, and vectors can be physically represented or drawn as arrows (Figure 2.3). The important characteristics of vectors (size and direction) are directly apparent on the figure. The length of the arrow represents the size or magnitude (500 N or 112 lbs) and the orientation in space represents its direction (15 degrees above horizontal).

The corresponding angular variable to force is a **moment of force** or **torque**. A moment is the rotating effect of a force and will be symbolized by an **M** for moment of force or **T** for torque. This book will use the term "torque" synonymously with "moment of force." This is a common English meaning for torque, although there is a more specific mechanics-of-materials meaning (a torsion or twisting moment) that leads some scientists to prefer the term "moment of force." When a force is applied to an object that is not on line with the center of the object, the force will create a torque that tends to rotate the object. In Figure 2.3 the impact force acts below the center of the ball and would create a torque that causes the soccer ball to acquire backspin. We will see later that the units of torque are pound-feet (lb·ft) and Newton-meters (N·m).

Let's look at an example of how kinematic and kinetic variables are used in a typical biomechanical measurement of **isometric** muscular strength. "Isometric" is a muscle research term referring to muscle actions performed in constant (*iso*) length (*metric*) conditions. The example of a spring is important for learning how mathematics and graphs can be used to understand the relationship between variables. This example will also help to understand how muscles, tendons, and ligaments can be said to

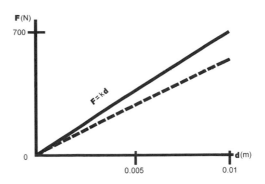

Figure 2.4. A graph (solid line) of the relationship between the force (**F**) required to stretch a spring a given displacement (**d**). The elasticity of the spring is the slope of the line. The slope is the constant (k) in Hooke's Law: (**F** = k · **d**).

Figure 2.3. Vector representation of the force applied by a foot to a soccer ball. The magnitude and direction properties of a vector are both apparent on the diagram: the length of the arrow represents 500 Newtons of force, while the orientation and tip of the arrow represent the direction (15° above horizontal) of the force.

have spring-like behavior. Figure 2.4 illustrates the force–displacement graph for the spring in a handgrip **dynamometer**. A **dynamometer** is a force-measuring device. As a positive force (**F**) pulls on the spring, the spring is stretched a positive linear distance (displacement = **d**). Displacement is a kinematic variable; force is a kinetic variable.

Therapists often measure a person's grip strength in essentially isometric conditions because the springs in hand dynamometers are very stiff and only elongate very small distances. The force–displacement graph in Figure 2.4 shows a very simple (predictable) and linear relationship between the force in the spring (**F**) and the resulting elongation (**d**). In other words, there is a uniform increase (constant slope

of the line) in force with increasing spring stretch. We will see later on in chapter 4 that biological tissues have much more complex (curved) mechanical behaviors when loaded by forces, but there will be linear regions of their load–deformation graphs that are representative of their elastic properties.

Let's extend our example and see how another mechanical variable can be derived from force and displacement. Many simple force measuring devices (e.g., bathroom and fishing scales) take advantage of the elastic behavior of metal springs that are stretched or compressed short distances. This relationship is essentially the mathematical equation (**F** = k · **d**) of the calibration line illustrated in Figure 2.4, and is called Hooke's Law. Hooke's Law is valid for small deformations of highly elastic materials like springs. The **stiffness** (elasticity) of the spring is symbolized as k, which represents the slope of the line. In chapter 4 we will look at the stiffness of biological tissues as the slope of the linear region of a graph like this. If we plug in the largest force and displacement (700 = k · 0.01), we can solve for the stiffness of the spring, and find it to be 70,000 N/m. This says that the spring force will increase 70,000 Newtons every meter it is stretched. This is about 15,730 pounds of tension if the spring were stretched to about 1.1 yards! Sounds pretty impressive, but remember that the springs are rarely elongated that much, and you might be surprised how stiff muscle-tendon units can get when strongly activated.

Engineers measure the stiffness or elasticity of a material with special machines that simultaneously record the force and deformation of the material. The slope of the load–deformation graph (force/length) in the linear region of loading is used to define stiffness. **Stiffness** is the measure of elasticity of the material, but this definition often conflicts with most people's common understanding of elasticity. People often incorrectly think elasticity means an object that is easily deformed with a low force, which is really compliance (length/force), the opposite of stiffness. An engineer would say that there was less stiffness or greater compliance in the second spring illustrated as a dashed line.

Can you find the stiffness (spring constant, k) that corresponds to the dashed calibration line in Figure 2.4? Remember that the stiffness, k, corresponds to the slope of the line illustrated in the figure and represents the change in force for a given change in length. The slope or rate of change of a variable or graph will be an important concept repeated again and again in biomechanics. Remember that forces and displacements are vectors, so directions are indicated by the sign (+ or –) attached to the number. What do you think the graph would look like if the force were reversed, i.e., to push and compress the spring rather than stretching it? What would happen to the sign of **F** and **d**?

It is also important to know that the previous example could also be measured using angular rather than linear measurements. There are isokinetic dynamometers

Activity: Elasticity

Take a rubber band and loop it between the index fingers of your hands. Slowly stretch the rubber band by moving one hand away from the other. The tension in the rubber band creates a torque that tends to abduct the metacarpophalangeal joints of your index finger. Does the tension your fingers sense resisting the torque from the rubber band uniformly increase as the band is stretched? Does a slightly faster stretch feel different? According to Hooke's Law, elastic materials like springs and rubber bands create forces directly proportional to the deformation of the material, but the timing of the stretch does not significantly affect the resistance. Chapter 4 will deal with the mechanical responses of biological tissues, which are not perfectly elastic, so the rate of stretch affects the mechanical response of the tissue.

that simultaneously measure the torque (**T**) and rotation (Figure 1.5). These angular measurements have been used to describe the muscular strength of muscle groups at various positions in the range of motion.

There are many other mechanical variables that help us understand how human movement is created. These variables (e.g., impulse, angular momentum, kinetic energy) often have special units of measurement. What all these mechanical variables and units have in common is that they can be expressed as combinations of only four base units. These base units are length, mass, and time. In the International System (SI) these units are the second (s), kilogram (kg), meter (m), and radian (rad). Scientific research commonly uses SI units because they are base 10, are used throughout the world, and move smoothly between traditional sciences. A Joule of mechanical energy is the same as a Joule of chemical energy stored in food. When this book uses mathematics to teach a conceptual understanding of mechanics in human movement (like in Figure 2.4), the SI system will usually be used along with the corresponding English units for a better intuitive feel for many students. The symbols used are based on the recommendations of the International Society of Biomechanics (ISB, 1987).

These many biomechanical variables are vitally important to the science of biomechanics and the integration of biomechanics with other kinesiological sciences. Application of biomechanics by kinesiology professionals does not have to involve quantitative biomechanical measurements. The next section will outline biomechanical principles based on the science and specialized terminology of biomechanics.

NINE FUNDAMENTALS OF BIOMECHANICS

Biomechanists measure all kinds of linear and angular mechanical variables to document and find the causes of human motion. While these variables and studies are extremely interesting to biomechanists, some kinesiology students and professionals may not find them quite so inherently stimulating. Most kinesiology professionals want to know the basic rules of biomechanics that they can apply in their jobs. This section proposes nine such principles of biomechanics and demonstrates how they relate to scientific laws. These biomechanical tools must be combined with other tools from your kinesiology toolbox to most effectively solve movement problems. Because these principles are the application rules for kinesiology professionals, they have usually been given less-scientific names so that we can communicate effectively with our clients.

Principles and Laws

The nine principles of biomechanics that follow take the form of general principles related to human movement. It is important to realize that principles for application are not the same as scientific laws. **Science** is a systematic method for testing hypotheses with experimental evidence for the purpose of improving our understanding of reality. Science uses a process, know as the scientific method, for testing a theory about a phenomenon with measurements, then reevaluating the theory based on the data. Ultimately, science is interested in finding the truth, facts, or laws of nature that provide the best understanding of reality. When experimentation shows data always consistent with a theory (given certain conditions), then the theory becomes a law. Scientists must always be open to new data and theories that may provide a more accurate description or improved understanding of a phenomenon. True scientific revolutions that throw out long-held and major theories are not as common as most people think. Though news reporters often herald scientific "breakthroughs," they are usually exaggerating the importance of a small step in what is a very slow process of weighing a great deal of evidence.

Note that science is not defined as a method for making practical applications of knowledge. **Technology** is the term usually used to refer to the tools and methods of applying scientific knowledge to solve problems or perform tasks. Remember that in chapter 1 we noted the belief of some scholars that studying academic disciplines and doing theoretical research are worthy enterprises without any need to show any practical application of knowledge. Even in "applied" fields like kinesiology, there is a long history of a theory-to-practice, or a science-to-profession gap (Harris, 1993). Why does this gap exist? It might exist because some scholars are hesitant to propose application based on what is often less-than-conclusive data, or they might be concerned about receiving less recognition for applied scholarship. Practitioners contribute to this gap as well by refusing to recognize the theoretical nature of science, by not reading widely to compile the necessary evidence for practice, and by demanding simple "how-to" rules of human movements when these simple answers often do not exist.

This text is based on the philosophy that the best use of the science of biomechanics is in its translation to principles for improving human movement. These principles are general rules for the application of biomechanics that are useful for most all human movements. Some of the principles are based on major laws of mechanics, many of which are hundreds of years old. For example, Newton's Laws of Motion are still used at NASA because they accurately model the motion of spacecraft, even though there are more recent advancements in theoretical physics that are only an improvement in very extreme conditions (high-energy or near the speed of light). Unfortunately, the human body is a much more complicated system than the space shuttle, and biomechanists have not had hundreds of years to make progress on theories of human movement. For these reasons, these nine principles of application should be viewed as general rules that currently fit what we currently know about the biomechanics of human movement.

Nine Principles for Application of Biomechanics

The nine principles of biomechanics proposed in this text were selected because they constitute the minimum number or core principles that can be applied to all human movements and because they provide a simple paradigm or structure to apply biomechanical knowledge. The names of

the principles are put in the common language of application; however, each can be directly linked to the concepts and laws of biomechanics. Special attention has been paid to make application of these principles both friendly and consistent with the specialized terminology of mechanics. As kinesiology professionals you will know the names of the biomechanical laws and theories behind these principles, but you will need to use more applied terminology when communicating with clients. This section will provide a description of each principle, and the application of these principles will be developed throughout the text. The principles can be organized (Figure 2.5) into ones dealing primarily with the creation of movement (process) and ones dealing with the outcome of various projectiles (product).

I want to point out that these principles are based primarily on work of several biomechanists (Norman, 1975; Hudson, 1995) who have developed generic biomechanical principles for all human movements. Many biomechanics books have proposed general principles for all movements (Meinel & Schnabel, 1998); various categories of human movements like throwing, catching, and running (e.g., Broer & Zernicke, 1979; Dyson, 1986; Kreighbaum & Barthels, 1996; Luttgens & Wells, 1982); or specific movements (e.g., Bunn, 1972; Groves & Camaione, 1975). Some biomechanists believe that general principles applicable to all sports are difficult to identify and have limited practical application due to unique goals and environmental contexts of skills (Hochmuch & Marhold, 1978). This book is based on the opposite philosophy. Kinesiology professionals should keep in mind the specific goals and contextual factors affecting a movement, but the nine principles of biomechanics are important tools for improving all human movements.

The first principle in biomechanics is the **Force–Motion principle**. Force–motion

Movement Principles

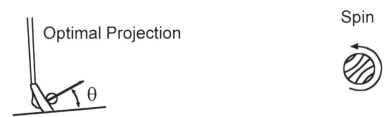

Balance

Inertia

Coordination Continuum

Range of Motion

1
2
3

Segmental Interaction

Force - Motion

Force -Time

Projectile Principles

Optimal Projection

θ

Spin

Figure 2.5. The nine principles of biomechanics can be classified into those related to movement of the body or a projectile. The human body can be a projectile, so all nine principles can be applied to the human body.

says that unbalanced forces are acting on our bodies or objects when we either create or modify movement. In quiet standing the force of gravity is balanced by ground reaction forces under our feet (Figure 2.6), so to move from this position a person creates larger horizontal and vertical forces with their legs. This simple illustration of the

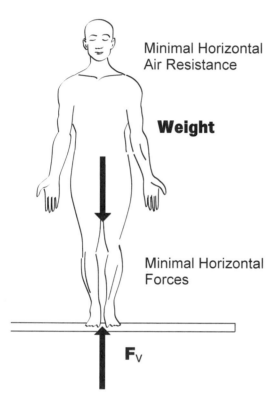

Figure 2.6. A free-body diagram of a person quietly standing. The major vertical forces acting on the person (gravity and ground reaction force) are illustrated, while horizontal forces are small enough to ignore.

body is our first example of what in mechanics is called a **free-body diagram**. A **free-body diagram** is a simplified model of any system or object drawn with the significant forces acting on the object. The complexity and detail of the free-body diagram depends on the purpose of the analysis. In-

spection of Figure 2.5 should make it qualitatively obvious that the addition of the two vertical forces illustrated would cancel each other out, keeping the person essentially motionless in the vertical direction. The Force–Motion principle here correctly predicts no change in motion, since there is no unbalanced force acting on the person. Later on in the text we will use free-body diagrams to actually calculate the effect of forces and torques on the motion of the human body, and we will study the effects of forces acting over time to change the motion of the human body. We will also come to see later that this principle is based on Newton's three laws of motion. The application of the Force–Motion principle in qualitative analysis will be explored throughout the text.

An important thing to notice in this principle is the sequence of events. Forces must act first, before changes in motion can occur. Detailed study of kinematics will illustrate when the motion occurred relative to the acceleration and force causing it. Suppose a person is running on a sidewalk and a small child darts directly in the runner's path to grab a bouncing ball. In order to avoid the child, the runner must change the state of motion. The Force–Motion principle tells the kinesiology professional that the runner's sideward movement (a change in direction and speed) had to be created by large forces applied by the leg to the ground. The force applied by the leg comes first and the sideward motion to avoid the collision was the result.

Substantial changes in motion do not instantly occur but are created over time, which leads us to the next principle of **Force–Time**. It is not only the amount of force that can increase the motion of an object; the amount of time over which force can be applied also affects the resulting motion. A person using a longer approach in bowling has more time to apply forces to increase ball speed. Increasing the time to apply force is also an important technique in slowing down objects (catching) and landing safely. The impulse–momentum relationship, the original language of Newton's second law, is the mathematical explanation of this important principle.

Another important principle to understand in the modification of motion is **Inertia**. **Inertia** can be defined as the property of all objects to resist changes in their state of motion. Newton's first law of motion outlines the principle of inertia. The Newtonian view of inertia as a fundamental property of motion was a major conceptual leap, rejecting the old Aristotelian view that constant application of force was required for motion. The linear and angular measures of inertia are mass (m) and moment of inertia (I). We will see that inertia can be viewed as a resistance to motion in the traditional sense, but this property can also be used to an advantage when modifying motion or transferring energy from one body segment to another.

The next principle involves the **Range of Motion** the body uses in movement. **Range of Motion** is the overall motion used in a movement and can be specified by linear or angular motion of the body segments. The purpose of some movements might require that some body segments limit range of motion, while others requiring maximum speed or force might require larger ranges of motion. Increasing the range of motion in a movement can be an effective way to increase speed or to gradually slow down from a high speed. A baseball pitcher taking a longer stride (Figure 2.7) is increasing the range of motion of the weight shift. Since moving through a range of motion takes time, this principle is related to the force–time principle.

The next biomechanical principle is **Balance**. **Balance** is a person's ability to control their body position relative to some base of support. Stability and mobility of body postures are inversely related, and

Figure 2.7. The forward stride of a pitcher increases the range of motion used to accelerate the body and eventually the baseball.

several biomechanical factors are involved in manipulating a person's stability and mobility. A handstand is a difficult gymnastic skill not only because of the muscular strength required, but also because of the small base of support in the anterior and posterior directions. Athletes in the starting blocks for sprints choose body postures with less stability in favor of increased mobility in the direction of the race.

How the muscle actions and body segment motions are timed in a human movement is usually referred to as coordination. The **Coordination Continuum** principle says that determining the optimal timing of muscle actions or segmental motions depends on the goal of the movement. If high forces are the goal of the movement, more simultaneous muscle actions and joints rotations are usually observed, while low-force and high-speed movements tend to have more sequential muscle and joint actions (Hudson, 1995; Kreighbaum & Barthels, 1996). These two strategies (simultaneous/sequential) can be viewed as a continuum, with the coordination of most motor skills falling somewhere between these two strategies.

The principle of **Segmental Interaction** says that the forces acting in a system of linked rigid bodies can be transferred through the links and joints. Muscles normally act in short bursts to produce torques that are precisely coordinated to complement the effects of torques created by forces at the joints. A wide variety of terms have been used to describe this phenomenon (transfer, summation, sequential) because there are many ways to study human movement. This variety of approaches has also created a confusing array of terminology classifying movements as either open or closed (kinematic or kinetic) chains. We will see that the exact mechanism of this principle of biomechanics is not entirely clear, and common classification of movements as open or closed chains is not clear or useful in analyzing movement (Blackard, Jensen, & Ebben, 1999; di Fabio, 1999; Dillman, Murray, & Hintermeister, 1994).

The biomechanical principle of **Optimal Projection** says that for most human movements involving projectiles there is an optimal range of projection angles for a specific goal. Biomechanical research shows that optimal angles of projection provide the right compromise between vertical velocity (determines time of flight) and horizontal velocity (determines range given the time of flight) within the typical conditions encountered in many sports. For example, in throwing most sport projectiles for horizontal distance, the typical air resistance and heights of release combine to make it beneficial for an athlete to use projection

angles below 45 degrees. Chapter 5 will give several examples of how biomechanical studies have determined desirable release angles for various activities. This research makes it easier for coaches to determine if athletes are optimizing their performance.

The last principle involves the **Spin** or rotations imparted to projectiles, and particularly sport balls. Spin is desirable on thrown and struck balls because it stabilizes flight and creates a fluid force called **lift**. This lift force is used to create a curve or to counter gravity, which affects the trajectory and bounce of the ball. A volleyball player performing a jump serve should strike above the center of the ball to impart topspin to the ball. The topspin creates a downward lift force, making the ball dive steeply and making it difficult for the opponent to pass. The spin put on a pass in American football (Figure 2.8) stabilizes the orientation of the ball, which ensures aerodynamically efficient flight. The natural application

Figure 2.8. The spin imparted to a football during a forward pass serves to stabilize ball flight, to provide aerodynamically efficient flight.

Interdisciplinary Issue: The Vertical Jump

Now that the principles are out of the bag, let's use them to look at a common sport movement, the vertical jump. Imagine an athlete is doing a standing vertical jump test. Which principles of biomechanics would be of most interest to scholars from motor development, motor learning, exercise physiology, or sport psychology studying the vertical jump test? What combinations of the sport sciences are most relevant to the concept of *skill* in vertical jumping? What sports science provides the most relevant information to the physical determinants of jumping ability? How could someone determine if the success of elite jumpers is more strongly related to genetics (nature/physical) than coaching (nurture/training)? How could a strength coach integrate jump training studies with biomechanical studies of jumping techniques?

of these biomechanical principles is in qualitative analysis of human movement.

QUALITATIVE ANALYSIS

The examples that illustrate the application of the principles of biomechanics in the solution of human movement problems in this book will be based on qualitative analyses. Research has shown that general principles of biomechanics provide a useful structure for qualitative analysis of human movement (Johnson, 1990; Matanin, 1993; Nielsen & Beauchamp, 1992; Williams & Tannehill, 1999; Wilkinson, 1996). Quantitative biomechanical analysis can also be used, but most kinesiology professionals will primarily be using qualitative analyses of movement rather than quantitative biomechanical analyses.

There are several models of qualitative analysis of human movement. Traditionally, kinesiology professionals have used a simple error detection and correction approach to qualitative analysis. Here the analyst relies on a mental image of the correct technique to identify "errors" in the performance and provide a correction. This approach has several negative consequences and is too simplistic a model for professional judgments (Knudson & Morrison, 2002). The application of the principles of biomechanics is illustrated in the present book using a more comprehensive vision of qualitative analysis than the simple error detection/correction of the past. This text uses the Knudson and Morrison (2002) model of qualitative analysis (Figure 2.9). This model provides a simple four-task structure: preparation, observation, evaluation/diagnosis, and intervention. This model of qualitative analysis is equally relevant to athletic or clinical applications of biomechanics to improving human movement.

In the preparation task of qualitative analysis the professional gathers relevant kinesiology knowledge about the activity, the performer, and then selects an observational strategy. In the observation task the analyst executes the observational strategy

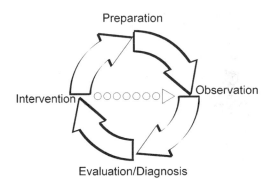

Figure 2.9. The four-task model of qualitative analysis. Adapted from Knudson and Morrison (2002).

to gather all relevant sensory information about the performance of the movement. The third task of qualitative analysis has two difficult components: evaluation and then diagnosis of performance. In evaluation the analyst identifies strengths and weaknesses of performance. Diagnosis involves the prioritizing of the potential interventions to separate causes of poor performance from minor or symptomatic weaknesses. Intervention is the last task of qualitative analysis. In this task the professional executes some action on behalf of the performer. Often in live qualitative analysis, the analyst will return immediately to the observation task to monitor the intervention and the mover's progress.

Application: Quantitative Analysis

An athletic trainer is planning a qualitative analysis of the lower-extremity muscular function of an athlete finishing up an anterior cruciate ligament (ACL) rehabilitation program. The trainer has run the athlete through the rehabilitation program, but wants a more functional evaluation of the athlete's ability and readiness for play. The athlete will be doing several drills, including multiple one-legged hops, shuttle runs, landings, jumps, and lateral cutting movements. For the preparation task of qualitative analysis, give examples of research or biomechanical principles that you think would be relevant to analyzing the athlete's ability to prevent damage to the ACL. Is there a task of qualitative analysis that more heavily relies on biomechanics than other sport sciences?

SUMMARY

Most biomechanical research has been based on rigid-body models of the skeletal system. Kinematics involves the description of the motion, while kinetics focuses on the forces that created the motion. There are many biomechanical variables and they can be classified as either scalars or vectors. Despite the precision of quantitative biomechanics, most kinesiology professionals apply biomechanics at a qualitative or conceptual level. The nine principles of biomechanics that can be used to apply biomechanics knowledge in professional practice are Force–Motion, Force–Time, Inertia, Range of Motion, Balance, Coordination Continuum, Segmental Interaction, Optimal Projection, and Spin. These nine principles can be applied using a comprehensive model (Knudson & Morrison, 2002) of qualitative analysis.

REVIEW QUESTIONS

1. What are major branches of mechanics, and which are most commonly used in performing biomechanical analyses of human movement?

2. What are the specific foci of kinematic and kinetic analyses, and provide some examples?

3. How are vector variables different from scalar variables?

4. How is a scientific principle different from a law?

5. The nine principles of biomechanics can be classified into which two areas of interest?

6. What are the nine principles of biomechanics?

7. What are some other factors that affect human movement and the application of the principles of biomechanics?

8. List as many reasons as possible for the apparent theory-to-practice gap between scholars and practitioners.

KEY TERMS

components
deformable body
dynamics
dynamometer
fluid
free-body diagram
isometric
kinematics
kinetics
mass
mechanics
resultant
scalar
science
strength (muscular)
stiffness
technology
torque/moment of force
vector
weight

SUGGESTED READING

Hudson, J. L. (1995). Core concepts in kinesiology. *JOPERD*, **66**(5), 54–55, 59–60.

Knudson, D., & Morrison, C. (2002). *Qualitative analysis of human movement* (2nd ed.). Champaign, IL: Human Kinetics.

Knuttgen, H. G., & Kraemer, W. J. (1987). Terminology and measurement in exercise performance. *Journal of Applied Sport Science Research*, **1**, 1–10.

Kreighbaum, E., & Bartels, K. M. (1996). *Biomechanics: A qualitative approach to studying human movement*. Boston: Allyn & Bacon.

Norman, R. (1975). Biomechanics for the community coach. *JOPERD*, **46**(3), 49–52.

Rogers, M. M., & Cavanagh, P. R. (1984). Glossary of biomechanical terms, concepts, and units. *Physical Therapy*, **64**, 82–98.

WEB LINKS

Physics and Mathematics Review provided by the physics department of the University of Guelph in Canada.

http://www.physics.uoguelph.ca/tutorials/tutorials.html

Knudson & Morrison (2002)—A link to the only book on the qualitative analysis of human movement.

http://www.humankinetics.com/products/showproduct.cfm?isbn=0736034625

BIOLOGICAL/STRUCTURAL BASES

The study of biomechanics requires an understanding of the structure of musculoskeletal systems and their mechanical properties. The three-dimensional computer model depicted here provides a good representation of the main structures of the ankle, but the response of these tissues to forces and the subsequent movement allowed requires an understanding of mechanics. The chapters in part II review key concepts of anatomy used in biomechanics and summarize key mechanical properties of the skeletal and neuromuscular systems. Part II lab activities show how biomechanics identifies the fascinating actions of muscles and joints in human movement.

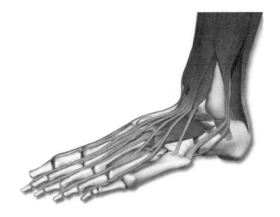

Image courtesy of Scott Barker, ATC.

CHAPTER 3

Anatomical Description and Its Limitations

In order to understand the origins of human movement, it is essential to understand anatomy. **Anatomy** is the study of the structure of the human body. Anatomy provides essential labels for musculoskeletal structures and joint motions relevant to human movement. Knowledge of anatomy also provides a common "language" of the human body and motions for kinesiology and medical professionals. Anatomy is an important prerequisite for kinesiology professionals trying to improve movement, prevent or treat injury. Anatomy is primarily a descriptive science and is not, by itself, enough to *explain the function* of the musculoskeletal system in movement. Knowledge of anatomy must be combined with biomechanics to accurately determine the musculoskeletal causes or the "how" human movement is created. This chapter reviews key anatomical concepts, shows how functional anatomy traditionally classifies muscle actions, shows how biomechanics is needed to determine muscle function in movement, and discusses the first two of the nine principles of biomechanics: **Range of Motion** and **Force–Motion**.

REVIEW OF KEY ANATOMICAL CONCEPTS

This section reviews several key concepts from human anatomy. A course in gross anatomy (macroscopic structures) is a typi-

cal prerequisite for the introductory biomechanics course. This section does not review all the bones, muscle, joints, and terms. Students and kinesiology professionals must continuously review and refresh their knowledge of anatomy. Anatomy describes the human body relative to the *anatomical position*. The anatomical position is approximated in Figure 3.1. The three spatial dimensions of the body correspond to the three anatomical planes: frontal, sagittal, and transverse. Recall that a plane of motion is a particular spatial direction or dimension of motion, and an axis is an imaginary line about which a body rotates. The anatomical axes associated with motion in each of these planes are the antero-posterior, medio-lateral, and longitudinal axes. Knowing these planes and axes is important to understanding medical descriptions of motion or movements. Even more important may be the functional implications of the orientation of these axes to the planes of motion they create. Note that motion in a particular plane (for example, sagittal) occurs by rotation about an axis oriented 90° (medio-lateral axis) to that plane. A person supinating their forearm to illustrate the anatomical position is creating motion in a transverse plane about a longitudinal axis roughly along the forearm. Functional anatomy applies knowledge of joint axes of rotation and muscle positions to hypothesize which muscles contribute to motion in an anatomical plane.

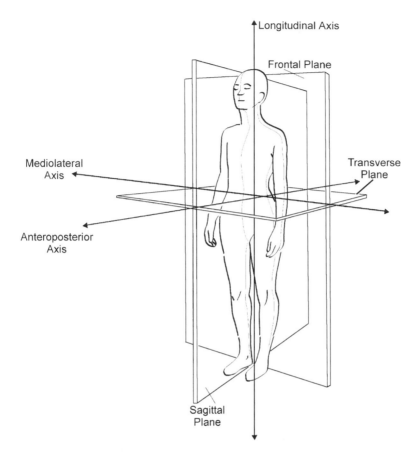

Figure 3.1. The major anatomical planes of motion, and axes of rotation.

Directional Terms

In addition to planes and axes, anatomy uses several directional terms to help describe the position of structures relative to the anatomical position. Toward the head is called superior, while toward the feet is inferior. Body parts toward the front of the body are anterior and objects to the back are in the posterior direction. Parts or motion toward the midline of the body are said to be medial, while motion or position toward the sides of the body are lateral. There are many other anatomical terms that have similar meanings as these but retain the original Latin or Greek form of classical anatomy. For example, superior is synonymous with cephalic, while inferior is the same a caudal. This book will use the more familiar English anatomical terms whenever possible.

Students with interests in sports medicine careers would do well to keep a medical dictionary handy and become familiar with the variety of classical anatomical terms used in medicine. Careful use of terminology is important in science and professions to prevent confusion. One example of the confusion that can occur with using unfamiliar Greek or Latin terms is the debate over the directional terms *valgus* and *varus*. The original Greek meanings of these terms (valgus [bowlegged] and varus [knock-kneed]) can be at odds with their typical use in orthopaedic medicine. Medicine usually defines genu (knee) valgus as an inward deviation of the knee joint, resulting in a knock-kneed appearance (Figure 3.2). Genu varus or varum usually corresponds to an outward deviating knee, which results in a bowlegged appearance. This leads to considerable confusion in describing anatomical abnormalities, and some have suggested that these terms be dropped or at least defined every time they are used (Houston & Swischuk, 1980). Some would look at Figure 3.2 and say the knee deviates medially, while others would say the lower leg deviates laterally. We will

see that this little problem of anatomical description is very similar to the multiple kinematic frames of reference (chapter 5) that are all correct descriptions of a single motion and the different units of measurement that can be used. Mechanics and anatomy both share the minor problem that there are several standards that have grown up with these sciences since people all over the world have been working on these same problems. Students should strive to read and write with special attention to the meaning of professional/scholarly terminology.

Joint Motions

Anatomy also has specific terminology describing the major rotations of bones at joints. "Flexion" refers to a decrease in joint angle in the sagittal plane, while "extension" is motion increasing joint angle (Figure 3.3a). Motion into the extremes of the range of motion are often noted as "hyper," as in hyperextension. Motion of a segment away from the midline in the frontal plane is "abduction," while movement back toward the midline is called "adduction" (Figure 3.3b). Joint motions in the transverse plane are usually called inward rotation (rotation of the anterior aspect of the segment toward the midline) and outward rotation (Figure 3.4). Some examples of special joint motion terms are "pronation," which refers to internal rotation of the forearm at the radioulnar joint, or "horizontal adduction," which is drawing the shoulder (glenohumeral joint) toward the midline in a transverse plane. Like the directional terms, anatomical terminology related to the rotations of joints is also used incorrectly. It is incorrect to say "a person is flexing a muscle" because flexion is a joint movement. It is important for kinesiology majors to use anatomical terms correctly. Refer to your anatomy book frequently to keep all

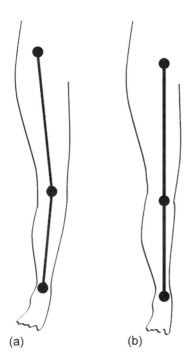

(a)   (b)

Figure 3.2. Orthopedic and pediatric medicine often calls the lower extremity deviation in (a) genu (knee) valgus because the distal segment (lower leg) deviates laterally from the midline of the body. Normal leg orientation in the frontal plane is illustrated in (b). The use of valgus and varus terminology is often inconsistent in the literature and should be clearly defined when used (Houston & Swischuk, 1980).

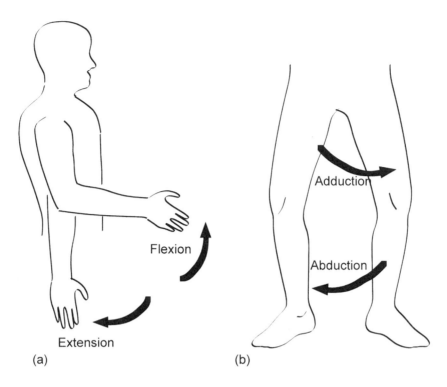

Figure 3.3. (a) Flexion and extension movements occur in a sagittal plane about a mediolateral axis; (b) adduction/abduction of the hip joint occurs in a frontal plane about an anteroposterior axis.

the joint motion terminology (this section does not review them all) fresh in your mind.

While there are attempts to standardize anatomical description throughout the world (Federative Committee on Anatomical Terminology, 1998), there remain regional inconsistencies in terminology. For example, some refer to the frontal plane as the "coronal" plane. Applied sciences such as medicine often develop specialized terms that are borrowed from anatomy, but that go against anatomical convention. A good example is related to how the foot acts during the stance phase of running. Medical and biomechanical studies have adopted the terms "pronation" and "supination" to refer to the complex triplanar actions of the subtalar joint. In normal running the foot strikes the ground on the lateral aspect

of the foot; the combined anatomical actions of eversion, plantar flexion, and abduction in the first part of stance is called pronation. This pronation serves to absorb the shock of the collision of the foot with the ground (Figure 3.5). The opposite motion (supination) stiffens the foot for the push off phase of stance. Here is another example of how anatomical terms are not always used in a consistent way. In your studies of biomechanics and other kinesiology disciplines, remember that adaptations and variations in anatomical terminology make it important to read carefully and often check background information. Modern biomechanical studies often assume quite a bit about reader expertise in the area and may cite sources giving necessary terminology and background information. This saves journal space but places a burden on

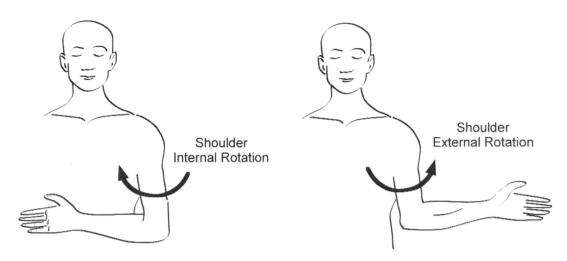

Figure 3.4. Inward and outward rotation of the shoulder joint occurs in a transverse plane about a longitudinal axis.

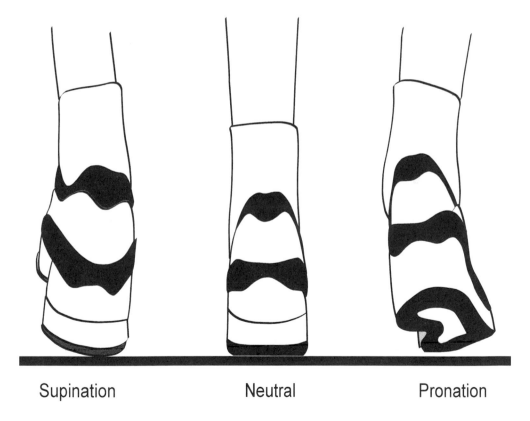

Supination   Neutral   Pronation

Figure 3.5. Frontal plane view of rear-foot motion in the first half of the stance phase of running. The foot lands in a supinated position. The motion of the foot and ankle to accommodate to the surface and absorb shock is called *pronation*.

the kinesiology professional to be knowledgeable about variations in descriptive terminology.

Review of Muscle Structure

The anatomical structure and microstructure of skeletal muscle has considerable functional importance. We will see later that the function of the complex structures of skeletal muscle can be easily modeled as coming from active and passive sources. This section will review a few of the structural components of skeletal muscle that are believed to be important in these active and passive tissue properties.

Careful dissection of skeletal muscle shows that muscles are composed of many distinct bundles of muscle fibers called *fascicles*. In cutting across a piece of beef or chicken you may have noticed the tissue is in small bundles. The connective tissue sheath that surrounds the whole muscle, bundling the fascicles together, is called *epimysium* (meaning over/above the muscle). Each fascicle is covered by connective tissue called *perimysium*, meaning "around the muscle." There are hundreds of muscle fibers within a fascicle, and an individual fiber is essentially a muscle cell. Muscle fibers are also covered with connective tissue called *endomysium* (within the muscle). The gradual blending of these connective tissue components of muscle forms a distinct tendon or fuses with the calcified connective tissue, the periosteum of bones. A schematic of the macrostructure of skeletal muscle is shown in Figure 3.6.

The specific arrangement of fascicles has a dramatic effect on the force and range-of-motion capability of the muscle

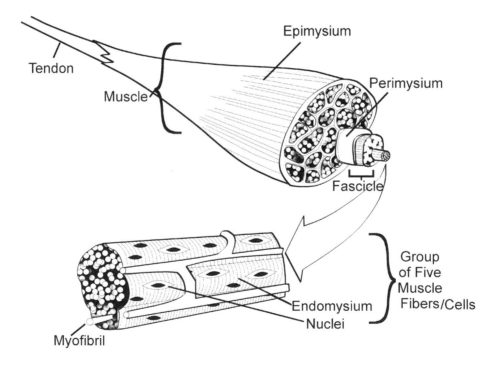

Figure 3.6. The macroscopic structure of muscle includes several layers of connective tissue and bundles of muscle fibers called fascicles. Muscle fibers (cells) are multinucleated and composed of many myofibrils.

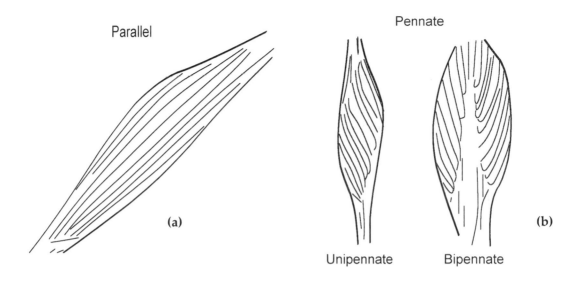

Figure 3.7. (a) Parallel arrangement of muscle fibers with the tendon favors range of motion over force. (b) Pennate arrangement of fibers are angled into the tendon and create greater force but less range of motion.

(Lieber & Friden, 2000). Anatomically, this fiber arrangement has been classified as either **parallel** or **pennate**. A **parallel** arrangement means that the muscle fascicles are aligned parallel to the long axis or line of pull of the muscle. Muscles like the rectus abdominis, sartorius, and biceps brachii have predominantly a parallel architecture (Figure 3.7a). **Pennate** muscles have fibers aligned at a small angle (usually less than 15°) to a tendon or **aponeurosis** running along the long axis of the muscle. An **aponeurosis** is a distinct connective tissue band within a muscle. This arrangement is called pennate because of the feathered appearance. The tibialis posterior and semimembranosus are primarily unipennate, while rectus femoris and gastrocnemius are bipennate (Figure 3.7b). An example of a multipennate muscle is the deltoid.

Muscles with parallel architecture favor range of motion over force development. The greater muscle excursion and velocity of parallel muscles comes from the greater number of sarcomeres aligned in se-

ries. The rectus abdominis can shorten from 1/3 to 1/2 of its length because of the parallel arrangement of fibers and fascicles. Small muscles may have a simple parallel design with fibers that run the length of the muscle, while larger parallel muscles have fibers aligned in series or end to end. These end-to-end connections and transverse connections within muscles make force transmission in muscle quite complex (Patel & Lieber, 1997; Sheard, 2000). Fiber architecture also interacts with the connective tissue within muscle to affect force or fiber shortening. The fibers in the center of the biceps do not shorten uniformly due to differences in the distal and proximal aponeurosis (Pappas, Asakawa, Delp, Zajac, & Draceet, 2002). The amount of tendon a muscle has and the ratio of tendon to fibers also affects the force and range-of-motion potential of a muscle.

In essence, pennate muscles can create a greater tension because of a greater physiological cross-sectional area per anatomical cross-sectional area, but have less range of shortening than a muscle with a parallel architecture. Physiological cross-sectional area is the total area of the muscle at right angles to the muscle fibers.

Muscle fibers are some of the largest cells in the body and are long cylindrical structures with multiple nuclei. A typical muscle cell is between 10 and 100 μm in diameter. The lengths of muscle fibers varies widely from a few centimeters to 30 cm long. Besides many nuclei there are hundreds to thousands of smaller protein filaments called **myofibrils** in every muscle fiber. If a muscle cell were to be imagined as a cylindrical straw dispenser, the myofibrils would be like the straws packed in this dispenser. Figure 3.8 illustrates the microstructure of a muscle fiber.

The microstructure of a muscle becomes even more fascinating and complex as you pull out a straw (myofibril), only to notice that there are even smaller threads or cylindrical structures within a myofibril. These many smaller fibers within each myofibril are all well organized and aligned with other adjacent myofibrils in a fiber. This is why looking at skeletal muscle under a light microscope gives the appearance of a consistent pattern of dark and light bands. This is how skeletal muscle came to be called *striated muscle* (Figure 3.8). These small sections of a myofibril between two Z lines (thin dark band) are called **sarcomeres**. Sarcomeres are the basic contractile structures of muscle.

Biomechanists model the active tension of whole muscles based on the behavior of the interaction of two contractile proteins in sarcomeres: **actin** and **myosin**. Actin is the thin protein filaments within the sarcomeres of a myofibril, and myosin the thicker protein filaments. Cross-bridges between myosin and actin are attached and detached with the chemical energy stored in adenosine triphosphate (ATP). You may be familiar with the names of the various zones (Z line, A band, and I band) and other substructures of a sarcomere.

While most biomechanists use simple models of the active tension of whole muscles, some biomechanists are interested in researching the mechanical behavior of the microstructures of myofibrils to increase our understanding of where active and passive forces originate. Considerable research is being done to understand muscle actions

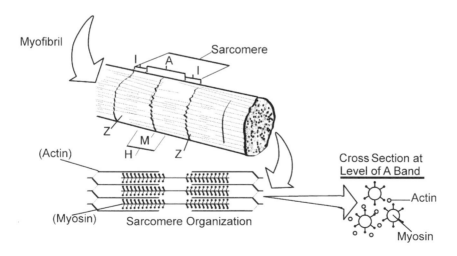

Figure 3.8. The microscopic structure of myofibril components of muscle fibers. Schematics of the sarcomere, as well as of the actin and myosin filaments are illustrated.

at this microscopic level from variations in myosin isoforms (Lutz & Lieber, 1999) to force transmission throughout the muscle fiber and muscle (Patel & Lieber, 1997; Sheard, 2000). Some muscle injuries could be due to this complex force production behavior and to nonuniform stresses in the sarcomeres of fibers (Morgan, Whitehead, Wise, Gregory, & Proske, 2000; Talbot & Morgan, 1996).

Many kinesiology students are familiar with muscular **hypertrophy** (increased muscle fiber diameter as a result of training), but they are unaware that chronic elongation of muscles (like in stretching) increases the number of sarcomeres in series within muscle fibers to increase their functional range of motion (Cox *et al.*, 2000; Williams & Goldspink, 1978). The number of sarcomeres and muscle fiber length are adaptable and strongly related to muscle performance (Burkholder, Fingado, Baron, & Lieber, 1994).

It is clear that biomechanics plays a role in understanding the functional significance of the gross and microstructural factors of the muscletendon unit. Most general concepts related to human movement, like muscular strength or range of motion, have many biomechanical factors and levels of structure that interact to determine how the concept actually affects movement. This is our first example of the paradox of learning: the more you know, the more you know what you don't know. Now that we have reviewed some of the major structural factors that affect muscle force and range of motion, let's define the kinds of actions muscles have.

MUSCLE ACTIONS

Muscle forces are the main internal motors and brakes for human movement. While gravity and other external forces can be used to help us move, it is the torques cre-

ated by skeletal muscles that are coordinated with the torques from external forces to obtain the human motion of interest. While some biomechanists are interested in the forces and motions created by smooth (visceral) or cardiac (heart) muscle, this text will focus on the actions of skeletal muscle that create human movement.

The activation of skeletal muscle has traditionally been called *contraction*. I will avoid this term because there are several good reasons why it is often inappropriate for describing what muscles actually do during movement (Cavanagh, 1988). Contraction implies shortening, which may only be accurate in describing the general interaction of actin and myosin in activated muscle. Contraction also conflicts with the many actions of muscles beyond shortening to overcome a resistance. Saying "eccentric contraction" is essentially saying "lengthening shortening"! Cavanagh suggests that the term "action" is most appropriate, and this book adopts this terminology. **Muscle action** is the neuromuscular activation of muscles that contributes to movement or stabilization of the musculoskeletal system. We will see that muscles have three major actions (eccentric, isometric, concentric) resulting from both active and passive components of muscle tension.

Mechanically, the three kinds of actions are based on the balance of the forces and torques present at any given instant (Figure 3.9). If the torque the activated muscles creates is exactly equal to the torque of the resistance, an isometric action results. A bodybuilder's pose is a good example of **isometric** muscle actions of opposing muscle groups. Recall that isometric literally means "same length."

A **concentric** action occurs when the torque the muscle group makes is larger than the torque of a resistance, resulting in muscle shortening. The upward lift of a dumbbell in an arm curl is the concentric phase of the exercise. In essence a

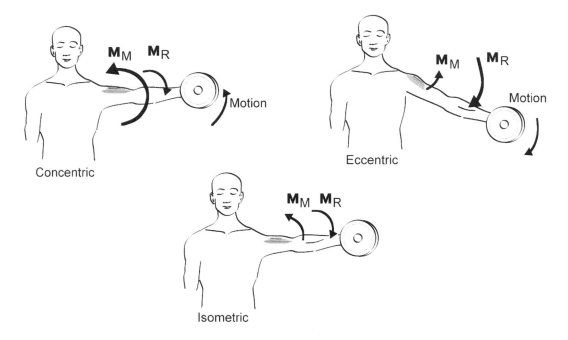

Figure 3.9. The three kinds of muscle action are determined by the balance of torques (moments of force: **M**). In concentric action the torque of the abductors ($\mathbf{M}_M$) is greater than the torque of the resistance ($\mathbf{M}_R$), so the arm rises. In isometric conditions the joint angle does not change because $\mathbf{M}_M$ and $\mathbf{M}_R$ are equal. In eccentric action $\mathbf{M}_M$ is less than $\mathbf{M}_R$, so the arm is lowered.

concentric action occurs when a muscle activation results in shortening of the muscle-tendon unit. When the lifter gradually lowers the weight in an arm curl, the torque the muscle group makes is less than the torque of the resistance. This lowering of the dumbbell is an **eccentric** muscle action or the lengthening of an activated muscle. In eccentric actions muscles are used as brakes on external forces or motion like the brakes of your car.

The importance of these different muscle actions cannot be overemphasized. Functional anatomical analysis and most people tend to focus primarily on the concentric actions of muscles. This overemphasis of what is usually in the minority of muscle actions for most movements gives a false impression of how muscles create human movement. The following section on the limits of functional anatomy will expand on this idea by showing that muscles

Application: Eccentric Actions and Muscle Injury

Eccentric actions are common to all muscles and virtually every human movement. Eccentric actions of high intensity, repetitive nature, or during fatigue are associated with muscle injury. When eccentrically active muscles are rapidly overcome by external forces, a *muscle strain* injury can occur. When people perform physical activity beyond typical levels, especially eccentric muscle actions, the result is usually *delayed-onset muscle soreness*. This is why it is important in conditioning to include both eccentric and concentric phases of exercises. Some athletic events would benefit from emphasis on eccentric training. For example, long jumpers and javelin throwers need strong eccentric strength in the takeoff and plant leg.

create movement in a variety of ways using all three muscle actions, not just concentric action.

Active and Passive Tension of Muscle

Activated muscles create forces by pulling about equally on all their attachments. This tensile force really has two sources: active and passive tension.

Active tension refers to the forces created between actin and myosin fibers in the sarcomeres of activated motor units. So active tension is the force created by the contractile proteins (actin and myosin) using chemical energy stored in ATP. This ability of muscles to create active tensile forces is unique compared to the connective tissue components (ligaments, tendons, bone) of the musculoskeletal system. The shape of this active tension potential of skeletal muscle is called the force–velocity relationship of muscle and is summarized in chapter 4.

Passive tension is the force that comes from an elongation of the connective tissue components of the muscletendon unit. When a person does a stretching exercise, the tension she feels in the muscles is the internal resistance of the muscletendon unit to the elongation of the stretch. This passive tension in stretching exercises can be quite large and may be responsible for the muscular weakness seen in muscles following stretching (Knudson, McHugh, & Magnusson, 2000). In the midranges of joint motion, passive tension does not significantly contribute to muscle forces in normal movement (Siegler & Moskowitz, 1984); however, it may be a factor in various neuromuscular disorders (Lamontagne, Malouin, & Richards, 2000). Muscle passive tension is a significant factor affecting movement at the extremities of joint range of motion. The increase in passive tension limiting

range of joint motion is quite apparent in multiarticular muscles and is called **passive insufficiency**. We will see in the following chapter that passive tension is an important component of the force–length relationship of muscle. The passive insufficiency of poor hamstring flexibility could lead to poor performance or risk of injury in activities that require combined hip flexion and knee extension, such as in a karate front kick (Figure 3.10). The passive tension in the hamstring muscles is high in Figure 3.10 because the muscle is simultaneously stretched across the hip and knee joint. We will learn later on in this chapter that the concept of range of motion is a complicated phenomenon that involves several mechanical variables.

Figure 3.10. The combined hip flexion and knee extension of a karate front kick may be limited by the passive insufficiency of the hamstring muscles. This technique requires excellent static and dynamic hamstring flexibility. Image courtesy of Master Steven J. Frey, 4th-Degree Black Belt.

Activity: Passive Tension

The effect of passive tension on joint motions can be felt easily in multi-joint muscles when the muscles are stretched across multiple joints. This phenomenon is called *passive insufficiency*. Lie down in a supine (face upwards) position and note the difference in hip flexion range of motion when the knee is flexed and extended. The hamstring muscle group limits hip flexion when the knee is extended because these muscles cross both the hip and the knee joints. Clinical tests like the straight-leg raise (Eksstrand, Wiktorsson, Oberg, & Gillquist, 1982), active knee extension (Gajdosik & Lusin, 1983), and the sit-and-reach (Wells & Dillon, 1952) all use passive insufficiency to evaluate hamstring static flexibility. Careful body positioning is required in flexibility tests because of passive insufficiency and other mechanical factors across several joints. Some aspects of this issue are explored in Lab Activity 3, which appears at the end of this volume.

Hill Muscle Model

One of the most widely used mechanical models of muscle that takes into account both the active and passive components of muscle tension is the three-component model developed by A. V. Hill in 1938 (Hill, 1970). Hill was an English physiologist who made substantial contributions to the understanding of the energetics (heat and force production) of isolated muscle actions. Hill was also interested in muscular work in athletics, and some of his experimental techniques represent ingenious early work in biomechanics (Hill, 1926, 1927). The **Hill muscle model** has two elements in series and one element in parallel (Figure 3.11). The contractile component (CC)

represents the active tension of skeletal muscle, while the parallel elastic component (PEC) and series elastic component (SEC) represent two key sources of passive tension in muscle. The Hill muscle model has been the dominant theoretical model for understanding muscle mechanics and is usually used in biomechanical computer models employed to simulate human movement.

We can make several functional generalizations about the mechanical behavior of muscle based on Figure 3.11. First, there is elasticity (connective tissue) in the production of active muscle tension modeled by the series elastic component. The source of this series elasticity is likely a mixture of the actin/myosin filaments, cross bridge stiffness, sarcomere nonuniformity, and other sarcomere connective tissue components. Second, the passive tension of relaxed muscle that is easily felt in stretching exercises or in passive insufficiency affects motion at the extremes of joint range of motion. The "p" in the parallel elastic component is a key for students to remember this as the primary source of passive tension in the Hill muscle model. Third, muscle tension results from a complex interaction of active and passive sources of tension. This third point can be generalized beyond the simple Hill muscle model as a result of recent research that has focused on the complex transmission of force within the connective tissue components of muscle (Patel & Lieber, 1997). Muscles may not create equal forces at their attachments because of force transmitted to extramuscular connective tissues (Huijing & Baan, 2001).

The separation of the passive tension into series and parallel components in the Hill model and the exact equations used to represent the elastic (springs) and contractile components are controversial issues. Whatever the eventual source and complexity of elastic tension, it is important to remember that the stretch and recoil of elas-

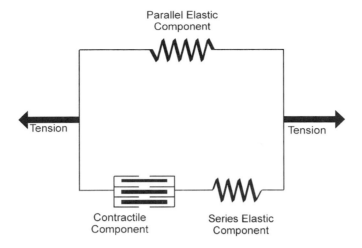

Figure 3.11. The Hill model of muscle describes the active and passive tension created by the MTU. Active tension is modeled by the contractile component, while passive tension is modeled by the series and parallel elastic components.

tic structures are an integral part of all muscle actions. It is likely that future research will increase our understanding of the interaction of active and passive components of muscle tension in creating human movement. There are many other complexities in how muscles create movement. The next section will briefly review the logic of functional anatomical analysis and how biomechanics must be combined with anatomy to understand how muscles create movement.

This section will show that such qualitative estimations of muscle actions are often incorrect. Similarly, many of the muscle actions hypothesized by coaches and therapists from subjective observation of movement are not correct (Herbert, Moore, Moseley, Schurr, & Wales, 1993). Kinesiology professionals can only determine the true actions of muscle by examining several kinds of biomechanical studies that build on anatomical information.

THE LIMITATIONS OF FUNCTIONAL ANATOMICAL ANALYSIS

Anatomy classifies muscles into functional groups (flexors/extensors, abductors/adductors, etc.) based on hypothesized actions. These muscle groups are useful general classifications and are commonly used in fitness education, weight training, and rehabilitation. These hypothesized muscle actions in movements and exercises are used to judge the relevance of various exercise training or rehabilitation programs.

Mechanical Method of Muscle Action Analysis

Functional anatomy, while not an oxymoron, is certainly a phrase that stretches the truth. Functional anatomy classifies muscles actions based on the mechanical method of muscle action analysis. This method essentially examines one muscle's line of action relative to one joint axis of rotation, and infers a joint action based on orientation and pulls of the muscle in the anatomical position (Figure 3.12). In the sagittal plane, the biceps brachii is classi-

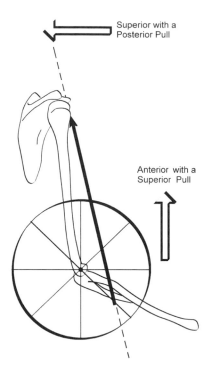

imately equally so that which end moves (if one does at all) depends on many biomechanical factors. Recall that there are three kinds of muscle actions, so that what the biceps brachii muscle does at the elbow in a particular situation depends on many biomechanical factors this book will explore.

While the biceps is clearly an elbow flexor, this analysis assumes quite a bit and does not take into consideration other muscles, other external forces, and the biarticular nature of the biceps. The long head of the biceps brachii crosses the shoulder joint. What if the movement of interest was the eccentric phase of the pull-over exercise (Figure 3.13), where the shoulder was the origin because the elbow angle essentially did not change while shoulder flexion and extension were occurring? It is not entirely clear if the long head of the biceps is in iso-

Figure 3.12. The mechanical method of muscle action analysis applied to biceps and elbow flexion in the sagittal plane. It is assumed that in the anatomical position the biceps pulls upward toward its anatomical origin from its anatomical insertion (radial tuberosity). The motion can be visualized using a bicycle wheel with the axle aligned on the joint axis. If the the muscle were pulling on the wheel from its illustrated direction and orientations relative to that joint axis (visualize where the line of action crosses the medial-lateral and superior-inferior axes of the sagittal plane), the wheel would rotate to the left, corresponding to elbow flexion. Unfortunately, the actions of other muscles, external forces, or other body positions are not accounted for in these analyses. More thorough and mathematical biomechanical analyses of the whole body are required to determine the true actions of muscles.

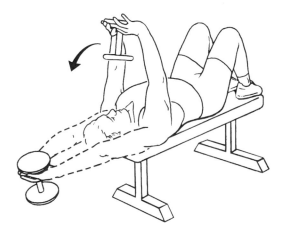

Figure 3.13. In the eccentric phase of the pullover exercise, the motion primarily occurs at the shoulder joint, with the elbow angle remaining unchanged. The isolated mechanical method of muscle action does not help in this situation to determine if the long head biceps (crossing both the elbow and shoulder joints) is isometrically active, concentrically active, or inactive. Do you think a biarticular muscle like the biceps can be doing two kinds of muscle actions at once? We will see later that extensive kinetic biomechanical models and EMG research must be combined to determine the actual action of muscles in many movements. Image courtesy of VHI Kits, Tacoma, WA.

fied as an elbow flexor because it is assumed that (1) the origins are at the shoulder joint, (2) the insertion is on the radial tuberosity, and (3) the anterior orientation and superior pull, as well as the superior orientation and posterior pull, would create elbow flexion. When a muscle is activated, however, it pulls both attachments approx-

metric or concentric action in this pull-over exercise example. Biomechanical data and analysis are necessary to determine the actual actions of muscles in movement. There are even cases where muscles accelerate a joint in the opposite direction to that inferred by functional anatomy (Zajac, 1991; Zajac & Gordon, 1989).

Rather invasive biomechanical measurements are usually required to determine exactly what muscle actions are occurring in normal movement. Many studies con-

ducted on animals have shown that muscles often have surprising and complex actions (see Biewener, 1998; Herzog, 1996a,b). One such study of the turkey (Roberts *et al.*, 1997) gastrocnemius (plantar flexor) found the muscle acted in essentially an isometric fashion in the stance phase of level running (Figure 3.14A), while concentric actions were used running uphill (Figure 3.14B). The invasive nature of these kinds of measurements and the interesting variations in the musculoskeletal structure of animals

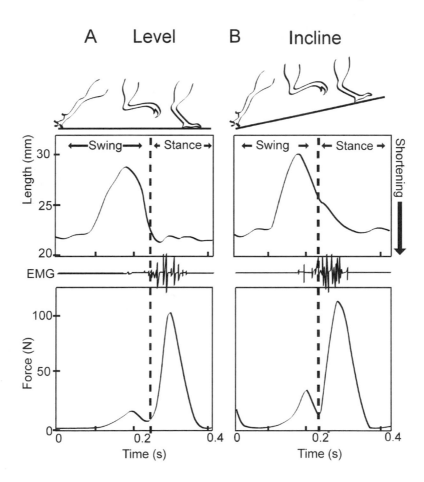

Figure 3.14. Simultaneous muscle force, length, and activation (EMG) measurements of the gastrocnemius of running turkeys. (A) Note that in level running the muscle creates considerable force but the fibers do not shorten, so the muscle is in isometric action and length changes are in the stretching and recoiling of the tendon. (B) in the stance phase of uphill running, the muscle fibers shorten (concentric action), doing mechanical work to lift the turkey's body. Reprinted with permission from Roberts *et al.* (1997). Copyright © 1997 American Association for the Advancement of Science.

(fish, kangaroo rats, wallabies; Biewener, 1998; Griffiths, 1989; Shadwick, Steffensen, Katz, & Knower, 1998) makes animal studies a major area of interest for the biomechanics of muscle function. Similar complex behavior of muscle actions has been observed in humans using recent improvements in ultrasound imaging (Finni, Komi, & Lepola, 2000; Finni *et al.*, 2001: Fukunaga, Ichinose, Ito, Kawakami, & Fukashiro, 1997; Fukunaga, Kawakami, Kubo, & Keneshisa, 2002; Kubo, Kawakami, & Fukunaga, 1999) and implantable fiberoptic force sensors (Komi, Belli, Huttunen, Bonnefoy, Geyssant, & Lacour, 1996). Recent studies of the human tibialis anterior have also documented nonlinear and nonisometric behavior of the muscle (lengthening of tendon and aponeurosis while fibers shorten) in isometric actions (Maganaris & Paul, 2000; Ito *et al.*, 1998). It is clear now that muscle actions in animal movements are more complicated than can be predicted by the concentric, single-joint analysis of functional anatomy.

Given these many examples of the complexity of muscle actions at the macro and microscopic levels, the hypothesized muscle actions from functional anatomy in many human movements should be interpreted with caution. Seemingly simple questions of what muscles contribute most to walking, jumping, or any movement represent surprisingly complex biomechanical issues. For example, should the word "eccentric" be used as an adjective to describe phases in weight training exercise (eccentric phase), when all the active muscles are clearly not in eccentric actions in the movement? If the active muscle group, body position, and resistance are well defined, this terminology is likely accurate. When the lifter "cheats" with other muscles in the exercise, modifies exercise technique, or performs a similar sporting movement, the eccentric adjective may not be accurate. The actions of other muscles, external forces like gravity, and the complexity of the musculoskeletal system can make the isolated analyses of functional anatomy in the anatomical position inaccurate for dynamic movement. Some biomechanical issues that illustrate this point are summarized here and developed throughout the book.

Interdisciplinary Issue: Anthropometry

Anthropometry is the science concerned with measurement of the physical properties (length, mass, density, moment of inertia, etc.) of a human body. Kinanthropometry is an area within kinesiology that studies how differences in anthropometry affect sport performance (see chapters 5 and 7 in Bloomfield, Ackland, & Elliott, 1994). The main organization in this area is the International Society for the Advancement of Kinanthropometry (ISAK). Since humans move in a wide variety of activities, many professionals use anthropometric data. Engineers use these measurements to design tools and workstations that fit most people and decrease risk of overuse injuries. Prosthetic and orthotic manufacturers often make anthropometric measurements on individuals to customize the device to the individual. Motor development scholars track the changes in anthropometric characteristics with growth and development. While people seem to have a wide variety of shapes and sizes, the relative (scaled to size) size of many anthropometric variables is more consistent. Biomechanists use many of these average physical measurements to make quite accurate kinetic or center-of-gravity calculations.

The Need for Biomechanics to Understand Muscle Actions

The traditional "kinesiological" analysis of movements of the early twentieth century essentially hypothesized how muscles contributed to motion in each phase of the skill by noting anatomical joint rotations and assuming muscles that create that joint rotation are active. Muscle actions in actual human movements, however, are not as simple as functional anatomy assumes (Bartlett, 1999). Several kinds of biomechanical research bear this out, and show that the combination of several kinds of quantitative biomechanical analysis are necessary to understand the functions of muscles in movements.

First, **electromyographic (EMG)** studies have documented general trends in activation of muscles in a particular muscle group, but with considerable potential variation in that trend or in activation between subjects (Basmajian & De Luca, 1985). The primary source of this variation may be the considerable redundancy (muscles with the same joint actions) of the muscular system. Nearly identical movements can be created by widely varying muscular forces or joint torques (Hatze, 2000; Patla, 1987; Winter, 1984).

EMG studies show that the activation patterns of individual muscles are not representative of all muscles in the same functional group (Arndt, Komi, Bruggemann, & Lukkariniemi, 1998; Bouisset, 1973), and muscles are quite sophisticated, with different motor unit activation depending on the task or muscle action (Babault, Pousson, Ballay, & Van Hoecke, 2001; Enoka, 1996; Gandevia, 1999; Gielen, 1999). Muscles within a muscle group can alternate periods of activity in low-level activities to minimize fatigue (Kouzaki, Shinohara, Masani, Kanehisa, & Fukunaga, 2002). Muscle activation can vary because of differences in joint angle, muscle action (Kasprisin & Grabiner, 2000; Nakazawa, Kawakami, Fukunaga, Yano, & Miyashita, 1993) or the degree of stabilization required in the task (Kornecki, Kebel, & Siemienski, 2001). For example, a manual muscle test for the bi-

ceps used by physical therapists uses isometric elbow flexion with the forearm in supination to minimize brachioradialis activity and maximize biceps activity (Basmajian and De Luca, 1985). Recent EMG studies, however, have also demonstrated that some of these procedures used to isolate specific muscles in physical therapy do not always isolate the muscle hypothesized as being tested (see Kelly, Kadrmas, & Speer, 1996; Rowlands, Wertsch, Primack, Spreitzer, Roberts, Spreitzer, & Roberts, 1995).

The activation of many muscles to create a specific force or action is called a muscle **synergy**. A muscle **synergy** is a combination of muscle actions that serves to optimally achieve a motor task. There is considerable recognition of the importance of muscle synergies and force sharing of muscles in biomechanical research (Arndt *et al.*, 1998; Herzog, 1996b, 2000) and in current rehabilitation and conditioning trends (see Interdisciplinary Issue on training muscles versus movements). How individual muscles share the load is complicated, depending on fiber type, contractile properties, cross-sectional area, moment arm, and antagonism (Ait-Haddou, Binding, & Herzog, 2000). Motor control uses the term *synergy* to refer to underlying rules of the neuromuscular system for using muscles to coordinate or create movements (Aruin, 2001; Bernstein, 1967).

Activity: Muscle Synergy

Make a tight fist in your dominant hand as forcefully and quickly as you can. Observe the actions of the superficial muscles of your arm. Why do you think biceps and triceps are isometrically activated in a power grip muscle synergy?

Recent EMG research has in addition begun to focus on different activation of intramuscular sections within a muscle beyond the traditional gross segmentation in classical anatomy (Mirka, Kelaher, Baker, Harrison, & Davis, 1997; Paton & Brown, 1994; Wickham & Brown, 1998). Wickham and Brown (1998) have confirmed different activation of seven distinct segments of the deltoid muscle, rather than the typical three sections (anterior, intermediate, posterior) of muscle fibers usually identified in anatomy. This line of research supports the EMG studies mentioned earlier which indicate that activation of muscles is much more complex than had been previously thought. Further microanatomy and EMG research on muscles, particularly those with large attachments, will most likely increase our understanding of how parts of the muscles are activated differently to create movement.

Second, the descriptions of musculoskeletal anatomy often do not account for variations in muscle attachment sites across individuals. The numbers and sites of attachments for the rhomboid and scalene muscles vary (Kamibayashi & Richmond, 1998). A person born with missing middle and lower fibers of trapezius on one side of their body must primarily rely on rhomboids for scapular retraction. Variations in skeletal structure are also hypothesized to contribute to risk of injury. For example, the shape of the acromion process of the scapula is believed to be related to a risk of impingement syndrome (Whiting & Zernicke, 1998). The role of anatomical variation in gross anatomy or in muscle architecture (Richmond, 1998) and their biomechanical effects of muscles actions and injury risk remain an important area of study.

Third, the linked nature of the human body makes the isolated functional anatomical analysis incomplete. This linking of body segments means that muscle actions have dramatic effects on adjacent and other joints quite distant from the ones the muscles cross (Zajac, 1991; Zajac & Gordon, 1989). This redistribution of mechanical energy at distant joints may be more important to some movements than the traditional joint action hypothesized by functional anatomy (Zajac, Neptune, & Kautz, 2002). Zajac and Gordon (1989), for example, showed how soleus activity in a sit-to-stand movement tends to extend the knee joint more than it plantar flexes the ankle joint. Physical therapists know that the pectoralis major muscle can be used to extend the elbow in closed kinetic chain (see chapter 6) situations for patients with triceps paralysis (Smith, Weiss, & Lehmkuhl, 1996). Functional anatomy does not analyze how forces and torques created by a muscle are distributed throughout all the joints of the skeletal system or how these loads interact between segments. Zajac and Gordon (1989) have provided a convincing argument that the classification of muscles as **agonists** or **antagonists** should be based on biomechanical models and joint accelerations, rather than torques the muscles create.

Dramatic examples of this wide variety of effects of muscles can be seen in multiarticular muscles (van Ingen Schenau *et al.*, 1989; Zajac, 1991). There is considerable interest in the topic of biarticular or multiarticular muscles, and it is known that they have different roles compared to similar monoarticular muscles (Hof, 2001; Prilutsky & Zatsiorsky, 1994; van Ingen Schenau *et al.*, 1995). Another example of the complexity of movement is how small differences in foot placement (angle of ankle plantar/dorsiflexion) dramatically affects which joint torques are used to cushion the shock in landing (DeVita & Skelly, 1992; Kovacs, Tihanyi, DeVita, Racz, Barrier, & Hortobagyi, 1999). A flat-footed landing minimizes a plantar flexor's ability to absorb shock, increasing the torque output of the hip and knee extensors. Small differences in foot angle in walking also affect the flexor

Interdisciplinary Issue: Training Muscles vs. Movements

In the strength and conditioning field a recent area of philosophical debate is related to a greater emphasis on training functional movements rather than training specific muscle groups (Gambetta, 1995, 1997). This debate is quite similar to the debate about the relative benefits of training with free weights or with machines. Training with free weights can more easily simulate the balance and stabilizing muscle actions in normal and sport movements. The advantage of machines is that they provide more muscle group-specific training with resistance that is not as dependent on position relative to gravity as free weights are. How might rehabilitation and conditioning professionals use biomechanics and EMG research to help match training to the demands of normal movement?

the model. The more simple the model, the easier the interpretation and application of results. For example, models of the motion of body segments in airborne skills in gymnastics and diving are quite effective in determining their effect on flight and rotation (Yeadon, 1998). As biomechanics models get more complicated and include more elements of the musculoskeletal system, the more difficult it is to validate the model. Interpretation is even complicated because of the many interrelated factors and variations in model parameters across subjects (Chow, Darling, & Ehrhardt, 1999; Hubbard, 1993).

Despite the many controversial issues in biomechanical modeling, these kinds of studies show that the actions of muscles in movements are quite complex and are related to segment and muscle geometry (Bobbert & van Ingen Schenau, 1988; Doorenbosch, Veeger, van Zandwij, & van Ingen Schenau, 1997), muscle elasticity (Anderson & Pandy, 1993), coordination (Bobbert & van Soest, 1994; Hatze, 1974; Nagano & Gerritsen, 2001), and accuracy or injury constraints (Fujii & Hubbard, 2002). It is possible that coordination in a movement even varies slightly across people because of differences in muscle mechanics (Chowdhary & Challis, 2001). Here we have the paradox of learning again. What muscles do to create movement is quite complex, so kinesiology scholars and professionals must decide what level of biomechanical system to study to best understand movement. The strength and conditioning field commonly groups muscles into functional groups like the knee extensors (quadriceps) or knee flexors (hamstrings). Whatever movements or level of analysis a kinesiology professional chooses, biomechanics needs to be added to anatomical knowledge to make valid inferences about human movement. The next section briefly shows how the sports medicine professions

or extensor dominance of the knee torque (Simonsen, Dyhre-Poulsen, Voigt, Aagaard, & Fallentin, 1997). The kinematics and kinetics chapters (5 and 6 & 7, respectively) will expand on the effects of the joints and segment actions in human movement.

The fourth line of biomechanical research documenting the complexity of muscular actions creating movement are **modeling** and **simulation**. **Modeling** involves the development of a mathematical representation of the biomechanical system, while **simulation** uses biomechanical models to examine how changes in various techniques and parameters affect the movement or body. Biomechanical models of the human body can be used to simulate the effects of changes in any of the parameters of

have integrated more biomechanical information into their professional practice.

Application: Muscle Groups

If muscles create movement in complex synergies that are adaptable, should kinesiology professionals abandon the common practice of naming muscle groups according to anatomical function (quads [knee extensors] or calf [ankle plantar flexors])? Such an extreme reaction to the complexity of biomechanics is not necessary. This common terminology is likely appropriate for prescribing general strength and conditioning exercises. It may even be an appropriate way to communicate anatomical areas and movements in working with athletes knowledgeable and interested in performance. Kinesiology professionals do need to qualitatively analyze movements at a deeper level than their clients, and remember that this simplified terminology does not always give an accurate picture of how muscles really act in human movement. Biomechanical and other kinesiology research must be integrated with professional experience in qualitatively analyzing movement.

Sports Medicine and Rehabilitation Applications

Musculoskeletal anatomy and its motion terminology are important in kinesiology and sports medicine, but it cannot be the sole basis for determining the function of muscles in human movement. Medical doctors specializing in sports medicine found that their extensive training in anatomy was not enough to understand injuries and musculoskeletal function in the athletes they treated (McGregor & Devereux, 1982).

This recognition by MDs that their strong knowledge of anatomy was incomplete to understand function and that they needed the science of kinesiology was a factor in the fusion of medical and kinesiology professionals that formed the American College of Sports Medicine (ACSM).

Today, many kinesiology students prepare for careers in medicine- and sports medicine-related careers (athletic training, physical therapy, orthotics, prosthetics, strength & conditioning). These professions are concerned with analyzing the actions of muscles in movement. Where can sports medicine professionals (athletic trainers, physical therapists, physical medicine, strength and conditioning) get the most accurate information on the biomechanical function of specific areas of the human body? Fortunately, there are several sources that strive to weigh the anatomical/clinical observations with biomechanical research. These sources focus on both normal and pathomechanical function of the human body. The following sources are recommended since they represent this balanced treatment of the subject, not relying solely on experience or research (Basmajian & Wolf, 1990; Kendall, McCreary, & Provance, 1993; Smith, Weiss, & Lehmkuhl, 1996).

It is important to remember that biomechanics is an indispensable tool for all kinesiology professionals trying to understand how muscles create movement, how to improve movement, and how problems in the musculoskeletal system can be compensated for. The last two sections of this chapter illustrate how biomechanical principles can be used to understand and improve human movement.

RANGE-OF-MOTION PRINCIPLE

One area where anatomical description is quite effective is in the area of the range of motion used in movement. Movement can be accurately described as combinations of joint angular motions. Remember that the biomechanical principle of range of motion, however, can be more generally defined as any motion (both linear or angular) of the body to achieve a certain movement goal. Specific joint motions can be of interest, but so too can the overall linear motions of the whole body or an extremity. Coaches can speak of the range of motion of a "stride" in running or an "approach" in the high jump. Therapists can talk about the range of motion for a joint in the transverse plane.

In human movement the performer can modify the number of joints, specific anatomical joint rotations, and amount of those rotations to tailor range of motion. Range of motion in movement can be imagined on a continuum from negligible motion to 100% of the physically possible motion. The **Range-of-Motion Principle** states that less range of motion is most effective for low-effort (force and speed) and high-accuracy movements, while greater range of motion favors maximum efforts related to speed and overall force production (Hudson, 1989). A person playing darts "freezes" or stabilizes most of the joints of the body with isometric muscle actions, and limits the dart throw to a small range of motion focused on elbow and wrist. The javelin thrower uses a long running approach and total body action to use considerable range of motion to maximize the speed of javelin release. The great accuracy required in golf putting favors limiting range of motion by using very few segments and limiting their motion to only what is needed to move the ball near the hole (Figure 3.15).

The application of the range-of-motion principle is more complicated when the effort of the movement is not maximal and when the load cannot be easily classified at the extreme of the continuum. A baseball or softball seems pretty light, but where on the range-of-motion continuum are these inter-

Figure 3.15. Very accurate movements like putting in golf limit range of motion by freezing most segments and using only a few segments. Photo courtesy of Getty Images.

mediate load activities? How much range of motion should you use when the load is a javelin, a shot, or your bodyweight (vertical jump)? Biomechanical studies can help kinesiology professionals decide how much range of motion is "about right." In the qualitative analysis of movement, this approach of identifying a range of correctness (like in range of motion) is quite useful because the professional can either reinforce the performer's good performance, or suggest less or more range of motion be used (Knudson & Morrison, 2002). The continuum of range of motion can also be qualitatively evaluated as a sliding scale (Knudson, 1999c) or volume knob (Hudson, 1995) where the performer can be told to fine tune range of motion by feedback (Figure 3.16). Let's look at how biomechanical research can help professionals evaluate the range of motion in a vertical jump.

The amount and speed of countermovement in a vertical jump is essential to a high jump. This range-of-motion variable

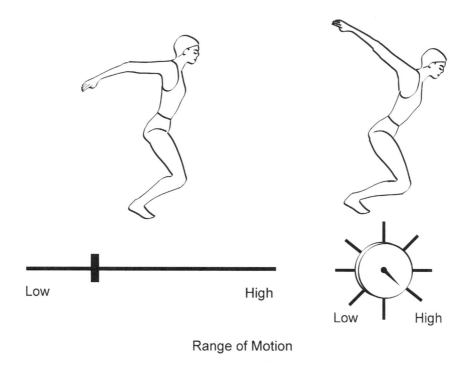

Range of Motion

Figure 3.16. Range of motion can be evaluated and pictured as an analog scale or a volume knob. If a change in range of motion is appropriate, the performer can be instructed to "increase" or "decrease" the range of motion in their movement.

can be expressed as a linear distance (drop in center of mass as percentage of height) or as body configuration, like minimum knee angle. We use the knee angle in this example because it is independent of a subject's height. One can hypothesize that maximizing the drop (range of motion) with a small knee angle in the countermovement would increase the height of the jump; however, this is not the case. Skilled jumpers tend to have minimum knee angles between 90 and 110° (Ross & Hudson, 1997). The potential benefits of range of motion beyond this point seems to be lost because of poorer muscular leverage, change in coordination, or diminishing benefits of extra time to apply force. The exact amount of countermovement will depend on the strength and skill of the jumper, but coaches can generally expect the knee angles in this range.

Another example of the complexity of applying the range-of-motion principle would be the overarm throw. In overarm throwing the athlete uses range of motion from virtually the entire body to transfer energy from the ground, through the body and to the ball. The range of motion (kinematics) of skilled overarm throwing has been extensively studied. Early motor development studies show that one range-of-motion variable (the length of the forward stride is usually greater than 50% of height) is important in a mature and forceful overarm throw (Roberton & Halverson, 1984). Stride length in throwing is the horizontal distance from the rear (push-off) foot to the front foot. This linear range of motion from leg drive tends to contribute 10 to 20% of the ball speed in skilled throwers (Miller, 1980). The skill of baseball pitching uses more stride range of motion, usually between 75 and 90% of standing height, and has been shown to significantly affect pitch speed (Montgomery & Knudson, 2002).

The axial rotations of the hips and trunk are the range-of-motion links between stride and arm action. The differentiation of hip and trunk rotation is believed to be an important milestone in mature throwing (Roberton & Halverson, 1984), and these movements contribute about 40 to 50% of ball speed in skilled throwers. In coaching high-speed throwing, coaches should look for hip and trunk opposition (turning the non-throwing side toward the target) in preparation for the throw. The optimal use of this range of motion is a coordination and segmental interaction issue that will be discussed later. Students interested in the skilled pattern of hip and trunk range of motion should look at the research on skilled pitchers (Fleisig, Barrentine, Zheng, Escamilla, & Andrews, 1999; Hong & Roberts, 1993).

Arm action is the final contributor to the range of motion used in overarm throwing. The complex joint actions of throwing contribute significantly (30–50% of ball velocity) to skilled throwing (Miller, 1980). To take advantage of the trunk rotation, the shoulder stays at roughly 90° of abduction to the spine (Atwater, 1979) and has been called the strong throwing position (Plagenhoef, 1971). With initiation of the stride, the elbow angle stays near 90° to minimize resistance to rotating the arm, so the major increase in ball speed is delayed until the last 75 ms (a millisecond [ms] is a thousandth of a second) before release (Roberts, 1991). Contrary to most coaching cues to "extend the arm at release," the elbow is typically 20° short of complete extension at release to prevent injury (Fleisig *et al.*, 1999). Inward rotation of the humerus, radioulnar pronation, and wrist flexion also contribute to the propulsion of the ball (Roberts, 1991), but the fingers usually do

not flex to add additional speed to the ball (Hore, Watts, & Martin, 1996).

In overarm throwing it appears that the range-of-motion principle can be easily applied in some motions like stride length using biomechanical research as benchmarks; however, it is much more difficult to define *optimal* amounts of joint motions or body actions in complex movements like overarm throwing. How range of motion might be changed to accommodate different level of effort throws, more specific tasks/techniques (e.g., curveball, slider), or individual differences is not clear. Currently, professionals can only use biomechanical studies of elite and skilled performers as a guide for defining desirable ranges of motion for movements. More data on a variety of performers and advances in modeling or simulation of movement are needed to make better recommendations on how modifications of range of motion may affect movement.

FORCE–MOTION PRINCIPLE

Another way to modify human movement is to change the application of forces. The **Force–Motion Principle** states that it takes unbalanced forces (and the subsequent torques they induce) to create or modify our motion. To know what size and direction of force to change, recall that a **free-body diagram** of the biomechanical system is usually employed. A major limitation of functional anatomical analysis was the limited nature of the forces and structures being considered. We are not in a position to perform quantitative calculations to determine the exact motion created at this point in the text, but this section will provide examples of the qualitative application of the Force–Motion Principle in improving human movement. Later on, in chapters 6 and 7, we will explore Newton's laws of motion and the major quantitative methods used in biomechanics to explore the forces that create human movement.

Kinesiology professionals often work in the area of physical conditioning to improve function. Function can be high-level sport performance or remediation of the effects of an injury, disuse, or aging. If muscle forces are the primary motors (hip extensors in running faster) and brakes (plantar flexors in landing from a jump), the Force–Motion Principle suggests that muscle groups that primarily contribute to the motion of interest should be trained. Remember that this can be a more complex task than consulting your anatomy book. How can we know what exercises, technique (speed, body position), or load to prescribe?

Imagine a physical education teacher working with students on their upper body muscular strength. A particular student is working toward improving his score on a pull-up test in the fitness unit. The forces in a pull-up exercise can be simplified into two vertical forces: the downward gravitation force of bodyweight and an upward force created by concentric muscle actions at the elbows, shoulders, and back. The considerable isometric actions of the grip, shoulder girdle, and trunk do not appear to limit this youngster's performance. You note that this student's bodyweight is not excessive, so losing weight is not an appropriate choice. The teacher decides to work on exercises that train the elbow flexors, as well as the shoulder adductors and extensors. The teacher will likely prescribe exercises like lat pulls, arm curls, and rowing to increase the student's ability to pull downward with a force larger than his bodyweight.

Suppose a coach is interested in helping a young gymnast improve her "splits" position in a cartwheel or other arm support stunt (Figure 3.17). The gymnast can easily overcome the passive muscular tension in the hip adductors to create a split in

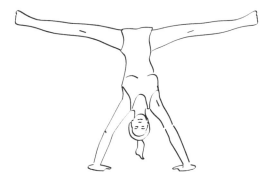

Figure 3.17. The Force–Motion Principle can be applied in a situation where a gymnast is having difficulty in performing inverted splits. The two forces that may limit the split are the passive tension resistance of the hip muscles or inadequate hip flexor/abductors. The coach must decide which forces limit this athlete's performance.

a seated position, but the downward force creating this static position is large (weight of the upper body) compared to the weight of the leg that assists the split in the inverted body position. The Force–Motion Principle suggests that the balance of forces at the hips must be downward to create the split in the dynamic action of the stunt. In other words, the forces of gravity and hip abductors must create a torque equal to the upward torque created by the passive tension in the hip adductors. If the gymnast is having trouble with this stunt, the two biomechanical solutions that could be considered are stretching the hip adductors (to decrease passive muscle tension resistance) and increase the muscular strength or activation of the hip abductors.

The examples of the Force–Motion Principle have been kept simple for several reasons. First, we are only beginning our journey to an understanding of biomechanics. Second, the Force–Motion Principle deals with a complex and deeper level of mechanics (kinetics) that explains the causes of motion. Third, as we saw in this chapter, the complexity of the biomechanical and neuromuscular system makes inference of muscle actions complicated. The variability in the forces and kinematics of human movement, therefore, have been of interest to a variety of scholars (see the Interdisciplinary Issue on Variability). The rest of the book will provide challenges to the perception that the causes of and solutions to human movement problems are simple and introduce you to the main areas of biomechanics that are used to answer questions about the causes of movement.

Interdisciplinary Issue: Variability

Scientists from a variety of disciplines have been interested in the variability of human movement performance. Biomechanical studies have often documented variability of kinematic and kinetic variables to determine the number of trails that must be analyzed to obtain reliable data (Bates, Osternig, Sawhill, & Janes, 1983; Rodano & Squadrone, 2002; Winter, 1984). Motor learning studies have focused on variability as a measure of neuromuscular control (e.g., Slifkin & Newell, 2000). The study of variability has also indicated that variability may play a role in potential injury (James, Dufek, & Bates, 2000). Multiple biomechanical measurements of kinematics and kinetics may provide important contributions to interdisciplinary studies of human movement variability.

SUMMARY

Anatomy is the descriptive study of the structure of the human body. This structural knowledge is an important prerequisite for the study of human movement, but must be combined with biomechanical knowledge to determine how muscles create human movement. Kinesiology professionals and students of biomechanics need to continually review their knowledge of musculoskeletal anatomy. Muscles tend to be activated in synergies to cooperate or coordinate with other forces to achieve movement goals. Muscle tension is created from active or passive components, and the action muscles create are either eccentric, concentric, or isometric. Biomechanical research has shown that the actions of muscles in normal movement are more complicated than what is hypothesized by functional anatomy. The Range-of-Motion Principle of biomechanics can be used to improve human movement. Modifying range of motion in the countermovement of the vertical jump, as well as the stride and body rotations in the overarm throw, were examples discussed. The Force–Motion Principle was applied to exercise training and how passive tension affects gymnastic performance.

REVIEW QUESTIONS

1. What are the major anatomical terms used in kinesiology and medicine to describe the position and motion of the body?

2. What structural and functional properties of muscle cells are different from other body cells?

3. How do fiber properties and arrangement affect force and range-of-motion potential of a muscle?

4. Name and define the three kinds of muscle actions.

5. What are the two major sources of muscle tension, and where in the range of motion are they most influential?

6. Explain the Hill three-component model of muscle and how the components relate to the sources of muscle tension.

7. What is an example of the Force–Motion Principle in human movement?

8. Why is the mechanical method of muscle action analysis used in functional anatomy inadequate to determine the actions of muscles in human movement?

9. How does biomechanics help kinesiology professionals understand the causes and potential improvement of human movement?

10. What factors should a kinesiologist consider when defining the appropriate range of motion for a particular movement?

KEY TERMS

active tension
agonist
antagonist
anatomy
anthropometry
actin
concentric
contractile component
eccentric
fascicle
Hill muscle model
hypertrophy
isometric
modeling
muscle action
myofibril
myosin
parallel elastic component
passive insufficiency
passive tension
pennation
sarcomere
series elastic component
simulation
synergy

SUGGESTED READING

Basmajian, J. V., & De Luca, C. J. (1985). *Muscles alive: Their functions revealed by electromyography* (5th. ed.). Baltimore: Williams & Wilkins.

Bloomfield, J., Ackland, T. R., & Elliott, B. C. (1994). *Applied anatomy and biomechanics in sport*. Melbourne: Blackwell Scientific Publications.

Cavanagh, P. R. (1988). On "muscle action" vs. "muscle contraction." *Journal of Biomechanics*, **21**, 69.

Gielen, S. (1999). What does EMG tell us about muscle function? *Motor Control*, **3**, 9–11.

Fitts, R. H., & Widrick, J. J. (1996). Muscle mechanics: Adaptations in muscle resulting from exercise training. *Exercise and Sport Sciences Reviews*, **24**, 427–473.

Hellebrandt, F. A. (1963). Living anatomy. *Quest*, **1**, 43–58.

Herbert, R., Moore, S., Moseley, A., Schurr, K., & Wales, A. (1993). Making inferences about muscles forces from clinical observations. *Australian Journal of Physiotherapy*, **39**, 195–202.

Herzog, W. (2000). Muscle properties and coordination during voluntary movement. *Journal of Sports Sciences*, **18**, 141–152.

Kleissen, R. F. M., Burke, J. H., Harlaar, J. & Zilvold, G. (1998). Electromyography in the biomechanical analysis of human movement and its clinical application. *Gait and Posture*, **8**, 143–158.

Lieber, R. L., & Bodine-Fowler, S. C. (1993). Skeletal muscle mechanics: Implications for rehabilitation. *Physical Therapy*, **73**, 844–856.

Lieber, R. L., & Friden, J. (2000). Functional and clinical significance of skeletal muscle architecture. *Muscle and Nerve*, **23**, 1647-1666.

Roberts, T. J., Marsh, R. L., Weyand, P. G., & Taylor, D. R. (1997). Muscular force in running turkeys: The economy of minimizing work. *Science*, **275**, 1113–1115.

Soderberg, G. L., & Knutson, L.M. (2000). A guide for use and interpretation of kinesiologic electromyographic data. *Physical Therapy*, **80**, 485-498.

Zajac, F. E. (2002). Understanding muscle coordination of the human leg with dynamical simulations. *Journal of Biomechanics*, **35**, 1011–1018.

WEB LINKS

Hypermuscle: Review of anatomical joint motion terminology from the University of Michigan.
> http://www.med.umich.edu/lrc/Hypermuscle/Hyper.html

PT Central Muscle Page: Comprehensive web muscle tables
> http://www.ptcentral.com/muscles/

Martindale's "Virtual" Medical Center-An electronic medical/anatomical library hosted by UC-Irvine.
> http://www-sci.lib.uci.edu/HSG/MedicalAnatomy.html

CHAPTER 4

Mechanics of the Musculoskeletal System

Many professionals interested in human movement function need information on how forces act on and within the tissues of the body. The deformations of muscles, tendons, and bones created by external forces, as well as the internal forces created by these same structures, are relevant to understanding human movement or injury. This chapter will provide an overview of the mechanics of biomaterials, specifically muscles, tendons, ligaments, and bone. The neuromuscular control of muscle forces and the mechanical characteristics of muscle will also be summarized. The application of these concepts is illustrated using the **Force–Time Principle** of biomechanics. An understanding of mechanics of musculoskeletal tissues is important in understanding the organization of movement, injury, and designing conditioning programs.

TISSUE LOADS

When forces are applied to a material, like human musculoskeletal tissues, they create **loads**. Engineers use various names to describe how loads tend to change the shape of a material. These include the principal or axial loadings of compression, tension, and shear (Figure 4.1). **Compression** is when an external force tends to squeeze the molecules of a material together. **Tension** is when the load acts to stretch or pull apart the material. For example, the weight of a body tends to compress the foot against the ground in the stance phase of running, which is resisted by tensile loading of the plantar fascia and the longitudinal ligament in the foot. **Shear** is a right-angle loading acting in opposite directions. A trainer creates a shearing load across athletic tape with scissor blades or their fingers when they tear the tape. Note that loads are not vectors (individual forces) acting in one direction, but are illustrated by two arrows (Figure 4.1) to show that the load results from forces from both directions.

When many forces are acting on a body they can combine to create combined loads called **torsion** and **bending** (Figure 4.2). In bending one side of the material is loaded in compression while the other side experiences tensile loading. When a person is in single support in walking (essentially a one-legged chair), the femur experiences bending loading. The medial aspect of the femur is in compression while the lateral aspect is in tension.

RESPONSE OF TISSUES TO FORCES

The immediate response of tissues to loading depends on a variety of factors. The size and direction of forces, as well as the mechanical strength and shape of the tissue, affect how the material structure will change. We will see in this section that mechanical strength and muscular strength are different concepts. This text will strive to use "muscular" or "mechanical" modifiers with the term strength to

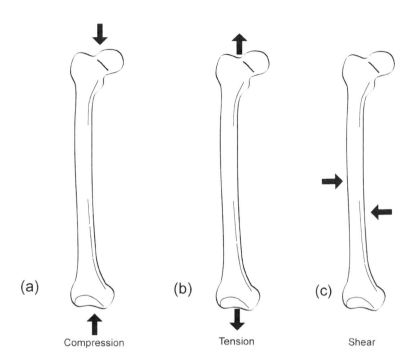

Figure 4.1. The principal axial loads of (a) compression, (b) tension, and (c) shear.

help avoid confusion. There are several important mechanical variables that explain how musculoskeletal tissues respond to forces or loading.

Stress

How hard a load works to change the shape of a material is measured by **mechanical stress**. **Mechanical stress** is symbolized with the Greek letter sigma (σ) and is defined as the force per unit area within a material ($\sigma = F/A$). Mechanical stress is similar to the concept of pressure and has the same units (N/m^2 and lbs/in^2). In the SI system one Newton per meter squared is one Pascal (Pa) of stress or pressure. As you read this book you are sitting in a sea of atmospheric gases that typically exert a pressure of 1 atm, 101.3 KPa (kilopascals), or 14.7 lbs/in^2 on your body. Note that mechanical stress is not vector quantity, but an even more complex quantity called a *tensor*. Tensors are generalized vectors that have multiple directions that must be accounted for, much like resolving a force into anatomically relevant axes like along a longitudinal axis and at right angles (shear). The maximum force capacity of skeletal muscle is usually expressed as a maximum stress of about 25–40 N/cm^2 or 36–57 lbs/in^2 (Herzog, 1996b). This force potential per unit of cross-sectional area is the same across gender, with females tending to have about two-thirds of the muscular strength of males because they have about two-thirds as much muscle mass a males.

Strain

The measure of the deformation of a material created by a load is called **strain**. This de-

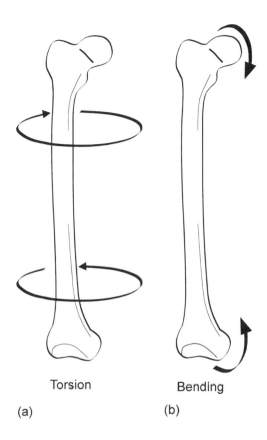

Torsion   Bending

(a)   (b)

Figure 4.2. Combined loads of (a) bending and (b) torsion. A bending load results in one side of the material experiencing tension and the other compression.

formation is usually expressed as a ratio of the normal or resting length (L_0) of the material. Strain (ϵ) can be calculated as a change in length divided by normal length: $(L - L_0)/L_0$. Imagine stressing a rubber band between two fingers. If the band is elongated to 1.5 times its original length, you could say the band experiences 0.5 or 50% tensile strain. This text will discuss the typical strains in musculoskeletal tissues in percentage units. Most engineers use much more rigid materials and typically talk in terms of units of microstrain. Think about what can withstand greater tensile strain: the shaft of a tennis racket, the shaft of a golf club, or the shaft of a fiberglass diving board?

Stiffness and Mechanical Strength

Engineers study the mechanical behavior of a material by loading a small sample in a materials testing system (MTS), which simultaneously measures the force and displacement of the material as it is deformed at various rates. The resulting graph is called a load-deformation curve (Figure 4.3), which can be converted with other measurements to obtain a stress–strain graph. Load-deformation graphs have several variables and regions of interest. The **elastic region** is the initial linear region of the graph where the slope corresponds to the **stiffness** or **Young's modulus** of elasticity of the material. **Stiffness** or Young's modulus is defined as the ratio of stress to strain in the elastic region of the curve, but is often approximated by the ratio of load to deformation (ignoring the change in dimension of the material). If the test were stopped within the elastic region the material would return to its initial shape. If the material were perfectly elastic, the force at a given deformation during restitution (unloading) would be the same as in loading. We will see later that biological tissues are not like a perfectly elastic spring, so they lose some of the energy in restitution that was stored in them during deformation.

Beyond the linear region is the plastic region, where increases in deformation occur with minimal and nonlinear changes in load. The yield point or elastic limit is the point on the graph separating the elastic and plastic regions. When the material is deformed beyond the yield point the material will not return to its initial dimensions. In biological materials, normal physiological loading occurs within the elastic region, and deformations near and beyond the elastic limit are associated with microstructural damage to the tissue. Another important variable calculated from these measurements is the mechanical strength of the material.

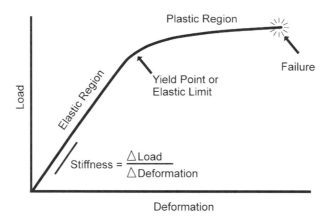

Figure 4.3. The regions and key variables in a load–deformation graph of an elastic material.

Activity: Failure Strength

Two strong materials are nylon and steel. Nylon strings in a tennis racket can be elongated a great deal (high strain and a lower stiffness) compared to steel strings. Steel is a stiff and strong material. Take a paper clip and apply a bending load. Did the paper clip break? Bend it back the opposite way and repeat counting the number or bends before the paper clip breaks. Most people cannot apply enough force in one shearing effort to break a paper clip, but over several loadings the steel weakens and you can get a sense of the total mechanical work/energy you had to exert to break the paper clip.

The **mechanical strength** of a material is the measurement of the maximum force or total **mechanical energy** the material can absorb before failure. Within the plastic region, the pattern of failure of the material can vary, and the definition of failure can vary based on the interest of the research. Conditioning and rehabilitation professionals might be interested in the yield strength (force at the end of the elastic region) of

healthy and healing ligaments. Sports medicine professionals may be more interested in the ultimate strength that is largest force or stress the material can withstand. Sometimes it is of interest to know the total amount of strain energy (see chapter 6) the material will absorb before it breaks because of the residual forces that remain after ultimate strength. This is failure strength and represents how much total loading the material can absorb before it is broken. This text will be specific in regards to the term *strength*, so that when used alone the term will refer to muscular strength, and the mechanical strengths of materials will be identified by their relevant adjective (yield, ultimate, or failure).

Viscoelasticity

Biological tissues are structurally complex and also have complex mechanical behavior in response to loading. First, biological tissues are **anisotropic**, which means that their strength properties are different for each major direction of loading. Second, the nature of the protein fibers and amount of calcification all determine the mechanical response. Third, most soft connective tissue components of muscle, tendons, and ligaments have another region in their load-deformation graph. For example, when a sample of tendon is stretched at a constant rate the response illustrated in Figure 4.4 is typical. Note that the response of the material is more complex (nonlinear) than the Hookean elasticity illustrated in Figure 2.4. The initial increase in deformation with little increase in force before the elastic region is called the *toe region*. The toe region corresponds to the straightening of the wavy collagen fiber in connective tissue (Carlstedt & Nordin, 1989). After the toe region, the slope of the elastic region will vary depending on the rate of stretch. This means that tendons (and other biological tissues) are not perfectly elastic but are **viscoelastic**.

Viscoelastic means that the stress and strain in a material are dependent on the rate of loading, so the timing of the force application affects the strain response of the material. Figure 4.5 illustrates the response of a ligament that is stretched to a set length at two speeds, slow and fast. Note that a high rate of stretch results in a higher stiffness than a slow stretch. Muscles and tendons also have increasing stiffness with increasing rates of stretch. The viscoelasticity of muscles and tendons has great functional significance. A slow stretch will result in a small increase in passive resistance (high compliance) from the muscle, while the muscle will provide a fast increase in passive resistance (high stiffness) to a rapid stretch. This is one of the reasons that stretching exercises should be performed slowly, to minimize the increase in force in the muscle–tendon unit (MTU) for a given amount of stretch. The solid lines of the

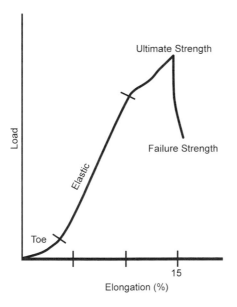

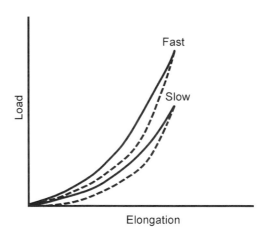

Figure 4.5. Load–deformation curves for tendon stretched at a fast and a slow rate to the same length. Force at any length in loading (solid line) are higher than in unloading (dashed line). A viscoelastic material has different stiffness at different rates of deformation. A faster stretch results in greater force in the tendon for a given length compared to a slow stretch. Hysteresis is the work and energy lost in the restitution of the tendon and can be visualized as the area between loading and unloading of the tendon.

Figure 4.4. The typical load–deformation (elongation) curve for human tendon is more complex than for many materials. Initial elongation is resisted by small force increases in the toe region, followed by the elastic region. Much of the physiological loading of tendons in normal movement are likely within the toe region (<5% strain: Maganaris & Paul, 2000), but elongation beyond the elastic limit damages the tendon.

Activity: Viscoelasticity

An extreme example of viscoelastic behavior that serves as a good demonstration for teaching stretching exercises is the behavior of Silly Putty™. Roll the putty into a cylinder, which serves as a model of muscle. The putty has low stiffness at slow stretching rates, so it gradually increases in length under low force conditions and has a plastic response. Now stretch the putty model quickly and note the much higher stiffness. This high stiffness makes the force in the putty get quite high at short lengths and often results in the putty breaking. You may be familiar with this complex (different) behavior of the material because the shape can be molded into a stable shape like a ball, but when the ball is loaded quickly (thrown at a wall or the floor) it will bounce rather than flatten out.

Application: Stress Relaxation

Guitar players will know that steel strings do not lose tension (consequently their tuning) as quickly as nylon strings; this phenomenon is not related to strength but to viscoelasticity. Steel guitar strings are much more elastic (stiffer) and have negligible viscoelastic properties compared to nylon strings. In a similar fashion, nylon tennis strings lose tension over time. Skilled players who prefer a higher tension to grab the ball for making greater spin will often need to cut out and replace nylon strings before they break. Gut strings are more elastic than nylon and tend to break before there is substantial stress relaxation. Similarly, when a static stretch holds a muscle group in an extended position for a long period of time the tension in the stretched muscle group decreases over time. This stress relaxation occurs quickly (most within the first 15 seconds), with diminishing amounts of relaxation with longer amounts of time (see Knudson, 1998). How might a coach set up a stretching routine that maximizes stress relaxation of the athlete's muscles? If many people dislike holding stretched positions for a long period to time, how might kinesiology professionals program stretching to get optimal compliance and muscle stress relaxation?

graph represent the loading response of the ligament, while the dashed lines represent the mechanical response of the tissue as the load is released (unloading).

There are other important properties of viscoelastic materials: **creep**, **stress relaxation**, and **hysteresis**. **Creep** is the gradual elongation (increasing strain) of a material over time when placed under a constant tensile stress. **Stress relaxation** is the decrease in stress over time when a material is elongated to a set length. For example, holding a static stretch at a specific joint position results in a gradual decrease in tension in the muscle from stress relaxation. If you leave a free weight hanging from a nylon cord, you might return several days later to find the elongation (creep) in the cord has stretched it beyond it initial length. Creep and stress relaxation are nonlinear responses and have important implications for stretching. These implications are dis-

cussed in the application box on "Flexibility on Stretching."

Hysteresis is the property of viscoelastic materials of having a different unloading response than its loading response (Figure 4.5). Hysteresis also provides a measure of the amount of energy lost because the material is not perfectly elastic. The area between the loading unloading of each stretch is the mechanical work lost from loading to unloading. We will learn in Chapter 6 that mechanical work is defined as force times displacement ($\mathbf{F} \cdot \mathbf{d}$), so work can be visualized as an area under a force-displacement graph. If you want to visualize the failure strength (work) of the material in Figure 4.3, imagine or shade in the total area above zero and below the load-deformation graph.

All these mechanical response variables of biological materials depend on precise measurements and characteristics of the samples. The example mechanical strengths and strains mentioned in the next section represent typical values from the literature. Do not assume these are exact values because factors like training, age, and disease all affect the variability of the mechanical response of tissues. Methodological factors like how the human tissues were stored, attached to the machine, or preconditioned (like a warm-up before testing) all affect the results. Remember that the rate of loading has a strong effect on the stiffness, strain, and strength of biological materials. The next section will emphasize more the strengths of tissues in different directions and how these are likely related to common injuries.

BIOMECHANICS OF THE PASSIVE MUSCLE–TENDON UNIT (MTU)

The mechanical response of the MTU to passive stretching is viscoelastic, so the response of the tissue depends on the time or rate of stretch. At a high rate of passive stretch the MTU is stiffer than when it is slowly stretched. This is the primary reason why slow, static stretching exercises are preferred over ballistic stretching techniques. A slow stretch results in less passive tension in the muscle for a given amount of elongation compared to a faster stretch. The load in an MTU during other movement conditions is even more complicated be-cause the load can vary widely with activation, previous muscle action, and kind of muscle action. All these variables affect how load is distributed in the active and passive components of the MTU. Keep in mind that the Hill model of muscle has a contractile component that modulates tension with activation, as well as two passive tension elements: the parallel elastic and the series elastic components. The mechanical behavior of activated muscle is presented in the upcoming section on the "Three Mechanical Characteristics of Muscle."

Tendon is the connective tissue that links muscle to bone and strongly affects how muscles are used or injured in movement. Tendon is a well-vascularized tissue whose mechanical response is primarily related to the protein fiber *collagen*. The parallel arrangement of collagen fibers in tendon and cross-links between fibers makes tendon about three times stronger in tension than muscle. The ultimate strength of tendon is usually about 100 MPa (megapascals), or 14,500 lbs/in^2 (Kirkendall & Garrett, 1997). Even though the diameter of tendons is often smaller than the associated muscle belly, their great tensile strength makes tendon rupture injuries rare. Acute overloading of the MTU usually results in **strains** (sports medicine term for overstretched muscle, not mechanical strain) and failures at the muscletendon junction or the tendon/bone interface (Garrett, 1996).

In creating movement, a long tendon can act as an efficient spring in fast bouncing movements (Alexander, 1992) because the stiffness of the muscle belly can exceed tendon stiffness in high states of activation. A muscle with a short tendon transfers force to the bone more quickly because there is less slack to be taken out of the tendon. The intrinsic muscles of the hand are well suited to the fast finger movements of a violinist because of their short tendons. The Achilles tendon provides shock absorption and compliance to smooth out the forces of the large calf muscle group (soleus and gastrocnemius).

BIOMECHANICS OF BONE

Unlike muscle, the primary loads experienced by most bones are compressive. The mechanical response of bone to compression, tension, and other complex loads depends on the complex structure of bones. Remember that bones are living tissues with blood supplies, made of a high per-

centage of water (25% of bone mass), and having considerable deposits of calcium salts and other minerals. The strength of bone depends strongly on its density of mineral deposits and collagen fibers, and is also strongly related to dietary habits and physical activity. The loading of bones in physical activity results in greater osteoblast activity, laying down bone. Immobilization or inactivity will result in dramatic decreases in bone density, stiffness, and mechanical strength. A German scientist is credited with the discovery that bones remodel (lay down greater mineral deposits) according to the mechanical stress in that area of bone. This laying down of bone where it is stressed and reabsorption of bone in the absence of stress is called **Wolff's Law**. Bone remodeling is well illustrated by the formation of bone around the threads of screws in the hip prosthetic in the x-ray in Figure 4.6.

Application: Osteoporosis

Considerable research is currently being directed at developing exercise machines as countermeasures for the significant bone density loss in extended space flight. A microgravity environment substantially decreases the loading of the large muscles and bones of the lower extremity, resulting in loss of bone and muscle mass. There is also interest in exercise as a preventative and remedial strategy for increasing the bone mass of postmenopausal women. The strong link between the positive stresses of exercise on bone density, however, is often complicated by such other things like diet and hormonal factors. In the late 1980s researchers were surprised to find that elite women athletes were at greater risk for **stress fractures** because they had the bone density of women two to three times their age. **Stress fractures** are very small breaks in the cortical (see below) bone that result from physical activity without adequate rest. What was discovered was that overtraining and the very low body fat that resulted in amenorrhea also affected estrogen levels that tended to decrease bone mass. This effect was stronger than the bone growth stimulus of the physical activity. High-level women athletes in many sports must be careful in monitoring training, diet, and body fat to maintain bone mass. Kinesiology professionals must be watchful for signs of a condition called the female athlete triad. The *female athlete triad* is the combination of disordered eating, amenorrhea, and osteoporosis that sometimes occurs in young female athletes.

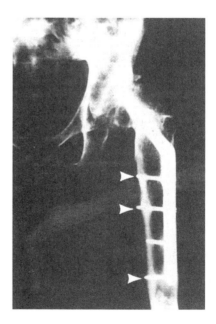

Figure 4.6. X-ray of a fractured femur with a metal plate repair. Note the remodeling of bone around the screws that transfer load to the plate. Reprinted with permission from Nordin & Frankel (2001).

The macroscopic structure of bone shows a dense, external layer called cortical (compact) bone and the less-dense internal cancellous (spongy) bone. The mechanical response of bone is dependent on this "sandwich" construction of cortical and cancellous bone. This design of a strong and stiff material with a weaker and more flexible interior (like fiberglass) results in a composite material that is strong for a given weight (Nordin & Frankel, 2001). This is much like a surf board constructed of fiberglass bonded over a foam core. Cortical bone is stiffer (maximum strain about 2%), while cancellous bone is less stiff and can withstand greater strain (7%) before failure. In general, this design results in ultimate strengths of bone of about 200 MPa (29,000 lbs/in²) in compression, 125 MPa (18,000 lbs/in²) in tension, and 65 MPa (9,500 lbs/in²) in shear (Hayes, 1986). This means that an excessive bending load on the femur like in Figure 4.2 would most likely cause a fracture to begin on the lateral aspect that is under tensile loading. Using sports rules to protect athletes from lateral blows (like blocking rules in American football) is wise because bone is weakest under shearing loads.

It is also important to understand that the ultimate strength of bone depends on nutritional, hormonal, and physical activity factors. Research done with an elite power-lifter found that the ultimate compressive strength of a lumbar vertebral body (more than 36,000 N or 4 tons) estimated from bone mineral measurements was twice that of the previous maximal value. More recent studies of drop jump training in pre-pubescent children has demonstrated that bone density can be increased, but it is unclear if peak forces, rates of loading, or repetitions are the training stimulus for the increases in bone mass (Bauer, Fuchs, Smith, & Snow, 2001). The following section will outline the mechanical response of ligaments to loading.

BIOMECHANICS OF LIGAMENTS

Ligaments are tough connective tissues that connect bones to guide and limit joint motion. Most joints are not perfect hinges with a constant axis of rotation, so they tend to have small accessory motions and moving axes of rotation that stress ligaments in several directions. The collagen fibers within ligaments are not arranged in parallel like tendons, but in a variety of directions. Normal physiological loading of most ligaments is 2–5% of tensile strain, which corresponds to a load of ≈500 N (112 lbs) in the human anterior cruciate ligament (Carlstedt & Nordin, 1989), except for "spring" ligaments that have a large percentage of elastin fibers (ligamentum flavum in the spine), which can stretch more than 50% of their resting length. The maximum strain of most ligaments and tendons is about 8–10% (Rigby, Hirai, Spikes, & Eyring, 1959).

Like bone, ligaments and tendons remodel according to the stresses they are subjected to. A long-term increase in the mechanical strength of articular cartilage with the loads of regular physical activity has also been observed (Arokoski, Jurvelin, Vaatainen, & Helminen, 2000). Inactivity, however, results in major decreases in the mechanical strength of ligaments and tendon, with reconditioning to regain this strength taking longer than deconditioning (Carlstedt & Nordin, 1989). The ability of the musculoskeletal system to adapt tissue mechanical properties to the loads of physical activity does not guarantee a low risk of injury. There is likely a higher risk of tissue overload when deconditioned individuals participate in vigorous activity or when trained individuals push the envelope, training beyond the tissue's ability to adapt during the rest periods between training bouts. We will see in the next section that muscle mechanical properties also change in response to activity and inactivity.

Application: Flexibility and Stretching

A common health-related fitness component is *flexibility*. Flexibility is defined as "the intrinsic property of body tissues, which determines the range of motion achievable without injury at a joint or group of joints" (Holt *et al.*, 1996:172). Flexibility can be mechanically measured as static and dynamic flexibility. Static flexibility refers to the usual linear or angular measurements of the actual limits of motion in a joint or joint complex. Static flexibility measurements have elements of subjectivity because of variations in testers and patient tolerance of stretch. Dynamic flexibility is the increase in the muscle group resistance to stretch (stiffness) and is a less subjective measure of flexibility (Knudson *et al.*, 2000). Inactivity and immobilization have been shown to decrease static range of motion (SROM) and increase muscle group stiffness (Akeson *et al.*, 1987; Heerkens *et al.*, 1986).

Stretching is a common practice in physical conditioning and sports. Stretching exercises must be carefully prescribed to focus tension on the MTUs and not the ligaments that maintain joint integrity (Knudson, 1998). Long-term stretching programs likely increase static range of motion by stimulating the production of new sarcomeres in muscle fibers (De Deyne, 2001). While much is known about the acute and chronic effects of stretching on static flexibility, less is know about its effect on dynamic flexibility (see Knudson *et al.*, 2000). One example of the complications in examining the effects of stretching by measuring muscle stiffness is the **thixotropic** property of muscle. **Thixotropy** is the variation in muscle stiffness because of previous muscle actions. If an active muscle becomes inactive for a long period of time, like sitting in a car or a long lecture, its stiffness will increase. Do your muscles feel tight after a long ride in the car? Enoka (2002) provides a nice review of this phenomenon and uses ketchup to illustrate its cause. A gel like ketchup if allowed to stand tends to "set" (like actin and myosin bound in a motionless muscle), but when shaken tends to change state and flow more easily. Most all of the increased stiffness in inactive muscles can be eliminated with a little physical activity or stretching. This does not, however, represent a long-term change in the stiffness of the muscles. Some of the most recent studies suggest that long-term effects of vigorous stretching are decreases in muscle viscosity and hysteresis, with no changes in tendon stiffness (Kubo *et al.*, 2001a, 2002). Another recent study combined MRI imaging and a biomechanical model to calculate the stress and strain in three hamstring muscles from isokinetic dynamometer measurements (Magnusson *et al.*, 2000). They found that the limits of SROM were within the toe region of the hamstring muscle stress–strain curve. Even if consistent stretching did create long-term decreases in muscle stiffness, it is unclear if this would translate to improved performance or lower risk of injury. More research is needed on the effects of stretching on muscle stiffness.

Interestingly, many studies have shown that the hypothesized performance-enhancing benefits of stretching prior to activity are incorrect. Stretching in the warm-up prior to activity has been shown to decrease muscular performance in a wide variety of tests (Knudson, 1999b). Muscle activation and muscular strength are significantly decreased for 15 and 60 minutes, respectively, following stretching (Fowles *et al.*, 2000a). The large stresses placed on MTUs in passive stretching create short-term weakening, but these loads have not been shown to increase protein synthesis (Fowles et al, 2000b).

Recent biomechanical and epidemiological research has also indicated that stretching during warm-up does not decrease the risk of injury (Knudson *et al.*, 2000; Shirier, 1999). Several lines of evidence now suggest that the best time to program stretching in conditioning programs is during the cool-down phase. Recommendations on stretching and flexibility testing have been proposed by Knudson (1999b) and Knudson *et al.* (2000).

Flexibility is also strongly related to variations in body position because the passive tension increases in each MTU, especially in multiarticular MTUs (passive insufficiency). This is why there are strict rules for body positioning in flexibility testing. What muscles of the leg and thigh are unloaded when students try to cheat by bending their knees in a sit-and-reach test? What calf muscle is unloaded in a seated toe-touch stretch when the ankle is plantar flexed?

THREE MECHANICAL CHARACTERISTICS OF MUSCLE

Previously we discussed the passive tension in an MTU as it is passively stretched. Now it is time to examine the tensile forces the MTU experiences in the wide variety actions, lengths, and other active conditions encountered in movement. The force potential of an MTU varies and can be described by three mechanical characteristics. These characteristics deal with the variations in muscle force because of differences in velocity, length, and the time relative to activation.

Force–Velocity Relationship

The **Force–Velocity Relationship** explains how the force of fully activated muscle varies with velocity. This may be the most important mechanical characteristic since all three muscle actions (eccentric, isomet-

ric, concentric) are reflected in the graph. We will see that the force or tension a muscle can create is quite different across actions and across the many speeds of movement. The discovery and formula describing this fundamental relationship in concentric conditions is also attributed to A. V. Hill. Hill made careful measurements of the velocity of shortening when a preparation of maximally stimulated frog muscle was released from isometric conditions. These studies of isolated preparations of muscle are performed in what we term *in vitro* (Latin for "in glass") conditions. Figure 4.7 illustrates the shape of the complete Force–Velocity Relationship of skeletal muscle. The Force–Velocity curve essentially states that the force the muscle can create decreases with increasing velocity of shortening (concentric actions), while the force the muscle can resist increases with increasing velocity of lengthening (eccentric actions). The force in isometric conditions is labeled P_0 in Hill's equation. The

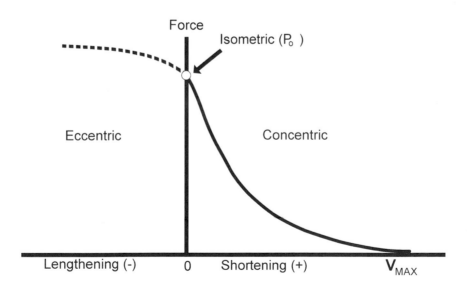

Figure 4.7. The *in vitro* Force–Velocity Relationship of muscle. Muscle force potential rapidly decreases with increasing velocity of shortening (concentric action), while the force within the muscle increases with increasing velocity of lengthening (eccentric action).

right side of the graph corresponds to how the tension potential of the muscle rapidly decreases with increases in speed of concentric shortening. Also note, however, that increasing negative velocities (to the left of isometric) show how muscle tension rises in faster eccentric muscle actions. In isolated muscle preparations the forces that the muscle can resist in fast eccentric actions can be almost twice the maximum isometric force (Alexander, 2002).

If the force capability of an *in vitro* muscle preparation varies with velocity, can this behavior be generalized to a whole MTU or muscle groups in normal movement? Researchers have been quite interested in this question and the answer is a strong, but qualified "yes." The torque a muscle group can create depends on the previous action, activation, rate of force development, and the combination of the characteristics of the muscles acting at that and nearby joints. Despite these complications, the *in vivo* (in the living animal) torque–angular velocity relationship of muscle groups usually matches the shape of the *in vitro* curve. These *in vivo* torque–angular velocity relationships are established by testing at many angular velocities on **isokinetic** and specialized dynamometers. These studies tend to show that in repeated isokinetic testing the peak eccentric torques are higher than peak isometric torques, but not to the extent of isolated muscle preparations (Holder-Powell & Rutherford, 1999), while concentric torques decline with varying slopes with increasing speed of shortening (De Koning *et al.*, 1985; Gulch, 1994; Pinniger *et al.*, 2000).

This general shape of a muscle's potential maximum tension has many implications for human movement. First, it is not possible for muscles to create large forces at high speeds of shortening. Muscles can create high tensions to initiate motion, but as the speed of shortening increases their ability to create force (maintain acceleration)

decreases. Second, the force potential of muscles at small speeds of motion (in the middle of the graph) depends strongly on isometric muscular strength (Zatsiorsky, 1995). This means that muscular strength will be a factor in most movements, but this influence will vary depending on the speed and direction (moving or braking) the muscles are used. Third, the inverse relationship between muscle force and velocity of shortening means you cannot exert high forces at high speeds of shortening, and this has a direct bearing on muscular power. In chapter 6 we will study mechanical power and look more closely at the right mix of force and velocity that creates peak muscular power output. This also means that isometric strength and muscle speed are really two different muscular abilities. Athletes training to maximize throwing speed will train differently based on the load and speed of the implements in their sport. Athletes putting the shot will do higher weight and low repetition lifting, compared to athletes that throw lighter objects like a javelin, softball, or baseball, who would train with lower weights and higher speeds of movement. One of the best books that integrates the biomechanics of movement and muscle mechanics in strength and conditioning is by Zatsiorsky (1995).

So there are major implications for human movement because of the functional relationship between muscle force and velocity. What about training? Does training alter the relationship between muscle force and velocity or does the Force–Velocity Relationship remain fairly stable and determine how you train muscle? It turns out that we cannot change the nature (shape) of the Force–Velocity Relationship with training, but we can shift the graph upward to improve performance (De Koning *et al.*, 1985; Fitts & Widrick, 1996). Weight training with high loads and few repetitions primarily shifts the force–velocity curve up near isometric conditions (Figure 4.8),

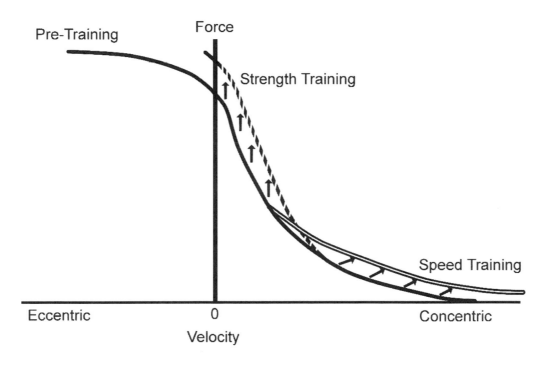

Figure 4.8. Training shifts the Force–Velocity curve upward and is specific to the kind of training. Heavy weight training primarily shifts the curve upward for isometric and slow concentric actions, while speed training improves muscle forces at higher concentric speeds.

while fast lifting of light loads shifts the curve up near V_{max}, which is the maximum velocity of shortening for a muscle.

Another area where the Force–Velocity Relationship shows dramatic differences in muscle performance is related to muscle fiber types. Skeletal muscle fibers fall on a continuum between slow twitch (Type I) and fast twitch (Type II). Type I are also called Slow-Oxidative (SO) because of their high oxidative glycolysis capacity (considerable mitochrondion, myoglobin, triglycerides, and capillary density). Type II fibers are also called Fast-Glycolytic (FG) because of their greater anaerobic energy capacity (considerable intramuscular ATP and glycolytic enzymes). Muscle fibers with intermediate levels are usually called FOG (Fast-Oxidative-Glycolytic) fibers. Muscle fibers type have been classified in many ways

(Scott, Stevens, & Binder-Macleod, 2001), but biomechanics often focuses on the twitch response and velocity of shortening characteristics of fiber types. This is because the force potential of fast and slow twitch fibers per given physiological cross-sectional area are about the same. The timing that the muscle fibers create force and speed of shortening, however, are dramatically different. This fact has major implications for high-speed and high-power movements.

The easiest way to illustrate these differences is to look at the **twitch** response of different fiber types. If an *in vitro* muscle fiber is stimulated one time, the fiber will respond with a twitch. The rate of tension development and decay of the twitch depends on the fiber type of the fiber. Figure 4.9 illustrates a schematic of the twitch re-

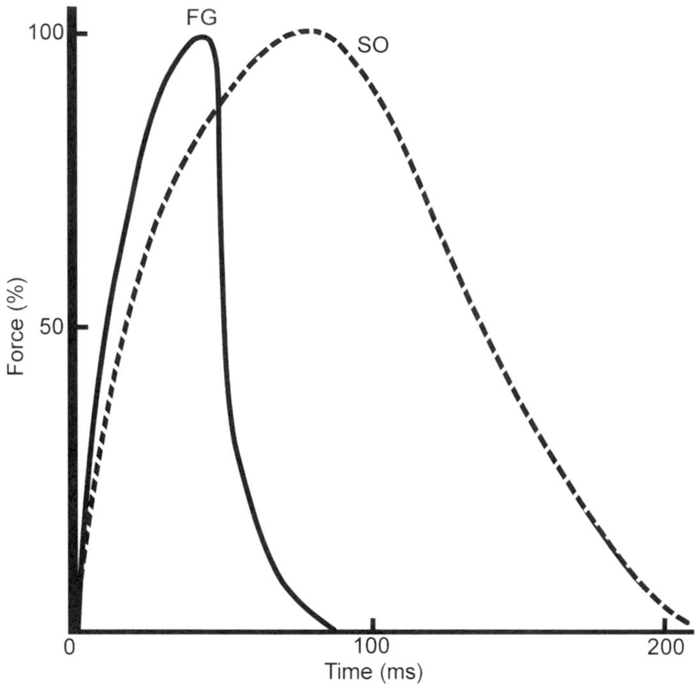

Figure 4.9. The twitch response of fast-twitch (FG) and slow-twitch (SO) muscle fibers. Force output is essentially identical for equal cross-sectional areas, but there are dramatic differences in the rise and decay of tension between fiber types that affect the potential speed of movement.

sponses of several fiber types. A fiber at the slow end of the fiber type continuum gradually rises to peak tension in between 60 and 120 ms (about a tenth of a second). A fiber at the high end of the continuum (FG) would quickly create a peak tension in 20 to 50 ms. This means that the muscle with a greater percentage of FG fibers can create a greater velocity of shortening than a similar (same number of sarcomeres) one with predominantly SO muscle fibers. Muscles with higher percentages of SO fibers will have a clear advantage in long-duration, endurance-related events.

Human muscles are a mix of fiber types. There are no significant differences in fiber types of muscles across gender, but within the body the antigravity muscles (postural muscles that primarily resist the torque created by gravity) like the soleus, erector spinae, and abdominals tend to have a higher percentage of slow fibers than fast fibers. The fiber type distribution of elite athletes in many sports has been well documented. There is also interest in the trainability and plasticity of fiber types (Fitts & Widrick, 1996; Kraemer, Fleck, & Evans, 1996). Figure 4.10 illustrates the Force–Velocity Relationship in the predominantly slow-twitch soleus and predominantly fast-twitch medial gastrocnemius muscles in a cat. This fiber distribution and mechanical behavior are likely similar to humans. If the gastrocnemius were a more significant contributor to high-speed movements in sport, how might exercise position and technique be used to emphasize the gastrocnemius over the soleus?

Interdisciplinary Issue: Speed

Running speed in an important ability in many sports. The force–velocity relationship suggests that as muscles shorten concentrically faster they can create less tension to continue to increase velocity. What are the main factors that determine sprinting speed? Do muscle mechanical properties dominate sprinting performance or can technique make major improvements in to running speed? Several lines of research suggest that elite sprinting ability may be more related to muscular and structural factors than technique. Near top running speed, stride rate appears to be the limiting factor rather than stride length (Chapman & Caldwell, 1983; Luthanen & Komi, 1978b; Mero, Komi, & Gregor, 1992). In the 100-meter dash running speed is clearly correlated with percentage of fast twitch fibers (Mero, Luthanen, Viitasalo, & Komi, 1981) and the length of muscle fascicles (Abe, Kumagai, & Brechue, 2000; Kumagai, Abe, Brechue, Ryushi, Takano, & Mizuno, 2000) in high-level sprinters. Athletes with longer fascicles are faster. Future research into genetic predisposition to fiber dominance and trainability might be combined with biomechanical research to help improve the selection and training of sprinters.

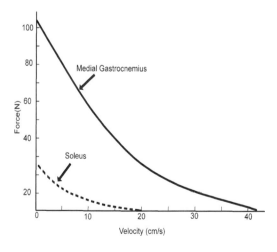

Figure 4.10. Differences in the Force–Velocity Relationship of the primarily fast-twitch medial gastrocnemius and primarily slow-twitch soleus of the cat. Reprinted, by permission, from Edgerton, Roy, Gregor, & Rugg, (1986).

Application: Domains of Muscular Strength

Therapists, athletes, and coaches often refer to a functional characteristic called **muscular strength**. While muscular strength is commonly measured in weight training with one-repetition maxima (1RM is the maximum weight a person can lift only one time), most researchers define muscular strength in isometric conditions at a specific joint angle to eliminate the many mechanical factors affecting muscle force (e.g., Atha, 1981; Knuttgen & Kraemer, 1987). Many fitness test batteries include tests for components called *muscular strength* and *muscular endurance*. Early physical education research demonstrated that muscular strength has several domains of functional expression. Statistical analysis of fitness testing demonstrated that muscular strength is expressed as static (isometric), dynamic (slow to moderate movements), and explosive for fast movement (Jackson & Frankiewicz, 1975). This corresponds closely to the major changes in force capability in the Force–Velocity Relationship. Others experts often include another domain of muscular strength related to eccentric actions: stopping strength (Zatsiorsky, 1995). Functional muscular strength is also complicated by the fact that the force a muscle group can express also depends on the inertia of the resistance. The peak force that can be created in a basketball chest pass is nowhere near peak bench press isometric strength because of the small inertia of the ball. The ball is easily accelerated (because of its low mass), and the force the muscles can create at high shortening velocities rapidly declines, so the peak force that can be applied to the ball is much less than with a more massive object. So the Force–Velocity property of skeletal muscle and other biomechanical factors is manifested in several functional "strengths." Kinesiology professionals need to be aware of how these various "muscular strengths" correspond to the movements of their clients. Professionals should use muscular strength terminology correctly to prevent the spread of inaccurate information and interpret the literature carefully because of the many meanings of the word *strength*.

Force–Length Relationship

The length of a muscle also affects the ability of the muscle to create tension. The **Force–Length Relationship** documents how muscle tension varies at different muscle lengths. The variation in potential muscle tension at different muscle lengths, like the Force–Velocity Relationship, also has a dramatic effect on how movement is created. We will see that the Force–Length Relationship is just as influential on the torque a muscle group can make as the geometry (moment arm) of the muscles and joint (Rassier, MacIntosh, & Herzog, 1999).

Remember that the tension a muscle can create has both active and passive sources, so the length–tension graph of muscle will have both of these components. Figure 4.11 illustrates the Force–Length Relationship for a skeletal muscle fiber. The active component of the Force–Length Relationship (dashed line) has a logical association with the potential numbers of cross-bridges between the actin and myosin filaments in the Sliding Filament Theory. Peak muscle force can be generated when there are the most cross-bridges. This is called resting length (L_0) and usually corresponds to a point near the middle of the range of motion. Potential active muscle tension decreases for shorter or longer muscle lengths because fewer cross-bridges are available for binding. The passive tension component (solid line) shows that passive tension increases very slowly near L_0 but dramatically increases as the muscle is elongated. Passive muscle tension usually does not contribute to movements in the middle portion of the range of motion, but does contribute to motion when muscles are stretched or in various neuromuscular disorders (Salsich, Brown, & Mueller, 2000). The exact shape of the Force–Length Relationship slightly varies between muscles because of differences in active (fiber area, angle of pennation) and passive (tendon) tension components (Gareis, Solomonow, Baratta, Best, & D'Ambrosia, 1992).

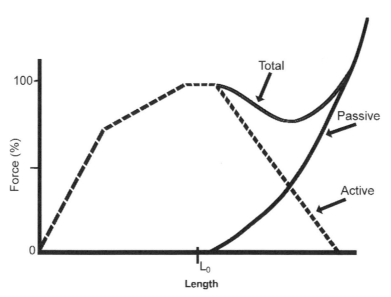

Figure 4.11. The Force–Length Relationship of human skeletal muscle. The active component follows an inverted "U" pattern according to the number of potential cross-bridges as muscle length changes. Passive tension increases as the muscle is stretched beyond its resting length (L_0). The total tension potential of the muscle is the sum of the active and passive components of tension.

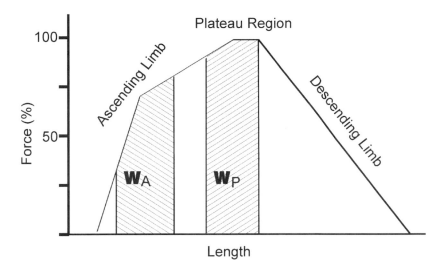

Figure 4.12. The three regions of the active component of the Length–Tension Relationship. Differences in the work ($W = F \cdot d$) the muscle can do in the ascending limb (W_A) versus the plateau region (W_P) are illustrated. Work can be visualized as the area under a force–displacement graph.

The active tension component of the Force–Length Relationship has three regions (Figure 4.12). The ascending limb represents the decreasing force output of the muscle as it is shortened beyond resting length. Movements that require a muscle group to shorten considerably will not be able to create maximal muscle forces. The plateau region represents the high muscle force region, typically in the midrange of the anatomical range of motion. Movements initiated near the plateau region will have the potential to create maximal muscle forces. The descending limb represents the decreasing active tension a muscle can make as it is elongated beyond resting length. At extremes of the descending limb the dramatic increases in passive tension provide the muscle force to bring a stretched muscle back to shorter lengths, even though there are virtually no potential cross-bridge attachment sites. Biomechanical research has begun to demonstrate that

muscles adapt to chronic locomotor movement demands and the coordination of muscles may be organized around muscles suited to work on the ascending, plateau, or descending limb of the force–length curve (Maganaris, 2001; Rassier *et al.*, 1999). It is clear that the length of muscles influences how the central nervous system coordinates their actions (Nichols, 1994).

Activity: Force–Length Relationship

Active insufficiency is the decreased tension of a multiarticular muscle when it is shortened across one or more of its joints. Vigorously shake the hand of a partner. Fully flex your wrist, and try to create a large grip force. What happened to your strength? Which limb of the Force–Length Relationship creates this phenomenon?

The implications for a muscle working on the ascending limb versus the plateau region of the force–length curve are dramatic. The mechanical work that a muscle fiber can create for a given range of motion can be visualized as the area under the graph because work is force times displacement. Note the difference in work (area) created if the muscle fiber works in the ascending limb instead of near the plateau region (Figure 4.12). These effects also interact with the force–velocity relationship to determine how muscle forces create movement throughout the range of motion. These mechanical characteristics also inter-act with the time delay in the rise and fall of muscle tension, the force–time relationship.

Force–Time Relationship

Another important mechanical characteristic of muscle is related to the temporal delay in the development of tension. The **Force–Time Relationship** refers to the delay in the development of muscle tension of the whole MTU and can be expressed as the time from the **motor action potential** (electrical signal of depolarization of the fiber that makes of the electromyogra-

Application: Strength Curves

The torque-generating capacity of a muscle primarily depends on its physiological cross-sectional area, moment arm, and muscle length (Murray, Buchanan, & Delp, 2000). The maximum torque that can be created by a muscle group through the range of motion does not always have a shape that matches the *in vitro* force–length relationship of muscle fibers. This is because muscles with different areas, moment arms, and length properties are summed and often over-come some antagonistic muscle activity in maximal exertions (Kellis & Baltzopoulos, 1997). There is also some evidence that the number of sarcomeres in muscle fibers may adapt to strength training and interact with muscle moment arms to affect were the peak torque oc-curs in the range of motion (Koh, 1995). "Strength curves" of muscle are often documented by multiple measurements of the isometric or isokinetic torque capability of a muscle group throughout the range of motion (Kulig, Andrews, & Hay, 1984). The torque-angle graphs creat-ed in isokinetic testing also can be interpreted as strength curves for muscle groups. The peak torque created by a muscle group tends to shift later in the range of motion as the speed of rotation increases, and this shift may be related to the interaction of active and passive sources of tension (Kawakami, Ichinose, Kubo, Ito, Imai, & Fukunaga, 2002). How the shape of these strength curves indicates various musculoskeletal pathologies is controversial (Perrin, 1993). Knowledge of the angles where muscle groups create peak torques or where torque output is very low is useful in studying movement. Postures and stances that put muscle groups near their peak torque point in the range of motion maximizes their potential contribution to mo-tion or stability (Zatsiorsky, 1995). In combative sports an opponent put in an extreme joint position may be easily immobilized (active insufficiency, poor leverage, or pain from the stretched position). Accommodating resistance exercise machines (Nautilus™ was one of the first) are usually designed to match the average strength curve of the muscle group or move-ment. These machines are designed to stay near maximal resistance (match the strength curve) throughout the range of motion (Smith, 1982), but this is a difficult objective because of indi-vidual differences (Stone, Plisk, & Collins, 2002). There will be more discussion of strength curves and their application in Chapter 7.

phy or EMG signal) to the peak or target muscle tension.

The time delay that represents the Force–Time Relationship can be split into two parts. The first part of the delay is related to the rise in muscle stimulation sometimes called active state or excitation dynamics. In fast and high-force movements the neuromuscular system can be trained to rapidly increase (down to about 20 ms) muscle stimulation. The second part of the delay involves the actual build-up of tension that is sometimes called contraction dynamics. Recall that the contraction dynamics of different fiber types was about 20 ms for FG and 120 ms for SO fibers. When many muscle fibers are repeatedly stimulated, the fusion of many twitches means the rise in tension takes even longer. The length of time depends strongly on the cognitive effort of the subject, training, kind of muscle action, and the activation history of the muscle group. Figure 4.13 shows a schematic of rectified electromyograph (measure the electrical activation of muscle) and the force of an isometric grip force. Note that peak isometric force took about 500 ms. Revisit Figure 3.14 for another example of the electromechanical delay (delay from raw EMG to whole muscle force). Typical delays in peak tension of whole muscle groups (the Force–Time Relationship) can be quite variable. Peak force can be developed in as little as 100 ms and up to over a second for maximal muscular strength efforts. The Force–Time Relationship is often referred to as the **electromechanical delay** in electromyographic (EMG) studies. This delay is an important thing to keep in mind when looking at EMG plots and trying to relate the timing of muscle forces to the movement. In chapter

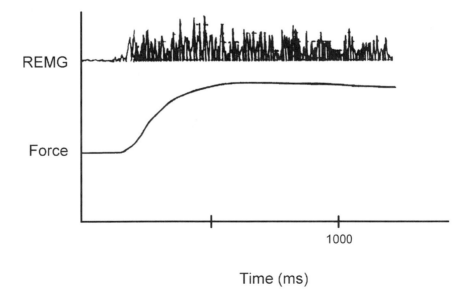

Figure 4.13. The rectified electromyographic (REMG) signal from the quadriceps and the force of knee extension in an isometric action. The delay between the activation (REMG) and the build-up of force in the whole muscle is the *electromechanical delay* and represents the Force–Time Relationship of the muscle. It takes 300 to 500 ms for peak force to be achieved after initial activation of the muscle group.

5 we will see that kinematics provides a precise description of how motion builds to a peak velocity and where this occurs relative to the accelerations that make it occur.

This delay in the development of muscle tension has implications for the coordination and regulation of movement. It turns out that deactivation of muscle (timing of the decay of muscle force) also affects the coordination of movements (Neptune & Kautz, 2001), although this section will limit the discussion of the Force–Time Relationship to a rise in muscle tension. Kinesiology professionals need to know about these temporal limitations so they understand the creation of fast movements and can provide instruction or cues consistent with what the mover's body does. For example, it is important for coaches to remember that when they see high-speed movement in the body, the forces and torques that created that movement preceded the peak speeds of motion they observed. The coach that provides urging to increase effort late in the movement is missing the greater potential for acceleration earlier in the movement and is asking the performer to increase effort when it will not be able to have an effect. Muscles are often preactivated before to prepare for a forceful event, like the activation of plantar flexors and knee extensors before a person lands from a jump. A delay in the rise of muscle forces is even more critical in movements that cannot be preprogrammed due to uncertain environmental conditions. Motor learning research shows that a couple more tenths of a second are necessary for reaction and processing time even before any delays for increases in activation and the electromechanical delay. To make the largest muscle forces at the initiation of an intended movement, the neuromuscular system must use a carefully timed movement and muscle activation strategy. This strategy is called the stretch-shortening cycle and will be discussed in the following section.

Application: Rate of Force Development

Biomechanists often measure force or torque output of muscle groups or movements with dynamometers. One variable derived from these force–time graphs is the rate of force development (F/t), which measures how quickly the force rises. A high rate of force development is necessary for fast and high-power movements. The vertical ground reaction forces of the vertical jump of two athletes are illustrated in Figure 4.14. Note that both athletes create the same peak vertical **ground reaction force**, but athlete A (dashed line) has a higher rate of force development (steeper slope) than athlete B (solid line). This allows athlete A to create a larger vertical impulse and jump higher. The ability to rapidly increase the active state and consequently muscle force has been demonstrated to contribute strongly to vertical jump performance (Bobbert & van Zandwijk, 1999). What kinds of muscle fibers help athletes create a quick rise in muscle force? What kinds of muscle actions allow muscles to create the largest tensions? The next two sections will show how the neuromuscular system activates muscles and coordinates movements to make high rates of muscle force development possible.

STRETCH-SHORTENING CYCLE (SSC)

The mechanical characteristics of skeletal muscle have such a major effect on the force and speed of muscle actions that the central nervous system has a preferred muscle action strategy to maximize performance in most fast movements. This strategy is most beneficial in high-effort events but is also usually selected in submaximal move-

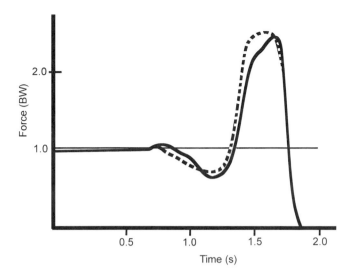

Figure 4.14. Vertical jump ground reaction forces in units of bodyweight (BW) for two athletes. Athlete A (dashed line) has a greater rate of force development compared to athlete B (solid line). The rate of force development is the slope of the graphs when vertical forces are building above 1BW. Do not interpret the up and down motion of the graph as motion of the jumper's body; it represents the sum of the vertical forces the athlete makes against the ground.

ments. Most normal movements unconsciously begin a **stretch-shortening cycle** (**SSC**): a countermovement away from the intended direction of motion that is slowed down (braked) with eccentric muscle action that is immediately followed by concentric action in the direction of interest. This bounce out of an eccentric results in potentiation (increase) of force in the following concentric action if there is minimal delay between the two muscle actions (Elliott, Baxter, & Besier, 1999; Wilson, Elliott, & Wood, 1991). In normal movements muscles are also used in shortening-stretch cycles where the muscle undergoes concentric shortening followed by eccentric elongation as the muscle torque decreases below the resistance torque (Rassier *et al.*, 1999).

Early research on frog muscle by Cavagna, Saibene, & Margaria (1965) demonstrated that concentric muscle work was potentiated (increased) when preceded by active stretch (eccentric action). This

phenomenon is know as the **stretch-shortening cycle** or stretch-shorten cycle and has been extensively studied by Paavo Komi (1984, 1986). The use of fiberoptic tendon force sensors and estimates of MTU length has allowed Komi and his colleagues to create approximate *in vivo* torque–angular velocity diagrams (Figure 4.15). The loop in the initial concentric motion shows the higher concentric tensions that are created following the rather less-than-maximal eccentric tensions. The performance benefit of SSC coordination over purely concentric actions is usually between 10 and 20% (see "Stretch-Shortening Cycle" activity), but the biomechanical origin of these functional benefits is still unclear. Many biomechanical variables have been examined to study the mechanism of the SSC, and the benefits of the SSC are dependent on when these variables are calculated (Bird & Hudson, 1998) and the resistance moved (Cronin, McNair, & Marshall, 2001b).

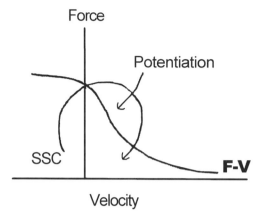

Figure 4.15. Schematic of the *in-vivo* muscle force–velocity behavior during an SSC movement, and force–velocity behavior derived from multiple isokinetic tests (see Barclay, 1997). Initial concentric shortening in an SSC creates higher forces than the classic Force–Velocity Curve. Muscle length is not measured directly but inferred from joint angle changes, so the true behavior of the muscle in human SSC motions is not known.

Activity: Stretch-Shortening Cycle

The benefits of stretch-shortening cycle muscle actions are most apparent in vigorous, full-effort movements. To see the size of the benefit for performance, execute several overarm throws for distance on a flat, smooth field. Measure the distance of your maximal-effort throw with your feet still and body facing the direction of your throw. Have someone help you determine about how far your trunk and arm backswing was in the normal throw. Measure the distance of a primarily concentric action beginning from a static position that matches your reversal trunk and arm position in the normal throw. Calculate the benefit of the SSC (prestretch augmentation) in the throw as (Normal − Concentric)/Concentric. Compare your results with those of others, the lab activity, and research on vertical jumping (Walshe, Wilson, & Murphy, 1996; Kubo *et al.*, 1999). The benefit of the SSC to other faster movements is likely to be even higher than in vertical jumping (Komi & Gollhofer, 1997). What factors might affect amount of prestretch augmentation? What might be the prestretch augmentation in other movements with different loads?

The mechanisms of the beneficial effects of SSC coordination is of considerable interest to biomechanics scholars. There are four potential sources of the greater muscle force in the concentric phase of an SSC: contractile potentiation, reflex potentiation, storage and reutilization of elastic energy, and the time available for force development (Komi, 1986; van Ingen Schenau, Bobbert, & de Haan, 1997). Contractile potentiation of muscle force is one of several variations in muscle force potential based on previous muscles actions. These phenomena are called history-dependent behaviors (see Herzog, Koh, Hasler, & Leonard, 2000; Sale, 2002). Shortening actions tend to depress force output of subsequent muscle actions, while eccentric actions tend to increase concentric actions that immediately follow. Force potentiation of muscle is also dependent on muscle length (Edman *et al.*, 1997).

Another mechanism for the beneficial effect of an SSC is a greater contribution from the myotatic or stretch reflex (see the section on "Proprioception of Muscle Action and Movement"). Muscle spindles are proprioceptors of muscle length and are particularly sensitive to fast stretch. When muscles are rapidly stretched, like in an SSC movement, muscle spindles activate a short reflex loop that strongly activates the muscle being stretched. Studies of athletes have shown greater activation of muscles in the concentric phase of an SSC movement compared to untrained subjects (Komi & Golhoffer, 1997). The lack of a precise value for the electromechanical delay (the Force–Time Relationship) makes it unclear if stretch reflexes contribute to greater muscle forces in the late eccentric phase or the following concentric phase. The contribution of reflexes to the SSC remains controversial and is an important area of study.

One of the most controversial issues is the role of elastic energy stored in the eccentric, which can be subsequently recovered in the concentric phase of an SSC. There has been considerable interest in the potential metabolic energy savings in the reutilization of stored elastic energy in SSC movements. It may be more accurate to say elastic mechanisms in the SSC are preventing energy loss or maintaining muscle efficiency (Ettema, 2001), rather than an energy-saving mechanism. Animal studies (e.g., wallabies and kangaroo rats) have been used to look at the extremes of evolutionary adaptation in muscletendon units related to SSC movement economy (Biewener, 1998; Biewener & Roberts, 2000; Griffiths, 1989; 1991). A special issue of the *Journal of Applied Biomechanics* was devoted to the role of stored elastic energy in the human vertical jump (Gregor, 1997). Recent *in vivo* studies of the human gastrocnemius muscle in SSC movements has shown that the compliant tendon allows the muscle fibers to act in near isometric conditions at joint reversal and while the whole muscle shortens to allow elastic recoil of the tendinous structures to do more positive work (Kubo, Kanehisa, Takeshita, Kawakami, Fukashiro, & Fukunaga, 2000b; Kurokawa, Fukunaga, & Fukashiro, 2001). The interaction of tendon and muscle must be documented to fully understand the benefits of the SSC action of muscles (Finni *et al.*, 2000).

Another mechanism for the beneficial effect of SSC coordination is related to the timing of force development. Recall that the rate of force development and the Force–Time Relationship have dramatic effect on high-speed and high-power movements. The idea is that if the concentric movement can begin with near-maximal force and the slack taken out of the elastic elements of the MTU, the initial acceleration and eventual velocity of the movement will be maximized. While this is logical, the interaction of other biomechanical factors (Force–Length Relationship, architecture, and leverage) makes it difficult to examine this hypothesis. Interested students should

see papers on this issue in the vertical jump (Bobbert, Gerritsen, Litjens, & van Soest, 1996; Bobbert & van Zandwijk, 1999) and sprint starts (Kraan, van Veen, Snijders, & Storm, 2001).

The most influential mechanism for the beneficial effect of an SSC will likely depend on the movement. Some events like the foot strike in sprinting or running jump (100 to 200 ms) require high rates of force development that are not possible from rest due to the Force–Time Relationship. These high-speed events require a well-trained SSC technique and likely have a different mix of the four factors than a standing vertical jump.

Plyometric (plyo=more metric=length) training will likely increase the athlete's ability to tolerate higher eccentric muscle forces and increase the potentiation of initial concentric forces (Komi, 1986). Plyometrics are most beneficial for athletes in high-speed and power activities. There has been considerable research on the biomechanics of lower-body drop jumping plyometrics (Bobbert, 1990). Early studies showed that jumpers tend to spontaneously adopt one of two techniques (Bobbert, Makay, Schinkelshoek, Huijing, & van Ingen Schenau, 1986) in drop jumping exercises. Recent research has focused on technique adaptations due to the compliance of the landing surface (Sanders & Allen, 1993), the effect of landing position (Kovacs *et al.*, 1999), and what might be the optimal drop height (Lees & Fahmi, 1994). Less research has been conducted on the biomechanics of upper body plyometrics (Newton, Kraemer, Hakkinen, Humphries, & Murphy, 1996; Knudson, 2001c). Loads for plyometric exercises are controversial, with loads ranging between 30 and 70% of isometric muscular strength because maximum power output varies with technique and the movement (Cronin, McNair, & Marshall, 2001a; Izquierdo, Ibanez, Gorostiaga, Gaurrues, Zuniga, Anton, Larrion, &

Hakkinen, 1999; Kaneko, Fuchimoto, Toji, & Suei, 1983; Newton, Murphy, Humphries, Wilson, Kraemer, & Hakkinen, 1997; Wilson, Newton, Murphy, & Humphries, 1993).

Plyometrics are not usually recommended for untrained subjects. Even though eccentric muscle actions are normal, intense unaccustomed eccentric activity is clearly associated with muscle damage. Eccentric-induced muscle fiber damage has been extensively studied and appears to be related to excessive strain in sarcomeres (see the review by Lieber & Friden, 1999) rather than the high forces of eccentric actions. Kinesiology professionals should carefully monitor plyometric technique and exercise intensity to minimize the risk of injury.

FORCE–TIME PRINCIPLE

The Force–Time Principle for applying biomechanics is not the same as the Force–Time Relationship of muscle mechanics. The **Force–Time Principle** states that the time available for force application is as important as the size of the forces used to create or modify movement. So the Force–Time Principle is concerned with the temporal strategy of force application in movements, while the Force–Time Relationship (electromechanical delay) states a fact that the tension build-up of muscle takes time. The electromechanical delay is clearly related to how a person selects the appropriate timing of force application. The Force–Time Principle will be illustrated in using forces to slow down an external object, and to project or strike an object.

Movers can apply forces in the opposite direction of the motion of an object to gradually slow down the object. Movements like catching a ball or landing from a jump (Figure 4.16) employ primarily eccentric

muscle actions to gradually slow down a mass over some period of time. Positioning the body to intercept the object early allows the mover to maximize the time the object can be slowed down. How does a gymnast maximize the time of force application to cushion the landing from a dismount from a high apparatus? Near complete extension of the lower extremities at touchdown on the mat allows near maximal joint range of motion to flex the joints and more time to bring the body to a stop.

The primary biomechanical advantage of this longer time of force application is safety, because the peak force experienced by the body (and consequently the stress in tissues) will be lower than during a short application of force. Moving the body and reaching with the extremities to maximize the time of catching also has strategic advantages in many sports. A team handball player intercepting the ball early not only has a higher chance of a successful catch, but they may prevent an opponent from intercepting. The distance and time the ball is in the air before contacting the catcher's hands is decreased with good arm extension, so there is less chance of an opponent intercepting the pass.

Imagine you are a track coach whose observations of a discus thrower indicate they are rushing their motion across the ring. The Force–Motion Principle and the force–velocity relationship make you think that slowing the increase in speed on the turns and motion across the circle might allow for longer throws. There is likely a limit to the benefit of increasing the time to apply because maximizing discus speed in an appropriate angle at release is the objective of the event. Are there timing data for elite discus throwers available to help with this athlete, or is there a little art in the application of this principle? Are you aware of other sports or activities where coaches focus on a controlled build-up in speed or unrushed rhythm?

Figure 4.16. The extension of the limbs before contact and increasing flexion in landing increases the time that forces can be applied to slow down the body. In chapter 6 we will see that this decreases the peak force on the body and decreases the risk of injury.

So there are sometimes limits to the benefit of increasing the time of force application. In movements with high demands on timing accuracy (baseball batting or a tennis forehand), the athlete should not maximize the time of force application because extra speed is of lower importance than temporal accuracy. If a tennis player preferred a large loop backswing where they used a large amount of time and the force of gravity to create racket head speed, the player will be vulnerable to fast and unpredictable strokes from an opponent. The wise opponent would mix up shot placements, spin, and increase time pressure to make it difficult for the player to get their long stroke in.

Suppose a patient rehabilitating from surgery is having difficulty using even the smallest weights in the clinic. How could a therapist use the Force–Time Principle to provide a therapeutic muscular overload? If bodyweight or assistive devices were available, could the therapist have the patient progressively increase the time they isometrically hold various positions? While this approach would tend to benefit muscular endurance more so than muscular strength, these two variables are related and tend to improve the other. Increasing the time of muscle activity in isometric actions or by modifying the cadence of dynamic exercises is a common training device used in rehabilitation and strength training. Timing is an important aspect of the application of force in all human movement.

Application: Force–Time and Range-of-Motion Principle Interaction

Human movements are quite complex, and several biomechanical principles often apply. This is a challenge for the kinesiology professional, who must determine how the task, performer characteristics, and biomechanical principles interact. In the sport of weight lifting, coaches know that the initial pull on the bar in snatch and clean-and-jerk lifts should not be maximal until the bar reaches about knee height. This appears counter to the Force–Time principle, where maximizing initial force over the time of the lift seems to be advantageous. It turns out that there are ranges of motion (postural) and muscle mechanical issues that are more important than a rigid application of the Force–Time Principle. The whole-body muscular strength curve for this lift is maximal near knee level, so a fast bar speed at the strongest body position compromises the lift because of the decrease in muscle forces in increasing speed of concentric shortening (Garhammer, 1989; Zatsiorsky, 1995). In other words, there is a Force–Motion advantage of a nearly-maximal start when the bar reaches knee level that tends to outweigh maximal effort at the start of the movement. Olympic lifts require a great deal of practice and skill. The combination of speed and force, as well as the motor skills involved in Olympic lifting, makes these whole-body movements popular conditioning exercises for high power sports (Garhammer, 1989).

In studying the kinetics of human movement (chapters 6 and 7), we will see several examples of how the human body creates forces over time. There will be many examples where temporal and other biomechanical factors make it a poor strategy to increase the time to apply force. In sprinting, each foot contact has to remain short (about 100 ms), so increasing rate of force development (Force–Motion Principle) is more appropriate than increasing the time of force application. It is important to realize that applying a force over a long time period can be a useful principle to apply, but it must be weighed with the other biomechanical principles, the environment, and subject characteristics that interact with the purpose of the movement.

NEUROMUSCULAR CONTROL

The mechanical response of muscles also strongly depends on how the muscles are activated. The neuromuscular control of movement is an active area of study where biomechanical research methods have been particularly useful. This section will summarize the important structures and their functions in the activation of muscles to regulate muscle forces and movement.

The Functional Unit of Control: Motor Units

The coordination and regulation of movement is of considerable interest to many scholars. At the structural end of the neuromuscular control process are the functional units of the control of muscles: motor units. A **motor unit** is one motor neuron and all the muscle fibers it innervates. A muscle may have from a few to several hundred motor units. The activation of a motor axon results in stimulation of all the fibers of that motor unit and the resulting twitch. All fibers of a motor unit have this synchronized, "all-or-nothing" response. Pioneering work in the neurophysiology of muscle activation was done in the early 20th century by Sherrington, Arian, and Denny-Brown (Burke, 1986).

Regulation of Muscle Force

If the muscle fibers of a motor unit twitch in unison, how does a whole muscle generate a smooth increase in tension? The precise regulation of muscle tension results from two processes: **recruitment** of different motor units and their **firing rate**.

Recruitment is the activation of different motor units within a muscle. Physiological research has determined three important properties of recruitment of motor units. First, motor units tend to be organized in pools or task groups (Burke, 1986). Second, motor units tend to be recruited in an asynchronous fashion. Different motor units are stimulated at slightly different times, staggering the twitches to help smooth out the rise in tension. There is evidence that some motor unit synchronization develops to increase rate of force development (Semmler, 2002), but too much synchronous recruitment results in pulses of tension/tremor that is associated with disease (Parkinson's) or extreme fatigue (final repetition of an exhaustive set of weight lifting). The recruitment of motor units is likely more complex than these general trends since serial and transverse connections between parallel architecture muscles allows active fibers to modify the tension and length of nearby fibers (Sheard, 2000).

The third organizational principle of recruitment has been called orderly recruitment or the **size principle** (Denny-Brown & Pennybacker, 1938; Henneman, Somjen, & Carpenter, 1965). It turns out that motor units tend to have specialized innervation and homogeneous fiber types, so motor units take on the characteristics of a particular fiber type. A small motor unit consists of a motor axon with limited myelination and primarily SO muscle fibers, while a large motor unit has a large motor axon (considerable myelination) and primarily FG fibers. The recruitment of a large motor unit by the brain results in the quickest message and build-up in tension, while recruitment of a small motor unit has a slower nerve conduction velocity and a gradual tension build-up (Figure 4.17).

In essence, the size principle says that motor units are recruited progressively from small (slow-twitch) to large (fast-twitch). A gradual increase in muscle force would result from recruitment of SO dominant motor units followed by FOG and FG dominant units, and motor units would be derecruited in reverse order if the force is to gradually decline. This holds true for most movements, but there is the ability to increase firing rate of large motor units within the size principle to move quickly or rapidly build up forces (Bawa, 2002; Burke, 1986). Have you ever picked up a light object (empty suitcase) when you expected a heavy one? If so, you likely activated many pools of large and small motor units immediately and nearly threw the object. Athletes in events requiring high rates of force development (jumping, throwing) will need to train their ability to override the size principle and activate many motor units rapidly. There are also many other factors that complicate the interpretation that the size principle is an invariant in motor control (Enoka, 2002). One example is that recruitment tends to be patterned for specific movements (Desmedt & Godaux, 1977; Sale, 1987), so the hamstring muscles may not be recruited the same in a jump, squat, or knee flexion exercise. Activation of muscles also varies across the kinds of muscle actions (Enoka, 1996; Gandevia, 1999; Gielen, 1999) and can be mediated by fatigue and sensory feedback (Enoka, 2002).

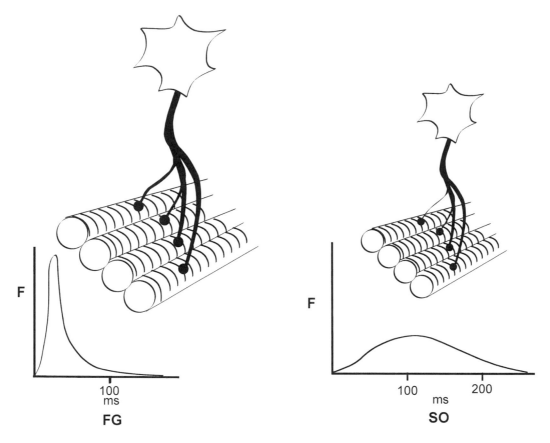

Figure 4.17. Schematic of the differences in the size of motor units. Typical twitch response, size of the motor nerve, and typical recruitment are illustrated.

Interdisciplinary Issue: The Control of Movement

With hundreds of muscles, each with hundreds of motor units that must be repeatedly stimulated, to coordinate in a whole-body movement, can the brain centrally control all those messages? If the brain could send all those messages in a preprogrammed fashion, could it also monitor and evaluate efferent sensory and proprioceptive information and adjust the movement? Early motor learning research and theory focused on the brain's central control of movement or a motor program. More recent research is based on a Bernstein or dynamical systems perspective (Feldman, Levin, Mitnitski, & Archambault, 1998; Schmidt & Wrisberg, 2000), where more general control strategies interact with sensory feedback. Since kinetic variables (torques, forces, EMG) can be measured or calculated using biomechanics, many motor control scholars are interested in looking at these variables to uncover clues as to how movement is coordinated and regulated. Some biomechanists are interested in the control of movement, so here is an ideal area for interdisciplinary research.

Firing rate or *rate coding* is the repeated stimulation of a particular motor unit over time. To create the muscle forces for normal movements, the frequency (Hz) that motor units are usually rate coded is between 10 and 30 Hz, while FG motor units have a faster relaxation time and can be rate coded between 30 and 60 Hz (Sale, 1992). The repeated stimulation of a motor unit increases the twitch force above the level of a single twitch (up to 10 times) because the tension in the fibers begins at a higher level, before the decay or relaxation in tension. Since recruitment tends to be asynchronous and firing rates vary with motor unit size, the twitches of the motor units in a whole muscle combine and cancel each other out, resulting in smooth changes in tension. When muscle is artificially stimulated for research or training purposes to elicit maximal force, the frequency used is usually higher than 60 Hz to make sure that motor unit twitches fuse into a **tetanus**. A **tetanus** is the summation of individual twitches into a smooth increase in muscle tension.

Both recruitment and firing rate have a dramatic influence on the range of muscle forces that can be created. How recruitment and firing rate interact to increase muscle forces is quite complex, but it appears that recruitment dominates for forces up to 50% of maximum with increasing importance of firing rate (Enoka, 2002). The combined effect of recruitment and firing rate of motor units is reflected in the size, density, and complexity of the eletromyographic (EMG) signal. Special indwelling EMG electrode techniques are used to study the recruitment of individual motor units (Basmajian & DeLuca, 1985).

Recall that in chapter 3 we learned how EMG research has shown that at the whole muscle level muscles are activated to in complex synergies to achieve movement or stabilization tasks. Physical medicine professionals often take advantage of this flexibility of the neuromuscular system by

Application: Neuromuscular Training

Unfortunately, athletes are often stereotyped as dumb jocks with gifted physical abilities. How much of movement ability do physical characteristics like muscular strength, speed, and coordination contribute to performance compared to neuromuscular abilities (a good motor brain)? Think about the ability your favorite athlete would have if he/she had a stroke that affected part of their motor cortex. In training and conditioning there are several areas of research where there is evidence that the effects of training on muscle activation by the central nervous system is underrated. First, it is well known that the majority of the initial gains in strength training (first month) are related to the neural drive rather than hypertrophy (see Sale 1992). Second, it is known that both normal and injured subjects are not usually able to achieve true maximum muscle force in a maximal voluntary contraction. This is called **muscle inhibition** and is studied using an electrical stimulation method called **twitch interpolation technique** (Brondino, Suter, Lee, & Herzog, 2002). Another area of neuromuscular research relates to the inability to express bilateral muscular strength (both arms or legs) equal to the sum of the unilateral strength of each extremity. This phenomenon has been called the bilateral deficit but the decrements (3–20%) are not always observed (Jakobi & Cafarelli, 1998). Interest in the biomechanics of the vertical jump has made this movement a good model for examining a potential bilateral deficit (Challis, 1998). How might a professional try to differentiate true differences in muscular strength between sides of the body and a bilateral deficit? If there truly is a bilateral deficit that limits the neuromuscular activation of two extremities, how should you train for bilateral movements (bilaterally, stronger or weaker limb)?

training muscle actions that compensate for physical limitations from disease or injury. Motor learning scholars are interested in EMG and the activation of muscles as clues to neuromuscular strategies in learning movements. While there has not been extensive research in this area, it appears that changes in EMG with practice/training depend on the nature of the task (Gabriel & Boucher, 2000). As people learn submaximal movements, the duration of EMG bursts decrease, there are decreases in extraneous and coactivation of muscles, and a reduction in EMG magnitude as the body learns to use other forces (inertial and gravitational) to efficiently create the movement (Englehorn, 1983; Moore & Marteniuk, 1986; Newell, Kugler, van Emmerick, & McDonald, 1989). Maximal-effort movements are believed to be more reliant on changes in the magnitude and rise time of activation, than the duration of muscle activation (Gottlieb, Corcos, & Agarwal, 1989). In maximal high-speed movements, the magnitude and rate of increase in activation tends to increase (Corcos, Jaric, Agarwal, & Gottlieb, 1993; Darling & Cooke, 1987; Gabriel & Boucher, 2000) with practice. The activation and cooperative actions of muscles to create skilled human movement are very complex phenomena.

Proprioception of Muscle Action and Movement

Considerable information about the body and its environment are used in the regulation of many movements. While persons use all their senses to gather information about the status or effectiveness of their movements, there are musculoskeletal receptors that provide information to the brain to help produce movement. These receptors of information about the motion and force in muscles and joints are called proprioceptors. While we usually do not consciously attend to this information, this information and the various reflexes they initiate are important in the organization of movement. A reflex is an involuntary response initiated by some sensory stimulus. Reflexes are only initiated if the sensory stimulus is above some threshold.

There are many proprioceptive receptors that monitor aspects of movement. Information about joint position is provided by four kinds of receptors. The vestibular system of the inner ear provides information about the head's orientation with respect to gravity. This section will summarize the important MTU proprioceptors that provide information on muscle length (muscle spindles) and force (Golgi tendon organs). Human movement performance relies on an integration of all sensory organs, and training can be quite effective in utilizing or overriding various sensory or reflex responses. A dancer spinning in the transverse plane prevents dizziness (from motion in inner ear fluid) by spotting—rotating the neck opposite to the spin to keep the eyes fixed on a point followed by a quick rotation with the spin so as to find that point again. Athletes in "muscular strength" sports not only train their muscle tissue to shift the Force–Velocity Relationship upward, they train their central nervous system to activate more motor units and override the inhibitory effect of Golgi tendon organs.

When muscle is activated, the tension that is developed is sensed by Golgi tendon organs. Golgi tendon organs are located at the musculotendinous junction and have an inhibitory effect on the creation of tension in the muscle. Golgi tendon organs connect to the motor neurons of that muscle and can relax a muscle to protect it from excessive loading. The intensity of this autogenic inhibition varies, and its functional significance in movement is controversial (Chalmers, 2002). If an active muscle were forcibly stretched by an external force, the

Golgi tendon organs would likely relax that muscle to decrease the tension and protect the muscle. Much of high speed and high muscular strength performance is training the central nervous system to override this safety feature of the neuromuscular system. The rare occurrence of a parent lifting part of an automobile off a child is an extreme example of overriding Golgi tendon organ inhibition from the emotion and adrenaline. The action of Golgi tendon organs is also obvious when muscles suddenly stop creating tension. Good examples are the collapse of a person's arm in a close wrist wrestling match (fatigue causes the person to lose the ability to override inhibition) or the buckling of a leg during the great loading of the take-off leg in running jumps.

Muscle spindles are sensory receptors located between muscle fibers that sense the length and speed of lengthening or shortening. Muscle spindles are sensitive to stretch and send excitatory messages to activate the muscle and protect it from stretch-related injury. Muscle spindles are sensitive to slow stretching of muscle, but provide the largest response to rapid stretches. The rapid activation of a quickly stretched muscle (100–200 ms) from muscle spindles is due to a short reflex arc. Muscle spindle activity is responsible for this **myotatic reflex** or stretch reflex. Use of a small rubber reflex hammer by physicians allows them to check the stretch reflex responses of patients. The large numbers of spindles and their innervation allows them to be sensitive and reset throughout the range of motion. Recall that a stretch reflex is one possible mechanism for the benefit of an SSC. Stretch reflexes may contribute to the eccentric braking action of muscles in follow-throughs. In performing stretching exercises, the rate of stretch should be minimized to prevent activation of muscle spindles.

The other important neuromuscular effect of muscle spindles is inhibition of the antagonist (opposing muscle action) mus-cle when the muscle of interest is shortening. This phenomenon is call **reciprocal inhibition**. Relaxation of the opposing muscle of a shortening muscle contributes to efficient movement. In lifting a drink to your mouth the initial shortening of the biceps inhibits triceps activity that would make the biceps work harder than necessary. Reciprocal inhibition is often overridden by the central nervous system when coactivation of muscles on both sides of a joint is needed to push or move in a specific direction. Reciprocal inhibition also plays a role in several stretching techniques designed to utilize neuromuscular responses to facilitate stretching. The contract–relax–agonist–contract technique of proprioceptive neuromuscular facilitation (PNF) is designed to use reciprocal inhibition to relax the muscle being stretched by contracting the opposite muscle group (Hutton, 1993; Knudson, 1998). For example, in stretching the hamstrings, the stretcher activates the hip flexors to help relax the hip extensors being stretched. The intricacies of proprioceptors in neuromuscular control (Enoka, 2002; Taylor & Prochazka, 1981) are relevant to all movement professionals, but are of special interest to those who deal with disorders of the neuromuscular system (e.g., neurologists, physical therapists).

SUMMARY

Forces applied to the musculoskeletal system create loads in these tissues. Loads are named based on their direction and line of action relative to the structure. Several mechanical variables are used to document the mechanical effect of these loads on the body. How hard forces act on tissue is mechanical stress, while tissue deformation is measured by strain. Simultaneous measurement of force applied to a tissue and its deformation allow biomechanists to determine the stiffness and mechanical strength of biological specimens. Musculoskeletal tissues are viscoelastic. This means that their deformation depends on the rate of loading and they lose some energy (hysteresis) when returning to normal shape. Bones are strongest in compression, while ligaments and tendon are strongest in tension. Three major mechanical characteristics of muscle that affect the tension skeletal muscles can create are the force–velocity, force–length, and force–time relationships. An important neuromuscular strategy used to maximize the initial muscle forces in most movements is the rapid reversal of a countermovement with an eccentric muscle action into a concentric action. This strategy is called the stretch-shortening cycle. The creation of muscular force is controlled by recruitment of motor units and modulating their firing rate. Several musculotendon proprioceptors provide length and tension information to the central nervous system to help regulate muscle actions. The Force–Time Principle is the natural application of the mechanical characteristics of muscle. The timing of force application is as important as the size of the forces the body can create. In most movement, increasing the time of force application can enhance safety, but kinesiology professionals must be aware of how this principle interacts with other biomechanical principles. The application of biomechanical principles is not easy, because they interact with each other and also with factors related to the task, individual differences, or the movement environment.

REVIEW QUESTIONS

1. What are the major kinds of mechanical loads experienced by muscle, tendon, and bone?

2. What are the mechanical variables that can be determined from a load-defor-mation curve, and what do they tell us about the response of a material to loading?

3. Compare and contrast the mechanical strength of muscle, tendon, ligaments and bone. How does this correspond to the incidence of various musculoskeletal injuries?

4. How does the passive behavior of the muscletendon unit affect the prescription of stretching exercises?

5. What are the functional implications of the Force–Velocity Relationship of skeletal muscle for strength training?

6. Use the Force–Length Relationship to describe how the active and passive components of muscle tension vary in the range of motion.

7. Compare and contrast the Force–Time Relationship of muscle with the Force–Time Principle of biomechanics.

8. When might increasing the time of force application not benefit the development of movement speed and why?

9. How does the brain control muscle force and how is muscle fiber type related?

10. What is the stretch-shortening cycle and in what kinds of movements is it most important?

11. What are the two major proprioceptors in muscle that monitor length and force?

12. How can range of motion in a movement be defined?

13. Explain how range of motion affects the speed, accuracy and force potential of movement. Give an example of when the Range of Motion Principle is mediated by other mechanical factors.

14. What biomechanical properties help contribute to the beneficial effect of continuous passive movement therapy (i.e., the slow, assisted motion of an injured limb) following surgery?

15. Write a complete description of a stretching or conditioning exercise. Identify the likely muscle actions and forces contributing to the movement.

KEY TERMS

bending
creep
compression
degrees of freedom
electromechanical delay
energy (mechanical)
firing rate(rate coding)
force–length relationship
force–time relationship
  (see electromechanical delay)
force–velocity relationship
hysteresis
isokinetic
load
motor unit
myotatic reflex
reciprocal inhibition
recruitment
shear
strain
strength (mechanical)
stress
stress fracture
stress relaxation
stretch-shortening cycle
tetanus
thixotropy
torsion
viscoelastic
Wolff's Law
Young's modulus

SUGGESTED READING

Alexander, R. M. (1992). *The human machine*. New York: Columbia University Press.

Biewener, A. A., & Roberts, T. J. (2000). Muscle and tendon contributions to force, work, and elastic energy savings: a comparative perspective. *Exercise and Sport Sciences Reviews*, **28**, 99–107.

Chapman, A. E. (1985). The mechanical properties of human muscle. *Exercise and Sport Sciences Reviews*, **13**, 443–501.

Enoka, R. (2002). *The neuromechanical basis of human movement* (3rd ed.). Champaign, IL: Human Kinetics.

De Luca, C. J. (1997). The use of surface electromyography in biomechanics. *Journal of Applied Biomechanics*, **13**, 135–163.

Fung, Y. C. (1981). *Biomechanics: Mechanical properties of living tissues*. New York: Springer-Verlag.

Gulch, R. W. (1994). Force–velocity relations in human skeletal muscle. *International Journal of Sports Medicine*, **15**, S2–S10.

Komi, P. V. (Ed.). (1992). *Strength and power in sport*. London: Blackwell Scientific Publications.

Latash, M. L., & Zatsiorsky, V. M. (Eds.) (2001). *Classics in movement science*. Champaign, IL: Human Kinetics.

Lieber, R. L., & Bodine-Fowler, S. C. (1993). Skeletal muscle mechanics: Implications for rehabilitation. *Physical Therapy*, **73**, 844–856.

Moritani, T., & Yoshitake, Y. (1998). The use of electromyography in applied physiology. *Journal of Electromyography and Kinesiology*, **8**, 363–381.

Mow, V. C., & Hayes, W. C. (Eds.) (1991). *Basic orthopaedic biomechanics*. New York: Raven Press.

Nordin, M., & Frankel, V. (2001). *Basic biomechanics of the musculoskeletal system* (3rd ed.). Baltimore: Williams & Wilkins.

Panjabi, M. M., & White, A. A. (2001). *Biomechanics in the musculoskeletal system*. New York: Churchill Livingstone.

Patel, T. J., & Lieber, R. L. (1997). Force transmission in skeletal muscle: From actomyosin to external tendons. *Exercise and Sport Sciences Reviews*, **25**, 321–364.

Rassier, D. E., MacIntosh, B. R., & Herzog, W. (1999). Length dependence of active force production in skeletal muscle. *Journal of Applied Physiology*, **86**, 1445–1457.

Scott, W., Stevens, J., & Binder-Macleod, S. A. (2001). Human skeletal muscle fiber type classifications. *Physical Therapy*, **81**, 1810–1816.

Whiting, W. C., & Zernicke, R. F. (1998). *Biomechanics of musculoskeletal injury*. Champaign, IL: Human Kinetics.

Zatsiorsky, V. (1995). *Science and practice of strength training*. Champaign, IL: Human Kinetics.

WEB LINKS

Electromyography (EMG) papers by Carlo DeLuca.
　　　http://www.delsys.com/library/tutorials.htm

MSRC—Musculoskeletal Research Center at the University of Pittsburgh focuses on the mechanical response of tissues to forces.
　　　http://www.pitt.edu/~msrc/

ISEK Standards—International Society for Electrophysiological Kinesiology standards for reporting EMG research.
　　　http://shogun.bu.edu/isek/publications/standards/index.html

Muscle Physiology Online—UC San Diego website with a comprehensive review of physiological and biomechanical aspects of muscle. See the macroscopic structure and muscle–joint interactions.
　　　http://muscle.ucsd.edu/index.html

Muscle structure, physiology, and mechanics of action slides by Paul Paolini at San Diego State.
　　　http://www.sci.sdsu.edu/Faculty/Paul.Paolini/ppp/lecture18/sld001.htm

Visible Human Project—National Institutes of Health human body imaging project.
　　　http://www.nlm.nih.gov/research/visible/visible_human.html

PART III

MECHANICAL BASES

Mechanics is the branch of physics that measures the motion of objects and explains the causes of that motion. The images here present illustrations of the kinematic and kinetic branches of biomechanics. Measurement of the three-dimensional motion of a golfer provides a precise kinematic description of the golf swing. The angular velocities of key body segments are plotted in the graph. The representation of the orientation of the key segments of the leg and the resultant force applied by the foot to the pedal of an exercise bike represents the kinetics, or the forces that cause human movement. A knowledge of the mechanics of exercise movements allows kinesiology professionals to understand those movements, develop specific training exercises, and change movement technique to improve performance. The chapters in part III introduce you to three key areas of this parent discipline of biomechanics: kinematics, kinetics, and fluid mechanics. The related lab activities explore qualitative and quantitative analyses of key mechanical variables important in understanding human movement.

Golf illustration provided courtesy of Skill Technologies Inc., Phoenix, AZ—www.skilltechnologies.com.

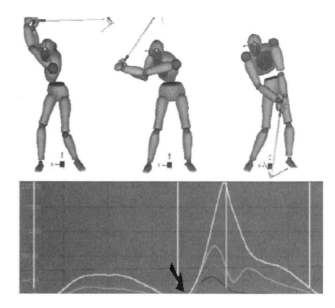

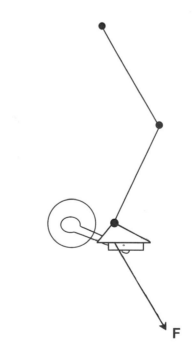

CHAPTER 5

Linear and Angular Kinematics

Kinematics is the accurate description of motion and is essential to understanding the biomechanics of human motion. Kinematics can range from anatomical descriptions of joint rotations to precise mathematical measurements of musculoskeletal motions. Recall from chapter 2 that kinematics is subdivided according to the kinds of measurements used, either linear or angular. Whatever the form of measurement, biomechanical studies of the kinematics of skilled performers provide valuable information on desirable movement technique. Biomechanics has a long history of kinematic measurements of human motion (Cappozzo, Marchetti, & Tosi, 1990). Accurate kinematic measurements are sometimes used for the calculation of more complex, kinetic variables. This chapter will introduce key kinematic variables in documenting both linear and angular human motions. The principles of biomechanics that apply kinematics to improving human movement are **Optimal Projection** and the **Coordination Continuum**.

LINEAR MOTION

Motion is change in position with respect to some frame of reference. In mathematical terms, linear motion is simple to define: final position minus initial position. The simplest linear motion variable is a scalar called **distance** (*l*). The use of the symbol *l* may be easy to remember if you associate it with the length an object travels irrespective of direction. Typical units of distance are meters and feet. Imagine an outdoor adventurer leaves base camp and climbs for 4 hours through rough terrain along the path illustrated in Figure 5.1. If her final position traced a 1.3-km climb measured relative to the base camp (0 km) with a pedometer, the distance she climbed was 1.3 km (final position – initial position). Note that 1.3 km (kilometers) is equal to 1300 meters. The odometer in your car works in a similar fashion, counting the revolutions (angular motion) of the tires to generate a measurement of the distance (linear variable) the car travels. Because distance is a scalar, your odometer does not tell you in what direction you are driving on the one-way street!

The corresponding vector quantity to distance is **displacement** (**d**). Linear displacements are usually defined relative to right-angle directions, which are convenient for the purpose of the analysis. For most two-dimensional (2D) analyses of human movements, like in Figure 5.1, the directions used are horizontal and vertical, so displacements are calculated as final position minus initial position in that particular direction. The usual convention is that motions to the right on the x-axis and upward along the y-axis are positive, with motion in the opposite directions negative. Since displacement is a vector quantity, if motion upward and to the right is defined as positive, motion downward and motion to the left is a negative displacement. Recall that the sign of a number in mechanics refers to direction.

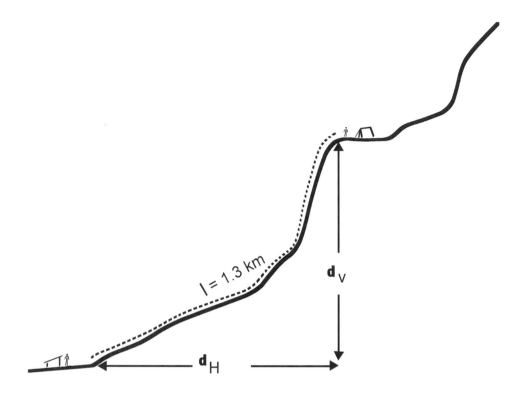

Figure 5.1. An outdoor adventurer climbs from base camp to a camp following the illustrated path. The distance the climber covers is 13 km. Her displacement is 0.8 km horizontally and 0.7 km vertically.

Assuming Figure 5.1 is drawn to the scale shown, it looks like the climber had a positive 0.8 km of horizontal displacement and 0.7 km of vertical displacement. This eyeballing of the horizontal and vertical components of the hike will be fairly accurate because displacement is a vector. Vectors can be conveniently represented by combinations of right-angle components, like the horizontal and vertical displacements in this example. If our adventurer were stranded in a blizzard and a helicopter had to lower a rescuer from a height of 0.71 km above base camp, what would be the rescuer's vertical displacement to the climber? The vertical displacement of the rescuer would be –0.01 km or 100 meters (final vertical position minus initial vertical position or 0.7 km – 0.71 km).

Biomechanists most often use measures of displacement rather than distance because they carry directional information that is crucial to calculation of other kinematic and kinetic variables. There are a couple of subtleties to these examples. First, the analysis is a simple 2D model of truly 3D reality. Second, the human body is modeled as a **point mass**. In other words, we know nothing about the orientation of the body or body segment motions; we just confine the analysis to the whole body mass acting at one point in space. Finally, an absolute frame of reference was used until the end, when we are interested in the displacement relative to a moving object—like the helicopter.

In other biomechanical studies of human motion the models and frames of reference can get quite complicated. The analysis might not be focused on whole-body movement but how much a muscle is shortening between two attachments. A three-dimensional (3D) analysis of the small accessory gliding motions of the knee joint motion would likely measure along anatomically relevant axes like proximal-distal, medio-lateral, and antero-posterior. Three-dimensional kinematic measurements in biomechanics require considerable numbers of markers, spatial calibration, and mathematical complexity for completion. **Degrees of freedom** represent the kinematic complexity of a biomechanical model. The degrees of freedom (dof) correspond to the number of kinematic measurements needed to completely describe the position of an object. A 2D point mass model has only 2 dof, so the motion of the object can be described with an x (horizontal) and a y (vertical) coordinate. The 3D motion of a body segment has 6 dof, because there are three linear coordinates (x, y, z) and three angles (to define the orientation of the segment) that must be specified. Good sources for a more detailed description of kinematics in biomechanics are available (Allard, Stokes, & Blanchi, 1995; Zatsiorsky, 1998). The field of biomechanics is striving to develop standards for reporting joint kinematics so that data can be exchanged and easily applied in various professional settings (Wu & Cavanagh, 1995).

The concept of **frame of reference** is, in essence, where you are measuring or observing the motion from. Reference frames in biomechanics are either absolute or relative. An absolute or global frame of reference is essentially motionless, like the apparent horizontal and vertical motion we experience relative to the earth and its gravitational field (as in Figure 5.1). A relative frame of reference is measuring from a point that is also free to move, like the motion of the foot relative to the hip or the plant foot relative to the soccer ball. There is

no one frame of reference that is best, because the biomechanical description that is most relevant depends on the purpose of the analysis.

This point of motion being relative to your frame of reference is important for several reasons. First, the appearance and amount of motion depends on where the motion is observed or measured from. You could always answer a question about a distance as some arbitrary number from an "unknown point of reference," but the accuracy of that answer may be good for only partial credit. Second, the many ways to describe the motion is much like the different anatomical terms that are sometimes used for the identical motion. Finally, this is a metaphor for an intellectually mature kinesiology professional who knows there is not one single way of seeing or measuring human motion because your frame of reference affects what you see. The next section will examine higher-order kinematic variables that are associated with the rates of change of an object's motion. It will be important to understand that these new variables are also dependent on the model and frame of reference used for their calculation.

Speed and Velocity

Speed is how fast an object is moving without regard to direction. Speed is a scalar quantity like distance, and most people have an accurate intuitive understanding of speed. **Speed** (s) is defined as the rate of change of distance ($s = l/t$), so typical units are m/s, ft/s, km/hr, or miles/hr. It is very important to note that our algebraic shorthand for speed (l/t), and other kinematic variables to come, means "the change in the numerator divided by the change in the denominator." This means that the calculated speed is an *average* value for the time interval used for the calculation. If you went jogging across town (5 miles) and arrived at the turn-around point in 30 minutes, your average speed would be (5 miles / 0.5 hours), or an average speed of 10 miles per hour. You likely had intervals where you ran faster or slower than 10 mph, so we will see how representative or accurate the kinematic calculation is depends on the size of time interval and the accuracy of your linear measurements.

Since biomechanical studies have used both the English and metric systems of measurements, students need to be able to convert speeds from one system to the other. Speeds reported in m/s can be converted to speeds that make sense to American drivers (mph) by essentially doubling them (mph = m/s · 2.23). Speeds in ft/s can be converted to m/s by multiplying by 0.30, and km/hour can be converted to m/s by multiplying by 0.278. Other conversion factors can be found in Appendix B. Table 5.1 lists some typical speeds in sports and other human movements that have been reported in the biomechanics literature. Examine Table 5.1 to get a feel for some of the typical peak speeds of human movement activities.

Table 5.1
TYPICAL PEAK SPEEDS IN HUMAN MOVEMENT

| | Speed | |
	m/s	mph
Bar in a bench press	0.25	0.6
Muscle shortening	0.5	1.1
Walking	1.1–1.8	2.5–4.0
Vertical jump	2.3	5.1
Free throw	7.0	15.7
Sprinting	12.0	26.8
Tennis forehand	20.0	44.6
Batting	31.3	70.0
Soccer kick	35.0	78.0
Baseball pitch	45.1	101
Tennis serve	62.6	140
Golf drive	66.0	148

Be sure to remember that speed is also relative to frame of reference. The motion of the runner in Figure 5.2 can correctly be described as 8 meters per second (m/s), 1 m/s, or –1.5 m/s. Runner A is moving 8 m/s relative to the starting line, 1 m/s faster than runner B, and 1.5 m/s slower than runner C. All are correct kinematic descriptions of the speed of runner A.

Application: Speed

One of the most important athletic abilities in many sports is speed. Coaches often quip that "luck follows speed" because a fast athlete can arrive in a crucial situation before their opponent. Coaches that often have a good understanding of speed are in cross-country and track. The careful timing of various intervals of a race, commonly referred to as pace, can be easily converted average speeds over that interval. Pace (time to run a specific distance) can be a good kinematic variable to know, but its value depends on the duration of the interval, how the athlete's speed changes over that interval, and the accuracy of the timing. How accurate do you think stopwatch measures of time and, consequently, speed are in track? What would a two-tenths-of-a-second error mean in walking (200 s), jogging (100 s), or sprinting (30 s) a lap on a 220 m track? Average speeds for these events are 1.1, 2.2, and 7.3 m/s, with potential errors of 0.1, 0.2, and 0.7%. Less than 1%! That sounds good, but what about shorter events like a 100-m dash or the hang time of a punt in football? American football has long used the 40-yard dash as a measure of speed, ability, and potential for athletes despite little proof of its value (see Maisel, 1998). If you measured time by freezing and counting frames of video (30 Hz), how much more accurate would a 40-meter dash timing be? If the current world record for 100 m is 9.79 seconds and elite runners cover 40 m (43.7 yards) out of the starting blocks in about 4.7 seconds, should you believe media guides that say that a certain freshman recruit at Biomechanical State University ran the forty (40-yard dash) in 4.3 seconds?

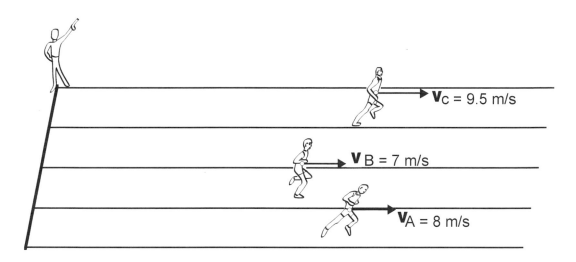

Figure 5.2. The speed of runner A depends on the frame of reference of the measurement. Runner A can be described as moving at 8 m/s relative to the track or –1.5 m/s relative to runner C.

Velocity is the vector corresponding to speed. The vector nature of velocity makes it more complicated that speed, so many people incorrectly use the words interchangeably and have incorrect notions about velocity. Velocity is essentially the speed of an object, *in a particular direction*. Velocity is the rate of change of displacement ($V = d/t$), so its units are the same as speed, and are usually qualified by a directional adjective (i.e., horizontal, vertical, resultant). Note that when the adjective "angular" is not used, the term *velocity* refers to linear velocity. If you hear a coach say a pitcher has "good velocity," the coach is not using biomechanical terminology correctly. A good question to ask in this situation is: "That's interesting. When and in what direction was the pitch velocity so good?"

The phrase "rate of change" is very important because velocity defines how quickly position is *changing* in the specified direction (displacement). Most students might recognize this phrase as the same one used to describe the derivative or the slope of a graph (like the hand dynamometer example in Figure 2.4).

Remember to think about velocity as a speed, but in a particular direction. A simple example of the velocity of human movement is illustrated by the path (dotted line in Figure 5.3) of a physical education student in a horizontal plane as he changes exercise stations in a circuit-training program. The directions used in this analysis are a fixed reference frame that is relevant to young students: the equipment axis and water axis.

The student's movement from his initial position (I) to the final position (F) can be vectorially represented by displacements along the equipment axis (d_E) and along the water axis (d_W). Note that the definition of these axes is arbitrary since the student must combine displacements in both directions to arrive at the basketballs or a drink. The net displacements for this student's movement are positive, because the final position measurements are larger than the initial positions. Let's assume that $d_E = 8$ m and $d_W = 2$ m and that the time it took this student to change stations was 10 seconds. The average velocity along the water axis would be $V_W = d_W/t = 2/10 =$

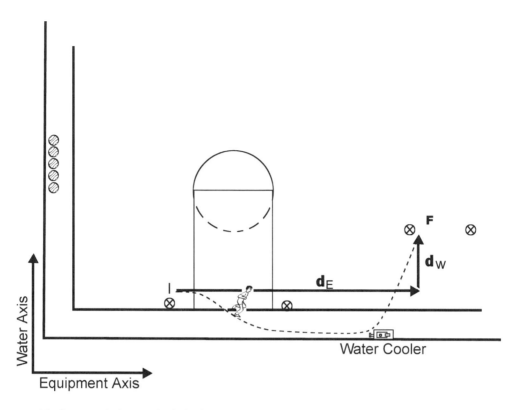

Figure 5.3. The horizontal plane path (dashed line) of a physical education student changing practice stations. The linear displacements can be measured along a water fountain axis and equipment axis.

0.2 m/s. Note that the motion of most interest to the student is the negative displacement ($-\mathbf{d}_W$), which permits a quick trip to the water fountain. The average velocity along the equipment axis would be 0.8 m/s ($\mathbf{V}_E = \mathbf{d}_L/t = 8/10$). Right-angle trigonometry can then be used to calculate the magnitude and direction of the resultant displacement ($\mathbf{d}_R$) and then the average velocity of the student. We will use right-angle trigonometry in chapter 6 to analyze the effect of force vectors. By the way, if your right-angle trigonometry is a little rusty, check out appendix D for a refresher.

Calculations of speed and velocity using algebra are *average* velocities over the time interval used. It is important to realize that the smaller the time interval the greater the potential accuracy of kinematic calcula-

tions. In the previous example, for instance, smaller time intervals of measurement would have detected the negative velocity (to get a drink) and positive velocity of the student in the water direction. Biomechanics research often uses high-speed film or video imaging (Gruen, 1997) to make kinematic measurements over very small time intervals (200 or thousands of pictures per second). The use of calculus allows for kinematic calculations ($\mathbf{v} = d_d/d_t$) to be made to *instantaneous* values for any point in time of interest. If kinematic calculations are based over a too large time interval, you may not be getting information much better than the time or pace of a whole race, or you may even get the unusual result of zero velocity because the race finished where it started.

Graphs of kinematic variables versus time are extremely useful in showing a pattern within the data. Because human movement occurs across time, biokinematic variables like displacement, velocity, and acceleration are usually plotted versus time, although there are other graphs that are of value. Figure 5.4 illustrates the horizontal displacement and velocity graphs for an elite male sprinter in a 100-m dash. Graphs of the speed over a longer race precisely document how the athlete runs the race. Notice that the athlete first approach their top speed at about the 40- to 50-meter mark. You can compare your velocity profile to Figure 5.4 and to those of other sprinters in Lab Activity 5.

Acceleration

The second derivative with respect to time, or the rate of change of velocity, is **acceleration**. Acceleration is how quickly velocity is changing. Remember that velocity changes when speed or direction change. This vec-

tor nature of velocity and acceleration means that it is important to think of acceleration as an unbalanced force in a particular direction. The acceleration of an object can speed it up, slow it down, or change its direction. It is incorrect to assume that "acceleration" means an object is speeding up. The use of the term "deceleration" should be avoided because it implies that the object is slowing down and does not take into account changes in direction.

Let's look at an example that illustrates why it is not good to assume the direction of motion when studying acceleration. Imagine a person is swimming laps, as illustrated in Figure 5.5. Motion to the right is designated positive, and the swimmer has a relatively constant velocity (zero horizontal acceleration) in the middle of the pool and as he approaches the wall. As his hand touches the wall there is a negative acceleration that first slows him down and then speeds him up in the negative direction to begin swimming again. Thinking of the acceleration at the wall as a push in the negative direction is correct throughout the

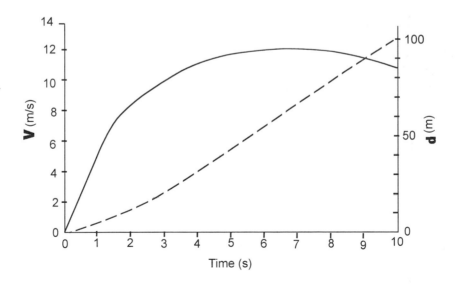

Figure 5.4. The displacement–time (dashed curve) and velocity–time (solid curve) graphs for the 100-m dash of an elite male sprinter.

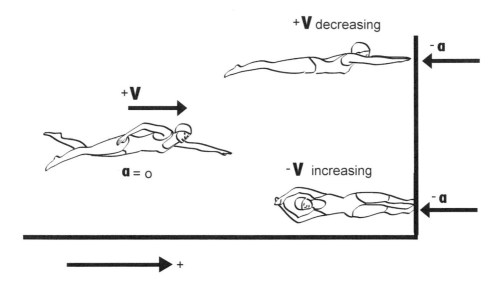

Figure 5.5. The motion and accelerations of swimmers as they change direction in lap swimming. If motion to the right is designated positive, the swimmer experiences a negative acceleration as they make the turn at the pool wall. The negative acceleration first slows positive velocity, and then begins to build negative velocity to start swimming in the negative direction. It is important to associate signs and accelerations with directions.

turn. As the swimmer touches the other wall there is a positive acceleration that decreases his negative velocity, and if he keeps pushing (hasn't had enough exercise) will increase his velocity in the positive direction back into the pool.

The algebraic definition of acceleration (**a**) is **V**/t, so typical units of acceleration are m/s² and ft/s². Another convenient way to express acceleration is in units of gravitational acceleration (*g*'s). When you jump off a box you experience (in flight) one *g* of acceleration, which is about −9.81 m/s/s or −32.2 ft/s/s. This means that, in the absence of significant air resistance, your vertical velocity will change 9.81 m/s every second in the negative direction. Note that this means you slow down 9.81 m/s every second on the way up and speed up 9.81 m/s every second on the way down. *G*'s are used for large acceleration events like a big change of direction on a roller coaster (4 *g*'s), the shockwaves in the lower leg following heel strike in running

(5 *g*'s), a tennis shot (50 *g*'s), or head acceleration in a football tackle (40–200 *g*'s). When a person is put under sustained (several seconds instead of an instant, like the previous examples) high-level acceleration like in jet fighters (5–9 *g*'s), pilots must a wear pressure suit and perform whole-body isometric muscle actions to prevent blacking out from the blood shifting in their body.

Acceleration due to gravity always acts in the same direction (toward the center of the earth) and may cause speeding up or slowing down depending on the direction of motion. Remember to think of acceleration as a push in a direction or a tendency to change velocity, not as speed or velocity. The vertical acceleration of a ball at peak flight in the toss of a tennis serve is 1 *g*, not zero. The vertical velocity may be instantaneously zero, but the constant pull of gravity is what prevents it from staying up there.

Let's see how big the horizontal acceleration of a sprinter is in getting out of the blocks. This is an easy example because the rules require that the sprinter have an initial horizontal velocity of zero. If video measurements of the sprinter showed that they passed the 10-m point at 1.9 seconds with a horizontal velocity of 7 m/s, what would be the runner's acceleration? The sprinter's change in velocity was 7 m/s (7 − 0), so the sprinter's acceleration was: $a = V/t = 7/1.9 = 3.7$ m/s/s. If the sprinter could maintain this acceleration for three seconds, how fast would he be running?

Close examination of the displacement, velocity, and acceleration graphs of an object's motion is an excellent exercise in qualitative understanding of linear kinematics. Examine the pattern of horizontal acceleration in the 100-m sprint mentioned earlier (see Figure 5.4). Note that there are essentially three phases of acceleration in this race that roughly correspond to the slope of the velocity graph. There is a positive acceleration phase, a phase of near zero acceleration, and a negative acceleration phase. Most sprinters struggle to prevent running speed from declining at the end of a race. Elite female sprinters have similar velocity graphs in 100-m races. What physiological factors might account for the inability of people to maintain peak speed in sprinting?

Note that the largest accelerations (largest rates of change of velocity) do not occur at the largest or peak velocities. Peak velocity must occur when acceleration is zero. Coaches often refer to quickness as the ability to react and move fast over short distances, while speed is the ability to cover moderate distances in a very short time. Based on the velocity graph in Figure 5.4, how might you design running tests to differentiate speed and quickness?

Acceleration is the kinematic (motion description) variable that is closest to a kinetic variable (explanation of motion). Kinesiology professionals need to remember that the pushes (forces) that create accelerations precede the peak speeds they eventually create. This delay in the development of motion is beyond the Fore–Time Principle mentioned earlier. Coaches observing movement cannot see acceleration, but they can perceive changes in speed or direction that can be interpreted as acceleration. Just remember that by the time the coach perceives the acceleration the muscular and body actions which created those forces occurred just before the motion changes you are able to see.

Uniformly Accelerated Motion

In rare instances the forces acting on an object are constant and therefore create a constant acceleration in the direction of the resultant force. The best example of this special condition is the force of earth's gravity acting on projectiles. A projectile is an object launched into the air that has no self-propelled propelling force capability (Figure 5.6). Many human projectile movements have vertical velocities that are sufficiently small so that the effects of air resistance in the vertical direction can be ignored (see chapter 8). Without fluid forces in the vertical direction, projectile motion is uniformly accelerated by one force, the force of gravity. There are exceptions, of course (e.g., skydiver, badminton shuttle), but for the majority of human projectiles we can take advantage of the special conditions of vertical motion to simplify kinematic description of the motion. The Italian Galileo Galilei is often credited with discovering the nearly constant nature of gravitational acceleration using some of the first accurate of measurements of objects falling and rolling down inclines. This section will briefly summarize these mathematical descriptions, but will emphasize several important facts about this kind of motion, and how this can help determine optimal angles of projection in sports.

When an object is thrown or kicked without significant air resistance in the ver-

Figure 5.6. Softballs (left) and soccer balls (right) are projectiles because they are not self-propelled when thrown or kicked.

tical direction, the path or **trajectory** will be some form of a parabola. The uniform nature of the vertical force of gravity creates a linear change in vertical velocity and a second-order change in vertical displacement. The constant force of gravity also assures that the time it takes to reach peak vertical displacement (where vertical velocity is equal to zero) will be equal to the time it takes for the object to fall to the same height that it was released from. The magnitude of the vertical velocity when the object falls back to the same position of release will be the same as the velocity of release. A golf ball tossed vertically at shoulder height at 10 m/s (to kill time while waiting to play through) will be caught at the same shoulder level at a vertical velocity of −10 m/s. The velocity is negative because the motion is opposite of the toss, but is the same magnitude as the velocity of release. Think about the 1 g of acceleration acting on this

golf ball and these facts about uniformly accelerated motion to estimate how many seconds the ball will be in flight.

This uniformly changing vertical motion of a projectile can be determined at any given instant in time using three formulas and the kinematic variables of displacement, velocity, acceleration, and time. My physics classmates and I memorized these by calling them VAT, SAT, and VAS. The various kinematic variables are obvious, except for "S," which is another common symbol for displacement. The final vertical velocity of a projectile can be uniquely determined if you know the initial velocity (V_i) and the time of flight of interest (VAT: $V_f^2 = V_i + at$). Vertical displacement is also uniquely determined by initial velocity and time of flight (SAT: $d = V_i t + 0.5at^2$). Finally, final velocity can be determined from initial velocity and a known displacement (VAS: $V_f^2 = V_i^2 + 2ad$).

Let's consider a quick example of using these facts before we examine the implications for the best angles of projecting objects. Great jumpers in the National Basketball Association like Michael Jordan or David Thompson are credited with standing vertical jumps about twice as high (1.02 m or 40 inches) as typical college males. This outstanding jumping ability is not an exaggeration (Krug & LeVeau, 1999). Given that the vertical velocity is zero at the peak of the jump and the jump height, we can calculate the takeoff velocity of our elite jumper by applying VAS. Solving for V_i in the equation:

$$V_f^2 = V_i^2 + 2ad$$

$$0 = V_i^2 + 2(-9.81)(1.002)$$

$$V_i = 4.47 \text{ m/s or } 9.99 \text{ mph}$$

We select the velocity to be positive when taking the square root because the initial velocity is opposite to gravity, which acts in the negative direction. If we wanted to calculate his hang time, we could calculate the time of the fall with SAT and double it because the time up and time down are equal:

$$d = V_i t + 0.5at^2$$

$$-1.02 = 0 + 0.5(-9.81)t^2$$

$$t = 0.456 \text{ s}$$

So the total fight time is 0.912 seconds. If you know what your vertical jump is, you can repeat this process and compare your takeoff velocity and hang time to that of elite jumpers. The power of these empirical relationships is that you can use the mathematics as models for simulations of projectiles. If you substitute in reasonable values for two variables, you get good predictions of kinematics for any instant in time. If you wanted to know when a partic-

ular height was reached, what two equations could you use? Could you calculate how much higher you could jump if you increased your takeoff velocity by 10%?

So we can see that uniformly accelerated motion equations can be quite useful in modeling the vertical kinematics of projectiles. The final important point about uniformly accelerated motion, which reinforces the directional nature of vectors, is that, once the object is released, the vertical component of a projectile's velocity is independent of its horizontal velocity. The extreme example given in many physics books is that a bullet dropped the same instant another is fired horizontally would strike level ground at the same time. Given constant gravitational conditions, the height of release and initial vertical velocity uniquely determine the time of flight of the projectile. The range or horizontal distance the object will travel depends on this time of flight and the horizontal velocity. Athletes may increase the distance they can throw by increasing the height of release (buying time against gravity), increasing vertical velocity, and horizontal velocity. The optimal combination of these depends on the biomechanics of the movement, not just the kinematics or trajectory of uniformly accelerated motion. The next section will summarize a few general rules that come from the integration of biomechanical models and kinematic studies of projectile activities. These rules are the basis for the Optimal Projection Principle of biomechanics.

OPTIMAL PROJECTION PRINCIPLE

For most sports and human movements involving projectiles, there is a range of angles that results in best performance. The **Optimal Projection Principle** refers to the angle(s) that an object is projected to achieve a particular goal. This section will outline some general rules for optimal projections that can be easily applied by coaches and teachers. These optimal angles are "rules of thumb" that are consistent with the biomechanical research on projectiles. Finding true optimal angles of projection requires integration of descriptive studies of athletes at all ability levels (e.g., Bartlett, Muller, Lindinger, Brunner, & Morriss, 1996), the laws of physics (like uniformly accelerated motion and the effects of air resistance; see chapter 8), and modeling studies that incorporate the biomechanical effects of various release parameters. Determining an exact optimal angle of projection for the unique characteristics of a particular athlete and in a particular environment has been documented using a combination of experimental data and modeling (Hubbard, de Mestre, & Scott, 2001). There will be some general trends or rules for teaching and coaching projectile events where biomechanical research has shown that certain factors dominate the response of the situation and favor certain release angles.

In most instances, a two-dimensional point-mass model of a projectile is used to describe the compromise between the height of release and the vertical and horizontal components of release velocity (Figure 5.7). If a ball was kicked and then landed at the same height, and the air resistance was negligible, the optimal angle of projection for producing maximum horizontal displacement would be 45°. Forty-five degrees above the horizontal is the perfect mix of horizontal and vertical velocity to maximize horizontal displacement. Angles above 45° create shorter ranges because the extra flight time from larger vertical velocity cannot overcome the loss in horizontal velocity. Angles smaller than 45° cause loss of flight time (lower vertical velocity) that cannot be overcome by the larger horizontal velocity. Try the activity below to explore optimal angles of projection.

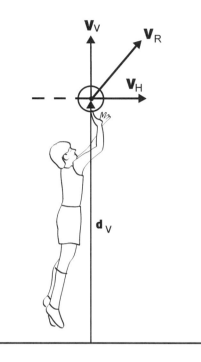

Figure 5.7. The three variables that determine the release parameters of a projectile in two-dimensions: height of release, and the horizontal and vertical velocities of release.

Activity: Angles of Projection

Use a garden hose to water the grass and try various angles of projection of the water. The air resistance on the water should be small if you do not try to project the water too far. Experiment and find the angle that maximizes the distance the water is thrown. First see if the optimal angle is about 45°, when the water falls back to the height that it comes out of the hose. What happens to the optimal angle during long-distance sprinkling as the height of release increases?

Note how the optimal angle of projection changes from 45° when the height of release is above and below the target.

Before we can look at generalizations about how these factors interact to apply the Optimal Projection Principle, the various goals of projections must be analyzed. The mechanical objectives of projectiles are displacement, speed, and a combination of displacement and speed. The goal of an archer is accuracy in displacing an arrow to the target. The basketball shooter in Figure 5.7 strives for the right mix of ball speed and displacement to score. A soccer goalie punting the ball out of trouble in his end of the field focuses on ball speed rather than kicking the ball to a particular location.

When projectile displacement or accuracy is the most important factor, the range of optimal angles of projection is small. In tennis, for example, Brody (1987) has shown that the vertical angle of projection (angular "window" for a serve going in) depends on many factors but is usually less than 4°. The goal of a tennis serve is the right combination of displacement and ball speed, but traditionally the sport and its statistics have emphasized the importance of consistency (accuracy) so as to keep the opponent guessing. In a tennis serve the height of projection above the target, the net barrier, the spin on the ball, the objective of serving deep into the service box, and other factors favor angles of projection at or above the horizontal (Elliott, 1983). Elite servers can hit high-speed serves 3° below the horizontal, but the optimal serving angle for the majority of players is between 0 and 15° above the horizontal (Elliott, 1983; Owens & Lee, 1969).

This leads us to our first generalization of the Optimal Projection Principle. *In most throwing or striking events, when a mix of maximum horizontal speed and displacement are of interest, the optimal angle of projection tends to be below 45°.* The higher point of release and dramatic effect of air resistance on most sport balls makes lower angles of release more effective. Coaches observing softball or baseball players throwing should look

for initial angles of release between 28 and 40° above the horizontal (Dowell, 1978). Coaches should be able to detect the initial angle of a throw by comparing the initial flight of the ball with a visual estimate of 45° angle (Figure 5.8). Note that there is a larger range of optimal or desirable angles that must accommodate differences in the performer and the situation. Increasing the height of release (a tall player) will tend to shift the optimal angle downward in the range of angles, while higher speeds of release (gifted players) will allow higher angles in the range to be effectively used. What do you think would happen to the optimal angles of release of a javelin given the height of release and speed of approach differences of an L5-disabled athlete compared to an able-bodied athlete?

There are a few exceptions to this generalization, which usually occur due to the special environmental or biomechanical conditions of an event. In long jumping, for example, the short duration of takeoff on the board limits the development of vertical velocity, so that takeoff angles are usually between 18 and 23° (Hay, Miller, & Canterna, 1986). We will see in chapter 8 that the effect of air resistance can quickly become dominant on the optimal release parameters for many activities. In football place-kicking, the lower-than-45° generalization applies (optimal angles are usually between 25 and 35°), but the efficient way the ball can be punted and the tactical importance of time during a punt make the optimal angle of release about 50°. With the wind at the punter's back he might kick above 50°, while using a flatter kick against a wind. The backspin put on various golf shots is another example of variations in the angle of release because of the desirable effects of spin on fluid forces and the bounce of the ball. Most long-distance clubs have low pitches, which agrees with our principle of a low angle of release, but a golfer might choose a club with more loft in

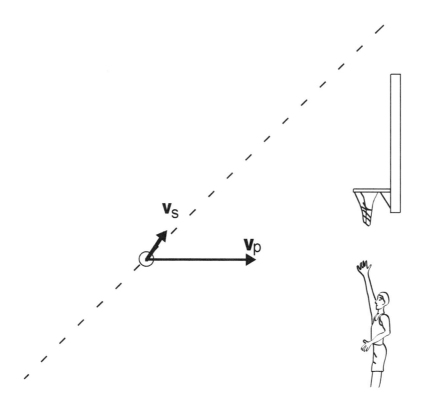

Figure 5.8. Coaches can visually estimate the initial path of thrown balls to check for optimal projection. The initial path of the ball can be estimated relative to an imaginary 45° angle. When throwing for distance, many small children select very high angles of release that do not maximize the distance of the throw.

situations where he wants higher trajectory and spin rate to keep a ball on the green.

The next generalization relates to projectiles with the goal of upward displacement from the height of release. *The optimal angle of projection for tasks emphasizing displacement or a mix of vertical displacement and speed tends to be above 45°.* Examples of these movements are the high jump and basketball shooting. Most basketball players (not the giants of the NBA) release a jump shot below the position of the basket. Considerable research has shown that the optimal angle of projection for basketball shots is between 49 and 55° (see Knudson, 1993). This angle generally corresponds to the arc where the minimum speed may be put on the ball to reach the goal, which is

consistent with a high-accuracy task. Ironically, a common error of beginning shooters is to use a very flat trajectory that requires greater ball speed and may not even permit an angle of entry so that the ball can pass cleanly through the hoop! Coaches that can identify appropriate shot trajectories can help players improve more quickly (Figure 5.9). The optimal angles of release in basketball are clearly not "high-arc" shots, but are slightly greater than 45° and match the typical shooting conditions in recreational basketball.

The optimal-angle-of-projection principle involves several generalizations about desirable initial angles of projection. These general rules are likely to be effective for most performers. Care must be taken in ap-

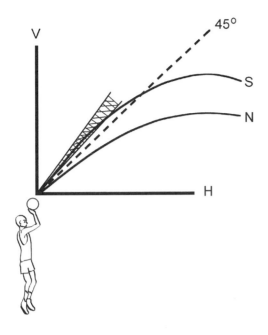

Figure 5.9. The optimal projection angles for most basketball jump shots are between 49 and 55° above the horizontal (hatched). These initial trajectories represent the right mix of low ball speed and a good angle of entry into the hoop. Novice shooters (N) often choose a low angle of release. Skilled shooters (S) really do not shoot with high arcs, but with initial trajectories that are in the optimal range and tailored to the conditions of the particular shot.

plying these principles in special populations. The biomechanical characteristics of elite (international caliber) athletes or wheelchair athletes are likely to affect the optimal angle of projection. Kinesiology professionals should be aware that biomechanical and environmental factors interact to affect the optimal angle of projection. For example, a stronger athlete might use an angle of release slightly lower than expected but which is close to optimal for her. Her extra strength allows her to release the implement at a higher point without losing projectile speed so that she is able to use a slightly lower angle of release. Professionals coaching projectile sports must keep

up on the biomechanical research related to optimal conditions for their athletes.

ANGULAR MOTION

Angular kinematics is the description of angular motion. Angular kinematics is particularly appropriate for the study of human movement because the motion of most human joints can be described using one, two, or three rotations. Angular kinematics should also be easy for biomechanics students because for every linear kinematic variable there is a corresponding angular kinematic variable. It will even be easy to distinguish angular from linear kinematics because the adjective "angular" or a Greek letter symbol is used instead of the Arabic letters used for linear kinematics.

Angular displacement (θ: theta) is the vector quantity representing the change in angular position of an object. Angular displacements are measured in degrees, radians (dimensionless unit equal to 57.3°), and revolutions (360°). The usual convention to keep directions straight and be consistent with our 2D linear kinematic calculations is to consider counterclockwise rotations as positive. Angular displacement measured with a **goniometer** is one way to measure **static flexibility**. As in linear kinematics, the frames of reference for these angular measurements are different. Some tests define complete joint extension as 0° while other test refer to that position as 180°. For a review of several physical therapy static flexibility tests, see Norkin & White (1995).

In analyzing the curl-up exercise shown in Figure 5.10, the angle between the thoracic spine and the floor is often used. This exercise is usually limited to the first 30 to 40° above the horizontal to limit the involvement of the hip flexors (Knudson, 1999a). The angular displacement of the thoracic spine in the eccentric phase would be –38° (final angle minus

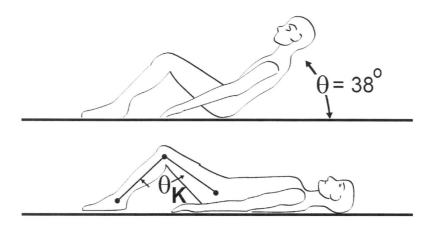

Figure 5.10. The angular kinematics of a curl-up exercise can be measured as the angle between the horizontal and a thoracic spine segment. This is an example of a absolute angle because it defines the angle of an object relative to external space. The knee joint angle (θ_K) is a relative angle because both the leg and thigh can move.

initial angle: $0 - 38 = -38°$). This trunk angle is often called an **absolute angle** because it is measured relative to an "unmoving" earth frame of reference. **Relative angles** are defined between two segments that can both move. Examples of relative angles in biomechanics are joint angles. The knee angle (θ_K) in Figure 5.10 is a relative angle that would tell if the person is changing the positioning of their legs in the exercise.

Angular Velocity

Angular velocity (ω: omega) is the rate of change of angular position and is usually expressed in degrees per second or radians per second. The formula for angular velocity is $\omega = \theta/t$, and calculations would be the same as for a linear velocity, except the displacements are angular measurements. Angular velocities are vectors are drawn by the right-hand rule, where the flexed fingers of your right hand represent the rotation of interest, and the extended thumb would be along the axis of rotation and would indicate the direction of the angular

velocity vector. This book does not give detailed examples of this technique, but will employ a curved arrow just to illustrate angular velocities and torques (Figure 5.11).

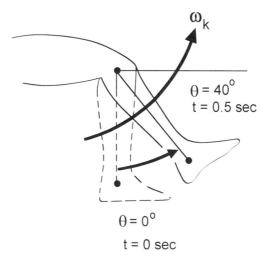

Figure 5.11. The average angular velocity of the first half of a knee extension exercise can be calculated from the change in angular displacement divided by the change in time.

The angular velocities of joints are particularly relevant in biomechanics, because they represent the angular speed of anatomical motions. If relative angles are calculated between anatomical segments, the angular velocities calculated can represent the speed of flexion/extension and other anatomical rotations. Biomechanical research often indirectly calculates joint angles from the linear coordinates (measurements) derived from film or video images, or directly from electrogoniometers attached to subjects in motion. It is also useful for kinesiology professionals to be knowledgeable about the typical angular velocities of joint movements. This allows professionals to understand the similarity between skills and determine appropriate training exercises. Table 5.2 lists typical peak joint angular speeds for a variety of human movements.

Table 5.2
TYPICAL PEAK ANGULAR SPEEDS IN HUMAN MOVEMENT

	Speed	
	deg/s	rad/s
Knee extension: sit-to-stand	150	2.6
Trunk extension: vertical jump	170	3.0
Elbow flexion: arm curl	200	3.5
Knee extension: vertical jump	800	14.0
Ankle extension: vertical jump	860	15.0
Wrist flexion: baseball pitching	1000	17.5
Radio/ulnar pronation: tennis serve	1400	24.4
Knee extension: soccer kick	2000	34.9
Shoulder flexion: softball pitch	5000	87.3
Shoulder internal rotation: pitching	7400	129.1

Let's calculate the angular velocity of a typical knee extension exercise and compare it to the peak angular velocity in the table. Figure 5.11 illustrates the exercise and the data for the example. The subject extends a knee, taking their leg from a vertical orientation to the middle of the range of motion. If we measure the angle of the lower leg from the vertical, the exerciser has moved their leg 40° in a 0.5-second period of time. The average knee extension angular velocity can be calculated as follows: $\omega_K = \theta/t = 40/0.5 = 80$ deg/s. The angular velocity is positive because the rotation is counterclockwise. So the exercise averages 80° per second of knee extension velocity over the half-second time interval, but the instantaneous angular velocity at the position shown in the figure is likely larger than that. The peak knee extension angular velocity in this exercise likely occurs in the midrange of the movement, and the knee extension velocity must then slow to zero at the end of the range of motion. This illustrates some limitations of free-weight exercises. There is a range of angular velocities (which have an affect on the linear Force–Velocity Relationship of the muscles), and there must be a decrease in the angular velocity of the movement at the end of the range of motion. This negative acceleration (if the direction of motion is positive) protects the joints and ligaments, but is not specific to many events where peak speed is achieved near the release of an object and other movements can gradually slow the body in the follow-through.

Angular Acceleration

The rate of change of angular velocity is **angular acceleration** ($\alpha = \omega/t$). Angular acceleration is symbolized by the Greek letter alpha (α). The typical units of angular acceleration are deg/s/s and radians/s/s. Like linear acceleration, it is best to think about

Interdisciplinary Issue: Specificity

One of the most significant principles discovered by early kinesiology research is the principle of specificity. Specificity applies to the various components of fitness, training response, and motor skills. Motor learning research suggests that there is specificity of motor skills, but there is potential transfer of ability between similar skills. In strength and conditioning the principle of specificity states that the exercises prescribed should be specific, as close as possible to the movement that is to be improved. Biomechanics research that measures the angular kinematics of various sports and activities can be used to assess the similarity and potential specificity of training exercises. Given the peak angular velocities in Table 5.2, how specific are most weight training or isokinetic exercise movements that are limited to 500° per second or slower? The peak speed of joint rotations is just one kinematic aspect of movement specificity. Could the peek speeds of joint rotation in different skills occur in different parts of the range of motion? What other control, learning, psychological, or other factors affect the specificity of an exercise for a particular movement?

angular acceleration as an unbalanced rotary effect. An angular acceleration of –200 rad/s/s means that there is an unbalanced clockwise effect tending to rotate the object being studied. The angular acceleration of an isokinetic dynamometer in the middle of the range of motion should be zero because the machine is designed to match or balance the torque created by the person, so the arm of the machine should be rotating at a constant angular velocity.

Angular kinematics graphs are particularly useful for providing precise descriptions of how joint movements occurred. Figure 5.12 illustrates the angular displacement and angular velocity of a simple elbow extension and flexion movement in the sagittal plane. Imagine that the data repre-

Interdisciplinary Issue: Isokinetic Dynamometers

Isokinetic (iso = constant or uniform, kinetic = motion) dynamometers were developed by J. Perrine in the 1960s. His Cybex machine could be set at different angular velocities and would accommodate the resistance to the torque applied by a subject to prevent angular acceleration beyond the set speed. Since that time, isokinetic testing of virtually every muscle group has become a widely accepted measure of muscular strength in clinical and research settings. Isokinetic dynamometers have been influential in documenting the balance of strength between opposing muscle groups (Grace, 1985). There is a journal (*Isokinetics and Exercise Science*) and several books (e.g., Brown, 2000; Perrin, 1993) that focus on the many uses of isokinetic testing. Isokinetic machines, however, are not truly isokinetic throughout the range of motion, because there has to be an acceleration to the set speed at the beginning of a movement that often results in a torque overshoot as the machine negatively accelerates the limb (Winter, Wells, & Orr, 1981) as well as another negative acceleration at the end of the range of motion. The effects of inertia (Iossifidou & Baltzopoulos, 2000) and muscular co-contraction (Kellis & Baltzopoulos, 1998) are other recent issues being investigated that affect the validity of isokinetic testing. It is important to note that the muscle group is not truly shortening or lengthening in an isokinetic fashion. Muscle fascicle-shortening velocity is not constant (Ichinose, Kawakami, Ito, Kanehisa, & Fukunaga, 2000) in isokinetic dynamometry even when the arm of the machine is rotating at a constant angular velocity. This is because linear motion of points on rotating segments do not directly correspond to angular motion in isokinetic (Hinson, Smith, & Funk, 1979) or other joint motions.

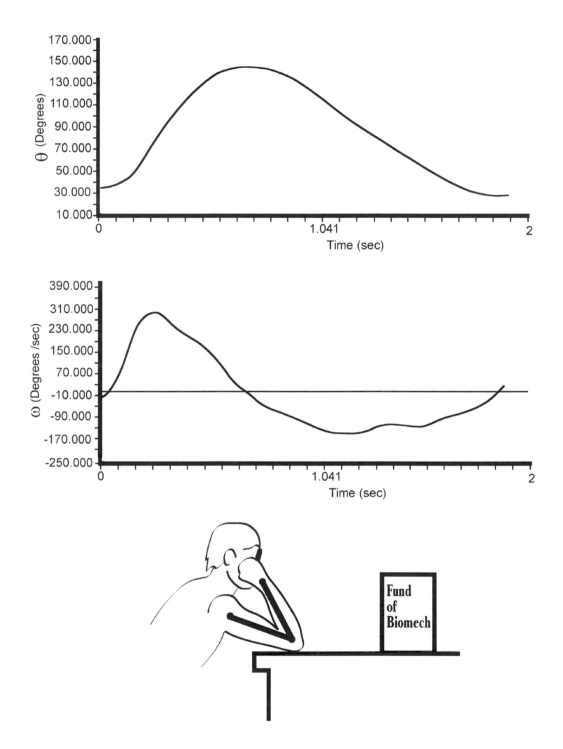

Figure 5.12. The angular displacement and angular velocity of a simple elbow extension/flexion movement to grab a book. See the text for an explanation of the increasing complexity of the higher-order kinematic variables.

sented a student tired of studying exercise physiology, who reached forward to grab a refreshing, 48-ounce *Fundamental of Biomechanics* text. Note as we look at the kinematic information in these graphs that the complexity of a very simple movement grows as we look at the higher-order derivatives (velocity).

The elbow angular displacement data show an elbow extended (positive angular displacement) from about a 37° to about an 146° elbow angle to grasp the book. The extension movement took about 0.6 seconds, but flexion with the book occurred more slowly. Since the elbow angle is defined on the anterior aspect of a subject's arm, larger numbers mean elbow extension. The corresponding angular velocity–time graph represents the speed of extension (positive ω) or the speed of flexion (negative ω). The elbow extension angular velocity peaks at about 300 deg/s (0.27 sec) and gradually slows. The velocity of elbow flexion increases and decreases more gradually than the elbow extension.

The elbow angular acceleration would be the slope of the angular velocity graph. Think of the elbow angular acceleration as an unbalanced push toward extension or flexion. Examine the angular velocity graph and note the general phases of acceleration. When are there general upward or downward trends or changes in the angular velocity graph? Movements like this often have three major phases. The extension movement was initiated by a phase of positive acceleration, indicated by an increasing angular velocity. The second phase is a negative acceleration (downward movement of the angular velocity graph) that first slows elbow extension and then initiates elbow flexion. The third phase is a small positive angular acceleration that slows elbow flexion as the book nears the person's head. These three phases of angular acceleration correspond to typical muscle activation in this movement. This move-

ment would usually be created by a triphasic pattern of bursts from the elbow extensors, flexors, and extensors. Accelerations (linear and angular) are the kinematic variables closest to the causes (kinetics) of the motion, and are more complex than lower-order kinematic variables like angular displacements.

Angular and linear kinematics give scientists important tools to describe and understand exactly how movement occur. Remember to treat the linear and angular measurements separately: like the old saying goes, "don't mix apples and oranges." A good example is your CD player. As the CD spins, a point near the edge travels a larger distance compared to a point near the center. How can two points make the same revolutions per minute and travel at different speeds? Easy, if you notice the last sentence mixes or compares angular and linear kinematic variables. In linear kinetics we will look at the trigonometric functions that allow linear measurements to be mapped to angular.

Biomechanists usually calculate angular kinematic variables from linear coordinates of body segments with trigonometry. There is another simple formula that converts linear to angular kinematics in special conditions. It is useful to illustrate why the body tends to extend segments prior to release events. The linear velocity of a point on a rotating object, *relative to its axis of rotation*, can be calculated as the product of its angular velocity and the distance from the axis to the point (called the radius): $V = \omega \cdot r$. The special condition for using this formula is to use angular velocity in radians/second. Using a dimensionless unit like rad/s, you can multiply a radius measured in meters and get a linear velocity in meters/second.

The most important point is to notice that the angular velocity and the radius are equally important in creating linear velocity. To hit a golf ball harder you can either use a longer club or rotate the club faster. We will see in chapter 7 that angular kinetic analysis can help us decide which of these two options is best for a particular situation. In most throwing events the arm is extended late in the throw to increase the linear velocity of a projectile. Angular kinetics is necessary to understand why this extension or increase in the radius of segments is delayed to just before release.

COORDINATION CONTINUUM PRINCIPLE

Many kinesiology professionals are interested in the coordination of movement. Coordination is commonly defined as the sequence and timing of body actions used in a movement. Unfortunately there is no universally agreed-upon definition or way to study coordination in the kinesiology literature. A wide variety of approaches has been proposed to describe the coordination of movement. Some approaches focus on the kinematics of the joint or segmental actions (Hudson, 1986; Kreighbaum & Bartels, 1996), while others are based on the joint forces and torques (kinetics) that create the movement (Chapman & Sanderson, 1990; Prilutsky, 2000; Putnam, 1991, 1993; Roberts, 1991; Zajac, 1991). This section presents the **Coordination Continuum Principle**, which is adapted from two kinematic approaches to defining coordination (Hudson, 1986; Kreighbaum & Bartels, 1996), because teachers and coaches most often modify the spatial and temporal aspects of movement. While teaching cues that focus on muscular effort may be used occasionally, the vast majority of teaching and coaching of movement remains in the positioning and motions of the body.

Kinematic coordination of movements can be pictured as a continuum ranging from simultaneous body actions to sequential actions. The **Coordination Continuum**

Principle suggests that movements requiring the generation of high forces tend to utilize simultaneous segmental movements, while lower-force and high-speed movements are more effective with more sequential movement coordination. A person lifting a heavy box simultaneously extends the hips, knees, and ankles (Figure 5.13). In overarm throwing, people usually use a more sequential action of the whole kinematic chain, beginning with the legs, followed by trunk and arm motions.

Figure 5.13. Coordination to move a heavy load usually involves simultaneous joint motions like in this squat lift.

Because coordination falls on a continuum and the speed and forces of movement vary widely, it is not always easy to determine what coordination pattern is best. In vertical jumping, resistance is moderate and the objective is to maximize height of takeoff and vertical velocity. While a vertical jump looks like a simultaneous movement, biomechanical studies show that the kinematics and kinetics of different jumpers have simultaneous and sequential characteristics (Aragon-Vargas & Gross, 1997a; Bobbert & van Ingen Schenau, 1988; Hudson, 1986). Kinesiology professionals need to remember that coordination is not an either/or situation in many activities. Until there is more research determining the most effective technique, there will be quite a bit of art to the coaching of movements not at the extremes of the continuum.

The motor development of high-speed throwing and striking skills tends to begin with restricted degrees of freedom and simultaneous actions. Children throwing, striking, or kicking tend to make initial attempts with simultaneous actions of only a few joints. Skill develops with the use of more segments and greater sequential action. In high-speed throwing, for example, the sequential or "differential" rotation of the pelvis and upper trunk is a late-developing milestone of high-skill throwing (Roberton & Halverson, 1984). It is critical that physical educators know the proper sequential actions in these low-force and high-speed movements. Kinematic studies help identify these patterns of motion in movement skills. Unfortunately, the youth of biomechanics means that kinematic documentation of coordination in the wide variety of human movements is not complete. Early biomechanics research techniques emphasized elite male performers, leaving little information on gender, special populations, lower skill levels, or age.

Suppose a junior high volleyball coach is working with a tall athlete on spiking. The potential attacker lacks a strong overarm pattern and cannot get much speed on the ball (Figure 5.14). The kinematics of the preparatory action lacks intensity, stretch, and timing. At impact the player's elbow and upper arm are well forward of her shoulder. The coach suspects that her overarm throwing pattern is still immature and must be developed before skilled spiking is possible. This coach has integrated biomechanical and motor development informa-

Figure 5.14. Poor sequential coordination in throwing and striking results in slow segment speeds at impact that can be visually identified by slow ball speeds, lack of eccentric loading of distal segments, or limited movement in the follow-through (like this volleyball spike).

tion to determine the best course of action to help this player improve. The lack of ball speed (kinematics), and muscle stretch-shortening cycles within a sequential coordination are biomechanical factors missing in this athlete. The forward elbow position at impact is a motor development indicator of an immature trunk and arm action within an overarm pattern. How coaches work on this problem may vary, but one good strategy would be to simplify the movement and work on throwing the volleyball. Sequential rotation of the trunk, arm, forearm, and wrist is the focus of training.

Strength and conditioning professionals closely monitor training technique, because body position and motion in exercises dramatically affect muscular actions and risk of injury. In strength training, resistances are near maximal, so coordination in most exercises tends to be simultaneous. Imagine someone performing a squat exercise with a heavy weight. Is the safest technique to simultaneously flex the hips and knees in the eccentric phase and then simultaneously extend in the concentric phase? If the resistance is lighter (bodyweight), like in standing up out of a chair, after a person leans forward to put their upper bodyweight over their feet, do the major joints of the body simultaneously act to stand? In the next chapter we will examine variations in conditioning for high-power and high-speed movements that are different than high-force (strength) movements. Do you think high-power movements will also have simultaneous coordination, or will the coordination shift a little toward sequential? Why?

SUMMARY

A key branch of biomechanics is kinematics, the precise description or measurement of human motion. Human motion is measured relative to some frame of reference and is usually expressed in linear (meters, feet) or angular (radians, degrees) units. Angular kinematics are particularly appropriate in biomechanics because these can be easily adapted to document joint rotations. There are many kinematic variables that can be used to document the human motion. Simple kinematic variables are scalars, while others are vector quantities that take into account the direction of motion. The time derivatives (rates of change) of position measurements are velocity and acceleration. The Optimal Projection Principle states that sporting events involving projectiles have a range of desirable initial angles of projection appropriate for most performers. The kinematic timing of segment motions falls on a Coordination Continuum

from simultaneous to sequential movement. High-force movements use more simultaneous joint rotations while high-speed movements use more sequential joint rotations.

REVIEW QUESTIONS

1. What is a frame of reference and why is it important in kinematic measurements?

2. Compare and contrast the scalar and vector linear kinematic variables.

3. Explain the difference between calculation of average and instantaneous velocities, and how does the length of the time interval used affect the accuracy of a velocity calculation?

4. Use the velocity graph in Figure 5.4 to calculate the average acceleration of the sprinter in the first and the last 10-m intervals. Compare you answers to the instantaneous values graph in Figure 5.5.

5. A patient lifts a dumbbell 1.2 m in 1.5 s and lowers it back to the original position in 2.0 s. Calculate the average vertical velocity of the concentric and eccentric phases of the lift.

6. Explain why linear and angular accelerations should be thought of as pushes in a particular direction rather than speeding up or slowing down.

7. Why are angular kinematics particularly well suited for the analysis of human movement?

8. From the anatomical position a person abducts their shoulder to 30° above the horizontal. What is the angular displacement of this movement with the usual directional (sign) convention?

9. A soccer player attempting to steal the ball from an opponent was extending her knee at 50 deg/s when her foot struck the opponent's shin pads. It the player's knee was stopped (0 deg/s) within 0.2 seconds, what angular acceleration did the knee experience?

10. A golfer drops a ball to replace a lost ball. If the ball had an initial vertical velocity of 0 m/s and had a vertical velocity before impact of –15.7 m/s exactly 1.6 seconds later, what was the vertical acceleration of the ball?

11. A softball coach is concerned that her team is not throwing at less than 70% speed in warm-up drills. How could she estimate or measure the speeds of the warm-up throws to make sure her players are not throwing too hard?

12. A biomechanist uses video images to measure the position of a box in the sagittal plane relative to a worker's toes during lifting. Which coordinate (x or y) usually corresponds to the height of the box and the horizontal position of the box relative to the foot?

13. Use the formula for calculating linear velocity from angular velocity ($V = \omega \cdot r$) to calculate the velocity of a golf club relative to the player's hands (axis of rotation). Assume the radius is 1.5 m and the angular velocity of the club was 2000 deg/s. Hint: remember to use the correct units.

14. Give an example of a fixed and a relative frame of reference for defining joint angular kinematics. Which frame of reference is better for defining anatomical rotations versus rotations in space?

15. What is the vertical acceleration of a volleyball at the peak of its flight after the ball is tossed upward in a jump serve?

KEY TERMS

absolute angle

acceleration

coordination continuum

degrees of freedom

displacement

distance

goniometer

isokinetic

point mass

relative angle

speed

static flexibility

trajectory

velocity

SUGGESTED READING

Cappozzo, A., Marchetti, M., & Tosi, V. (Eds.) (1990). *Biolocomotion: A century of research using moving pictures*. Rome: Promograph.

Hudson, J. L. (1986). Coordination of segments in the vertical jump. *Medicine and Science in Sports and Exercise*, **18**, 242–251.

Kreighbaum, E., & Barthels, K. M. (1996). *Biomechanics: A qualitative approach to studying human movement*. Boston: Allyn & Bacon.

Lafortune, M. A., & Hennig, E. M. (1991). Contribution of angular motion and gravity to tibial acceleration. *Medicine and Science in Sports and Exercise*, **23**, 360–363.

Mero, A., Komi, P. V., & Gregor, R. J. (1992). Biomechanics of sprint running: A review. *Sports Medicine*, **13**, 376–392.

Plagenhoef, S. (1971). *Patterns of human motion: A cinematographic analysis*. Englewood Cliffs, NJ: Prentice-Hall.

Zatsiorsky, V. M. (1998). *Kinematics of human motion*. Champaign, IL: Human Kinetics.

WEB LINKS

Projectiles—page on the path of the center of gravity of a skater by Debra King and others from Montana State University.

    http://btc.montana.edu/olympics/physbio/biomechanics/pm-intro.html

Kinematics simulations on golf and freefall—part of mechanics simulations at explorescience.com.

    http://www.explorescience.com/activities/activity_list.cfm?categoryID=10

Free kinematic analysis software by Bob Schleihauf at San Francisco State.

    http://www.kavideo.sfsu.edu/

Kwon3D—Theoretical background information for kinematic and kinetic software developed by Young-Hoo Kwon of Texas Woman's University.

    http://kwon3d.com/theories.html

CHAPTER **6**

Linear Kinetics

In the previous chapter we learned that kinematics or descriptions of motion could be used to provide information for improving human movement. This chapter will summarize the important laws of kinetics that show how forces overcome inertia and how other forces create human motion. Studying the causes of linear motion is the branch of mechanics known as **linear kinetics**. Identifying the causes of motion may be the most useful kind of mechanical information for determining what potential changes could be used to improve human movement. The biomechanical principles that will be discussed in this chapter are **Inertia**, **Force–Time**, and **Segmental Interaction**.

LAWS OF KINETICS

Linear kinetics provides precise ways to document the causes of the linear motion of all objects. The specific laws and mechanical variables a biomechanist will choose to use in analyzing the causes of linear motion often depends on the nature of the movement. When instantaneous effects are of interest, Newton's Laws of Motion are most relevant. When studying movements over intervals of time is of interest, the Impulse–Momentum Relationship is usually used. The third approach to studying the causes of motion focuses on the distance covered in the movement and uses the Work–Energy Relationship. This chapter summarizes these concepts in the context of

human movement. Most importantly, we will see how these laws can be applied to human motion in the biomechanical principles of Force–Motion, Force–Time, and Coordination Continuum Principles.

NEWTON'S LAWS OF MOTION

Arguably, some of the most important discoveries of mechanics are the three laws of motion developed by the Englishman, Sir Isaac Newton. Newton is famous for many influential scientific discoveries, including developments in calculus, the Law of Universal Gravitation, and the Laws of Motion. The importance of his laws cannot be overemphasized in our context, for they are the keys to understanding how human movement occurs. The publication of these laws in his 1686 book *De Philosophiae Naturalis Principia Mathematica* marked one of the rare occasions of scientific breakthrough. Thousands of years of dominance of the incorrect mechanical views of the Greek philosopher Aristotle were overturned forever.

Newton's First Law and First Impressions

Newton's first law is called the **Law of Inertia** because it outlines a key property of matter related to motion. Newton stated that all objects have the inherent property to resist a change in their state of motion.

His first law is usually stated something like this: objects tend to stay at rest or in uniform motion unless acted upon by an unbalanced force. A player sitting and "warming the bench" has just as much inertia as a teammate of equal mass running at a constant velocity on the court. It is vitally important that kinesiology professionals recognize the effect inertia and Newton's first law have on movement technique. The linear measure of inertia (Figure 6.1) is mass and has units of kg in the SI system and slugs in the English system. This section is an initial introduction to the fascinating world of kinetics, and will demonstrate how our first impressions of how things work from casual observation are often incorrect.

Understanding kinetics, like Newton's first law, is both simple and difficult: simple because there are only a few physical laws that govern all human movement, and these laws can be easily understood and demonstrated using simple algebra, with only a few variables. The study of biomechanics can be difficult, however, because the laws of mechanics are often counterintuitive for most people. This is because the observations of everyday life often lead to incorrect assumptions about the nature of the world and motion. Many children and adults have incorrect notions about inertia,

and this view of the true nature of motion has its own "cognitive inertia," which is hard to displace. The natural state of objects in motion is to slow down, right? Wrong! The natural state of motion is to continue whatever it is doing! Newton's first law shows that objects tend to resist changes in motion, and that things only seem to naturally slow down because forces like friction and air or water resistance that tend to slow an object's motion. Most objects around us appear at rest, so isn't there something natural about being apparently motionless? The answer is yes, if the object is initially at rest! The same object in linear motion has the same natural or inertial tendency to keep moving. In short, the mass (and consequently its linear inertia) of an object is the same whether it is motionless or moving.

We also live in a world where most people take atmospheric pressure for granted. They are aware that high winds can create very large forces, but would not believe that in still air there can be hundreds of pounds of force on both sides of a house window (or a person) due to the pressure of the atmosphere all around us. The true nature of mechanics in our world often becomes more apparent under extreme conditions. The pressure of the sea of air we live in becomes real when a home explodes or implodes from a passing tornado, or a

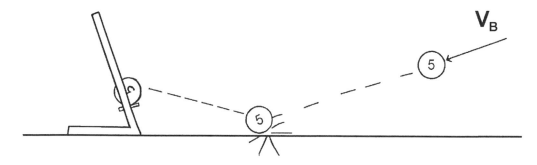

Figure 6.1. All objects have the inherent property of inertia, the resistance to a change in the state of motion. The measure of linear motion inertia is mass. A medicine ball has the same resistance to acceleration (5 kg of mass) in all conditions of motion, assuming it does not travel near the speed of light.

fast-moving weather system brings a change in pressure that makes a person's injured knee ache. People interested in scuba diving need to be knowledgeable about pressure differences and the timing of these changes when they dive.

So casual observation can often lead to incorrect assumptions about the laws of mechanics. We equate forces with objects in contact or a collision between two objects. Yet we live our lives exercising our muscles against the consistent force of gravity, a force that acts at quite a distance whether we are touching the ground or not. We also tend to equate the velocity (speed and direction) of an object with the force that made it. In this chapter we will see that the forces that act on an object do not have to be acting in the direction of the resultant motion of the object (Figure 6.2). It is the skilled person that creates muscle forces to precisely combine with external forces to balance a bike or throw the ball in the correct direction.

Casual visual observation also has many examples of perceptual illusions about the physical realities of our world. Our brains work with our eyes to give us a mental image of physical objects in the world, so that most people routinely mistake this constructed mental image for the actual object. The color of objects is also an illusion based on the wavelengths of light that are reflected from an object's surface. So what about touch? The solidity of objects is also a perceptual illusion because the vast majority of the volume in atoms is "empty" space. The forces we feel when we touch things are the magnetic forces of electrons on the two surfaces repelling each other, while the material strength of an object we bend is related to its physical structure and chemical bonding. We also have a distorted perception of time and the present. We rely on light waves bouncing off objects and toward our eyes. This time delay is not a problem at all, unless we want to observe

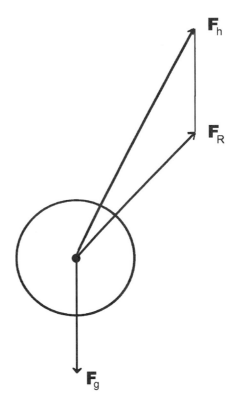

Figure 6.2. Force and motion do not always act in the same direction. This free-body diagram of the forces and resultant force ($\mathbf{F}_R$) on a basketball before release illustrates how a skilled player applies a force to an object ($\mathbf{F}_h$) that combines with the force of gravity ($\mathbf{F}_g$) to create the desired effect. The motion of the ball will be in the direction of $\mathbf{F}_R$.

very high-speed or distant objects like in astronomy. There are many other examples of our molding or construction of the nature of reality, but the important point is that there is a long history of careful scientific measurements which demonstrate that certain laws of mechanics represent the true nature of object and their motion. These laws provide a simple structure that should be used for understanding and modifying motion, rather than erroneous perceptions about the nature of things. Newton's first law is the basis for the Inertia Principle in applying biomechanics.

Interdisciplinary Issue: Body Composition

A considerable body of kinesiology research has focused on the percentage of fat and lean mass in the human body. There are metabolic, mechanical, and psychological effects of the amount and location of fat mass. In sports performance, fat mass can be both an advantage (increased inertia for a football lineman or sumo wrestler) and a disadvantage. Increasing lean body mass usually benefits performance, although greater mass means increasing inertia, which could decrease agility and quickness. When coaches are asked by athletes "How much should I weigh?" they should answer carefully, focusing the athlete's attention first on healthy body composition. Then the coach can discuss with the athlete the potential risks and benefits of changes in body composition. How changes in an athlete's inertia affect their sport performance should not be evaluated without regard to broader health issues.

Newton's Second Law

Newton's second law is arguably the most important law of motion because it shows how the forces that create motion (kinetics) are linked to the motion (kinematics). The second law is called the **Law of Momentum** or **Law of Acceleration**, depending on how the mathematics is written. The most common approach is the famous **F = ma**. This is the law of acceleration, which describes motion (acceleration) for any instant in time. The formula correctly written is $\Sigma F = m \cdot a$, and states that the acceleration an object experiences is proportional to the resultant force, is in the same direction, and is inversely proportional to the mass. The larger the unbalanced force in a particular direction, the greater the acceleration of the object in that direction. With increasing mass, the inertia of the object will decrease the acceleration if the force doesn't change.

Let's look at an example using skaters in the push-off and glide phases during ice skating (Figure 6.3). If the skaters have a mass of 59 kg and the horizontal forces are known, we can calculate the acceleration of the skater. During push-off the net horizontal force is +200 N because air resistance is negligible, so the skater's horizontal acceleration is: $\Sigma F = m \cdot a$, 200 = 59a, so a = 3.4 m/s/s. The skater has a positive acceleration and would tend to speed up 3.4 m/s every second if she could maintain her push-off force this much over the air resistance. In the glide phase, the friction force is now a resistance rather than a propulsive force. During glide the skater's acceleration is –0.08 m/s/s because: $\Sigma F = m \cdot a$, –5 = 59a, so a = –0.08 m/s/s.

The kinesiology professional can qualitatively break down movements with Newton's second law. Large changes in the speed or direction (acceleration) of a person means that large forces must have been applied. If an athletic contest hinges on the agility of an athlete in a crucial play, the coach should select the lightest and quickest player. An athlete with a small mass is easier to accelerate than an athlete with a larger mass, provided they can create sufficient forces relative to body mass. If a smaller player is being overpowered by a larger opponent, the coach can substitute a larger more massive player to defend against this opponent. Note that increasing force or decreasing mass are both important in creating acceleration and movement.

Newton's second law plays a critical role in quantitative biomechanics. Biomechanists wanting to study the net forces that create human motion take acceleration and body segment mass measurements and apply **F = ma**. This working backward from kinematics to the resultant kinetics is called

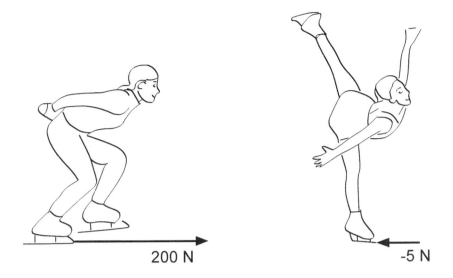

Figure 6.3. Friction forces acting on ice skaters during push-off and gliding. Newton' Second Law of Motion applied in the horizontal direction (see text) will determine the horizontal acceleration of the skater.

inverse dynamics. Other scientists build complex computer models of biomechanical systems and use **direct dynamics**, essentially calculating the motion from the "what-if" kinetics and body configurations they input.

Newton's Third Law

Newton's third law of motion is called the **Law of Reaction**, because it is most often translated as: for every action there is an equal and opposite reaction. For every force exerted, there is an equal and opposite force being exerted. If a patient exerts a sideways force of +150 N on an elastic cord, there has to be −150-N reaction force of the cord on the patient's hand (Figure 6.4). The key insight that people often miss is that a force is really a mutual interaction between two bodies. It may seem strange that if you push horizontally against a wall, the wall is simultaneously pushing back toward you, but it is. This is *not* to say that a force on a

free-body diagram should be represented by two vectors, but a person must understand that the effect of a force is not just on one object. If someone ever did not seem to kiss you back, you can always take some comfort in the fact that at least in mechanical terms they did.

An important implication of the law of reaction is how reaction forces can change the direction of motion opposite to our applied force when we exert our force on objects with higher force or inertia (Figure 6.5a). During push-off in running the athlete exerts downward and backward push with the foot, which creates a **ground reaction force** to propel the body upward and forward. The extreme mass of the earth easily overcomes inertia, and the ground reaction force accelerates our body in the opposite direction of force applied to the ground. Another example would be eccentric muscle actions where we use our muscles as brakes, pushing in the opposite direction to another force. The force exerted by the tackler in Figure 6.5b ends up being an eccentric

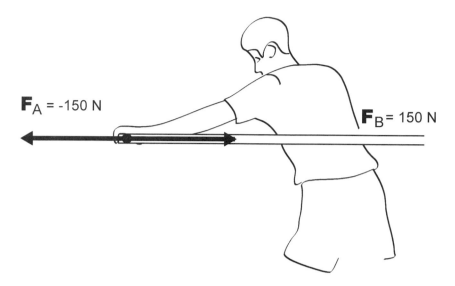

Figure 6.4. Newton's third law states that all forces have an equal and opposite reaction forces on the other object, like in this elastic exercise. The –150-N (**F**$_A$) force created by the person on the elastic cord coincides with a 150-N reaction force (**F**$_B$) exerted on the person by the cord.

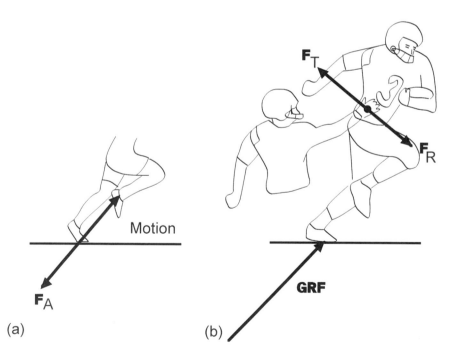

Figure 6.5. A major consequence of Newton's third law is that the forces we exert on an object with larger inertia often create motion in the direction opposite of those forces. In running, the downward backward push of the foot on the ground (**F**$_A$) late in the stance (a) creates a ground reaction force which acts forward and upward, propelling the runner through the air. A defensive player trying to make a tackle from a poor position (b) may experience reaction forces that create eccentric muscle actions and injurious loads.

muscle action as the inertia and ground reaction forces created by the runner are too great. Remember that when we push or pull, this force is exerted on some other object and the object pushes or pulls back on us too!

There are several kinds of force-measuring devices used in biomechanics to study how forces modify movement. Two important devices are the **force platform** (or force plates) and pressure sensor arrays. A force plate is a rigid platform that measures the forces and torques in all three dimensions applied to the surface of the platform (Schieb, 1987). Force plates are often mounted in a floor to measure the ground reaction forces that are equal and opposite to the forces people make against the ground (see Figure 6.5). Since the 1980s, miniaturization of sensors has allowed for rapid development of arrays of small-force sensors that allow measurement of the distribution of forces (and pressure because the area of the sensor is known) on a body. Several commercial shoe insoles with these sensors are available for studying the pressure distribution under a person's foot (see McPoil, Cornwall, & Yamada, 1995). There are many other force-measuring devices (e.g., load cell, strain gauge, isokinetic dynamometer) that help biomechanics scholars study the kinetics of movement.

INERTIA PRINCIPLE

Newton's first law of motion, or the **Law of Inertia**, describes the resistance of all objects to a change in their state of linear motion. In mechanics, the measure of inertia is an object's mass. Application of Newton's first law in biomechanics is termed the **Inertia Principle**. This section will discuss how teachers, coaches, and therapists adjust movement inertia to accommodate the task. Our focus will be on the linear inertia (mass) of movement, so the inertial resistance to rotation will be summarized in chapter 7.

The first example of application of the inertia principle is to reduce mass in order to increase the ability to rapidly accelerate. Obvious examples of this principle in track are the racing flats/shoes used in competition versus the heavier shoes used in training. The heavier shoes used in training provide protection for the foot and a small inertial overload. When race day arrives, the smaller mass of the shoes makes the athlete's feet feel light and quick. We will see in chapter 7 that this very small change in mass, because of its position, makes a much larger difference in resistance to rotation (angular inertia). Let's add a little psychology and conditioning to the application of lowering inertia. Warm-up for many sports involves a gradual increase in intensity of movements, often with larger inertia. In baseball or golf, warm-up swings are often taken with extra weights, which when taken off make the "stick" feel very light and fast (Figure 6.6).

In movements where stability is desired over mobility, the Inertia Principle suggests that mass should be increased. Linemen in football and centers in basketball have tasks that benefit more from increasing muscle mass to increase inertia, than from decreasing inertia to benefit quickness. Adding mass to a golf club or tennis racket will make for faster and longer shots if the implement can be swung with the same velocity at impact. If an exercise machine tends to slide around in the weight room, a short-term solution might be to store some extra weights on the base or legs of the machine. If these new weights are not a safety risk (in terms of height or potential for tripping people), the increased inertia of the station would likely make the machine safer.

Another advantage of increased inertia is that the added mass can be used to modify the motion of another body segment.

Figure 6.6. Mass added to sporting implements in warm-up swings makes the inertia of the regular implement (when the mass is removed) feel very light and quick. Do you think this common sporting ritual of manipulating inertia is beneficial? If so, is the effect more biomechanical or psychological?

The preparatory leg drives and weight shifts in many sporting activities have several benefits for performance, one being putting more body mass in motion toward a particular target. The forward motion of a good percentage of body mass can be transferred to the smaller body segments just prior to impact or release. We will be looking at this transfer of energy later on in this chapter when we consider the **Segmental Interaction Principle**. The defensive moves of martial artists are often designed to take advantage of the inertia of an attacker. An opponent striking from the left has inertia that can be directed by a block to throw to the right.

An area where modifications in inertia are very important is strength and conditioning. Selecting masses and weights for training and rehabilitation is a complicated

issue. Biomechanically, it is very important because the inertia of an external object has a major influence on amount of muscular force and how those forces can be applied (Zatsiorsky, 1995). Baseball pitchers often train by throwing heavier or lighter than regulation baseballs (see, e.g., Escamilla, Speer, Fleisig, Barrentine, & Andrews, 2000). Think about the amount of force that can be applied in a bench press exercise versus a basketball chest pass. The very low inertia of the basketball allows it to accelerate quickly, so the peak force that can be applied to the basketball is much lower than what can be applied to a barbell. The most appropriate load, movement, and movement speed in conditioning for a particular human movement is often difficult to define. The principle of specificity says the movement, speed, and load should be similar to the actual activity; therefore, the overload should only come from moderate changes in these variables so as to not adversely affect skill.

Suppose a high school track coach has shot put athletes in the weight room throwing medicine balls. As you discuss the program with the coach you find that they are using loads (inertia) substantially lower than the shot in order to enhance the speed of upper extremity extension. How might you apply the principle of inertia in this situation? Are the athletes fully using their lower extremities in a similar motion to shot putting? Can the athletes build up large enough forces before acceleration of the medicine ball, or will the force–velocity relationship limit muscle forces? How much lower is the mass of the medicine ball than that of the shot? All these questions, as well as technique, athlete reaction, and actual performance, can help you decide if training is appropriate. The biomechanical research on power output in multi-segment movements suggests that training loads should be higher than the 30 to 40% of 1RM seen in individual muscles and muscle groups (see the following section on muscle power; Cronin *et al.*, 2001a,b; and Funato, Matsuo, & Fukunaga, 1996). Selecting the inertia for weight training has come a long way from "do three sets of 10 reps at 80% of your maximum."

MUSCLE ANGLE OF PULL: QUALITATIVE AND QUANTITATIVE ANALYSIS OF VECTORS

Before moving on to the next kinetic approach to studying the causes of movement, it is a good time to review the special mathematics required to handle vector quantities like force and acceleration. The linear kinetics of a biomechanical issue

called *muscle angle of pull* will be explored in this section. While a qualitative understanding of adding force vector is enough for most kinesiology professionals, quantifying forces provides a deeper level of explanation and understanding of the causes of human movement. We will see that the linear kinetics of the pull of a muscle often changes dramatically because of changes in its geometry when joints are rotated.

Qualitative Vector Analysis of Muscle Angle of Pull

While the attachments of a muscle do not change, the angle of the muscle's pull on bones changes with changes in joint angle. The angle of pull is critical to the linear and angular effects of that force. Recall that a force can be broken into parts or components. These pulls of a muscle's force in two dimensions are conveniently resolved into longitudinal and rotational components. This local or relative frame of reference helps us study how muscle forces affect the body, but do not tell us about the orientation of the body to the world like absolute frames of reference do. Figure 6.7 illustrates typical angles of pull and these components for the biceps muscle at two points in the range of motion. The linear kinetic effects of the biceps on the forearm can be illustrated with arrows that represent force vectors.

The component acting along the longitudinal axis of the forearm (F_L) does not create joint rotation, but provides a load that stabilizes or destabilizes the elbow joint. The component acting at right angles to the forearm is often called the rotary component (F_R) because it creates a torque that contributes to potential rotation. Remember that vectors are drawn to scale to show their magnitude with an arrowhead to represent their direction. Note that in the extended position, the rotary component is similar to the stabilizing component.

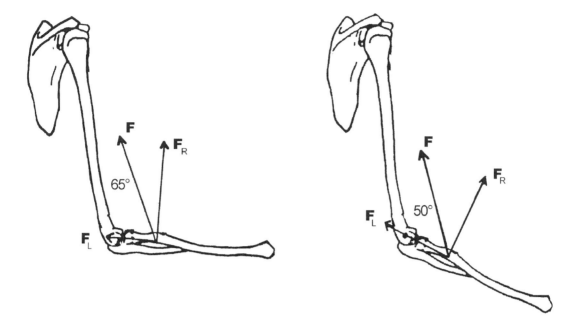

Figure 6.7. Typical angles of pull of the biceps brachii muscle in an arm curl. The angular positions of the shoulder and elbow affect the angle of pull of the muscle, which determines the size of the components of the muscle force. Muscle forces (**F**) are usually resolved along the longitudinal axis of the distal segment (**F**$_L$) and at right angles to the distal segment to show the component that causes joint rotation (**F**$_R$).

In the more flexed position illustrated, the rotary component is larger than the smaller stabilizing component. In both positions illustrated, the biceps muscle tends to flex the elbow, but the ability to do so (the rotary component) varies widely.

This visual or qualitative understanding of vectors is quite useful in studying human movement. When a muscle pulls at a 45° angle, the two right-angle components are equal. A smaller angle of pull favors the longitudinal component, while the rotary component benefits from larger angles of pull. Somewhere in the midrange of the arm curl exercise the biceps has an angle of pull of 90°, so all the bicep's force can be used to rotate the elbow and there is no longitudinal component.

Vectors can also be qualitatively added together. The rules to remember are that the forces must be drawn accurately (size and direction), and they then can be added together in tip-to-tail fashion. This graphical method is often called drawing a *parallelogram of force* (Figure 6.8). If the vastus lateralis and vastus medialis muscle forces on the right patella are added together, we get the resultant of these two muscle forces. The resultant force from these two muscles can be determined by drawing the two muscle forces from the tip of one to the tail of the other, being sure to maintain correct length and direction. Since these diagrams can look like parallelograms, they are called a parallelogram of force. Remember that there are many other muscles, ligaments, and joint forces not shown that affect knee function. It has been hypothesized that an imbalance of greater lateral forces in the quadriceps may contribute to patellofemoral pain syndrome (Callaghan & Oldham, 1996). Does the resultant force (**F**$_R$)

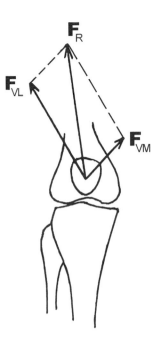

Figure 6.8. Any vectors acting on the same object, like the vastus medialis ($\mathbf{F}_{VM}$) and vastus lateralis ($\mathbf{F}_{VL}$) of the right knee, can be added together to find a resultant ($\mathbf{F}_R$). This graphical method of adding vectors is called a *parallelogram of forces*.

in Figure 6.8 appear to be directed lateral to the longitudinal axis of the femur?

Quantitative Vector Analysis of Muscle Angle of Pull

Quantitative or mathematical analysis provides precise answers to vector resolution (in essence subtraction to find components) or vector composition. Right-angle trigonometry provides the perfect tool for this process. A review of the major trigonometric relationships (sine, cosine, tangent) is provided in Appendix D. Suppose an athlete is training the isometric stabilization ability of their abdominals with leg raises in a Roman chair exercise station. Figure 6.9a illustrates a typical orientation and magnitude of the major hip flexors (the iliopsoas

group) that hold their legs elevated. The magnitude of the weight of the legs and the hip flexor forces provide a large resistance for the abdominal muscles to stabilize. This exercise is not usually appropriate for untrained persons.

If an iliopsoas resultant muscle force of 400 N acts at a 55° angle to the femur, what are the rotary ($\mathbf{F}_R$) and longitudinal ($\mathbf{F}_L$) components of this force? To solve this problem, the rotating component is moved tip to tail to form a right triangle (Figure 6.9b). In this triangle, right-angle trigonometry says that the length of the adjacent side to the 55° angle ($\mathbf{F}_L$) is equal to the resultant force times cos 55°. So the stabilizing component of the iliopsoas force is: $\mathbf{F}_L = 400(\cos 55°) = 229$ N, which would tend to compress the hip joint along the longitudinal axis of the femur. Likewise, the rotary component of this force is the side opposite the 55° angle, so this side is equal to the resultant force times sin 55°. The component of the 400-N iliopsoas force that would tend to rotate the hip joint, or in this example isometrically hold the legs horizontally, is: $\mathbf{F}_R = 400 \sin 55° = 327$ N upward. It is often a good idea to check calculations with a qualitative assessment of the free body diagram. Does the rotary component look larger than the longitudinal component? If these components are the same at 45°, does it make sense that a higher angle would increase the vertical component and decrease the component of force in the horizontal direction?

When the angle of pull (θ) or push of a force can be expressed relative to a horizontal axis (2D analysis like above), the horizontal component is equal to the resultant times the cosine of the angle θ. The vertical component is equal to the resultant times the sine of the angle θ. Consequently, how the angle of force application affects the size of the components is equal to the shape of a sine or cosine wave. A qualitative understanding of these functions helps one

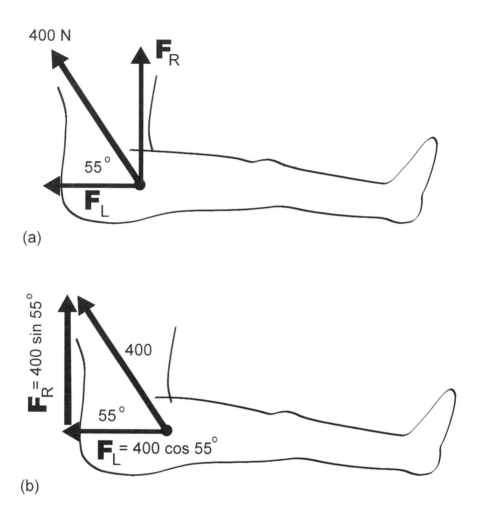

Figure 6.9. Right-angle trigonometry is used to find the components of a vector like the iliopsoas muscle force illustrated. Notice the muscle force is resolved into components along the longitudinal axis of the femur and at right angles to the femur.

understand where the largest changes occur and what angles of force application are best. Let's look at a horizontal component of a force in two dimensions. This is analogous to our iliopsoas example, or how any force applied to an object will favor the horizontal over the vertical component. A cosine function is not a linear function like our spring example in chapter 2. Figure 6.10 plots the size of the cosine function as a percentage of the resultant for angles of pull from 0 to 90°.

A 0° (horizontal) angle of pull has no vertical component, so all the force is in the horizontal direction. Note that, as the angle of pull begins to rise (0 to 30°), the cosine or horizontal component drops very slowly, so most of the resultant force is directed horizontally. Now the cosine function begins to change more rapidly, and from 30 to 60° the horizontal component has drops from 87 to 50% the size of the resultant force. For angles of pull greater than 60°, the cosine drops off very fast, so there is a dra-

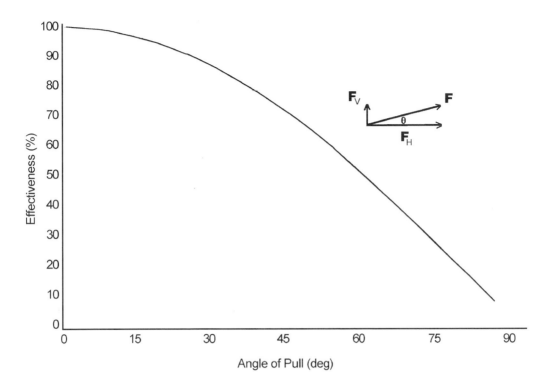

Figure 6.10. Graph of the cosine of angle **θ** between 0 and 90° (measured from the right horizontal) shows the percentage effectiveness of a force (**F**) in the horizontal direction (**F**$_H$). This horizontal component is equal to **F** cos(**θ**), and angle **θ** determines the tradeoff between the size of the horizontal and vertical components. Note that the horizontal component stays large (high percentage of the resultant) for the first 30° but then rapidly decreases. The sine and cosine curves are the important nonlinear mathematical functions that map linear biomechanical variables to angular.

matic decrease in the horizontal component of the force, with the horizontal component becoming 0 when the force is acting at 90° (vertical). We will see that the sine and cosine relationships are useful in angular kinetics as well. These curves allow for calculation of several variables related to angular kinetics from linear measurements. Right-angle trigonometry is also quite useful in studying the forces between two objects in contact, or precise kinematic calculations.

CONTACT FORCES

The linear kinetics of the interaction of two objects in contact is also analyzed by resolving the forces into right-angle components. These components use a local frame of reference like the two-dimensional muscle angle of pull above, because using horizontal and vertical components are not always convenient (Figure 6.11). The forces between two objects in contact are resolved into the **normal reaction** and **friction**. The normal reaction is the force at right angles to the surfaces in contact, while friction is the force acting in parallel to the surfaces. Friction is the force resisting the sliding of the surfaces past each other.

When the two surfaces are dry, the force of friction (**F**) is equal to the product of the coefficient of friction (μ) and the normal reaction (**F**$_N$), or **F** = $\mu \cdot$ **F**$_N$. The coefficient

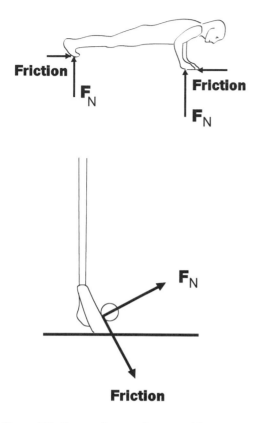

Figure 6.11. Forces of contact between objects are usually resolved into the right-angle components of *normal reaction* (F_N) and *friction*.

of friction depends on the texture and nature of the two surfaces, and is determined by experimental testing. There are coefficients of static (non-moving) friction (μ_S) and kinetic (sliding) friction (μ_K). The coefficients of kinetic friction are typically 25% smaller than the maximum static friction. It is easier to keep an object sliding over a surface than to stop it and start the object sliding again. Conversely, if you want friction to stop motion, preventing sliding (like with anti-lock auto brakes) is a good strategy. Figure 6.12 illustrates the friction force between an athletic shoe and a force platform as a horizontal force is increased. Please note that the friction grows in a linear fashion until the limiting friction is reached ($\mu_S \cdot F_N$), at which point the shoe begins to slide across the force platform. If the weight on the shoe created a normal reaction of 300 N, what would you estimate the μ_S of this rubber/metal interface?

Typical coefficients of friction in human movement vary widely. Athletic shoes have coefficients of static friction that range from 0.4 to over 1.0 depending on the shoe and sport surface. In tennis, for example, the linear and angular coefficients of friction range from 0.4 to over 2.0, with shoes responding differently to various courts

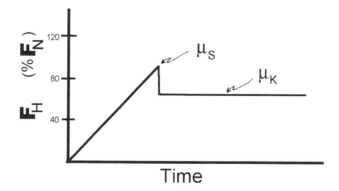

Figure 6.12. The change in friction force between an athletic shoe and a force platform as a horizontal force is applied to the shoe. The ratio of the friction force (F_H) on this graph to the normal force between the shoe and force platform determine the coefficient of friction for these two surfaces.

(Nigg, Luthi, & Bahlsen, 1989). Epidemiological studies have shown that playing on lower-friction courts (clay) had a lower risk of injury (Nigg *et al.*, 1989). Many teams that play on artificial turf use flat shoes rather than spikes because they believe the lower friction decreases the risk of severe injury. The sliding friction between ice and a speed skating blade has been measured, demonstrating coefficients of kinetic friction around 0.005 (van Ingen Schenau, De Boer, & De Groot, 1989).

IMPULSE–MOMENTUM RELATIONSHIP

Human movement occurs over time, so many biomechanical analyses are based on movement-relevant time intervals. For example, walking has a standardized gait cycle (Whittle, 1996), and many sport movements are broken up into phases (usually, preparatory, action, and follow-through). The mechanical variables that often used in these kinds of analyses are impulse (**J**) and momentum (**p**). These two variables are related to each other in the original language of Newton's second law: the change in momentum of an object is equal to the impulse of the resultant force in that direction. The impulse–momentum relationship is Newton's Second law written over a time interval, rather than the instantaneous (**F** = m**a**) version.

Impulse is the effect of force acting over time. Impulse (**J**) is calculated as the product of force and time (**J** = **F** · t), so the typical units are N·s and lb·s. Impulse can be visualized as the area under a force–time graph. The vertical ground reaction force during a foot strike in running can be measured using a force platform, and the area under the graph (integral with respect to time) represents the vertical impulse (Figure 6.13). A person can increase the motion of an object by applying a greater im-

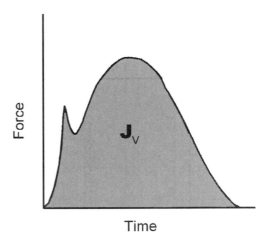

Figure 6.13. The vertical impulse (J_V) of the vertical ground reaction force for a footstrike in running is the area under the force–time graph.

pulse, and both the size of the force and duration of force application are equally important. Impulse is the mechanical variable discussed in the following section on the "Force–Time Principle." In movement, the momentum a person can generate, or dissipate in another object, is dependent on how much force can be applied and the amount of time the force is applied.

Newton realized that the mass of an object affects its response to changes in motion. Momentum is the vector quantity that Newton said describes the quantity of motion of an object. Momentum (**p**) is calculated as the product of mass and velocity (**p** = m · **v**). The SI unit for momentum is kg·m/s. Who would you rather accidentally run into in a soccer game at a 5-m/s closing velocity: a 70- or 90-kg opponent? We will return to this question and mathematically apply the impulse–momentum relationship later on in this chapter once we learn about a similar kinetic variable called *kinetic energy*.

The association between impulse (force exerted over time) and change in momentum (quantity of motion) is quite useful in gaining a deeper understanding of many sports. For example, many impacts create very large forces because the time interval of many elastic collisions is so short. For a golf ball to change from zero momentum to a very considerable momentum over the 0.0005 seconds of impact with the club requires a peak force on the golf ball of about 10,000 N, or greater than 2200 pounds (Daish, 1972). In a high-speed soccer kick, the ball is actually on the foot for about 0.016 seconds, so that peak forces

on the foot are above 230 pounds (Tol, Slim, van Soest, & van Dijk, 2002; Tsaousidis & Zatsiorsky, 1996). Fortunately, for many catching activities in sport an athlete can spread out the force applied to the ball over longer periods of time. The Impulse–Momentum Relationship is the mechanical law that underlies the **Force–Time Principle** introduced earlier in chapters 2 and 4. Let's revisit the application of the **Force–Time Principle** with our better understanding of linear kinetics.

Interdisciplinary Issue: Acute and Overuse Injuries

A very important area of research by many kinesiology and sports medicine scholars is related to musculoskeletal injuries. Injuries can be subclassified into acute injuries or overuse injuries. Acute injuries are single traumatic events, like a sprained ankle or breaking a bone in a fall from a horse. In an acute injury the forces create tissue loads that exceed the ultimate strength of the biological tissues and cause severe physical disruption. Overuse injuries develop over time (thus, chronic) from a repetitive motion, loading, inadequate rest, or a combination of the three. Injuries from repetitive vocational movements or work-related musculoskeletal disorders (WMSDs) are examples of chronic injuries (Barr & Barbe, 2002). Stress fractures and anterior tibial stress syndrome (shin splits) are classic examples of overuse injuries associated with running. Runners who overtrain, run on very hard surfaces, and are susceptible can gradually develop these conditions. If overuse injuries are untreated, they can develop into more serious disorders and injuries. For example, muscle overuse muscles can sometimes cause inflammation of tendons (tendinitis), but if the condition is left untreated degenerative changes begin to occur in the tissue that are called tendinoses (Khan, Cook, Taunton, & Bonar, 2000). Severe overuse of the wrist extensors during one-handed backhands irritates the common extensor tendon attaching at the lateral epicondyle, often resulting in "tennis elbow."

The etiology (origin) of overuse injuries is a complex phenomenon that requires interdisciplinary research. The peak force or acceleration (shock) of movements is often studied in activities at risk of acute injury. It is less clear if peak forces or total impulse are more related to the development of overuse injuries. Figure 6.14 illustrates the typical vertical ground reaction forces measured with a force platform in running, step aerobics, and walking. Note that the vertical forces are normalized to units of bodyweight. Notice that step aerobics has peak forces near 1.8 BW because of the longer time of force application and the lower intensity of movement. Typical vertical ground reaction forces in step aerobics look very much like the forces in walking (peak forces of 1.2 BW and lower in double support) but tend to be a bit larger because of the greater vertical motion. The peak forces in running typically are about 3 BW because of the short amount of time the foot is on the ground. Do you think the vertical impulses of running and step aerobics are similar? Landing from large heights and the speed involved in gymnastics are very close to injury-producing loads. Note the very high peak force and rate of force development (slope of the F–t curve) in the landing ground reaction force. Gymnastic coaches should limit the number of landings during practice and utilize thick mats or landing pits filled with foam rubber to reduce the risk of injury in training.

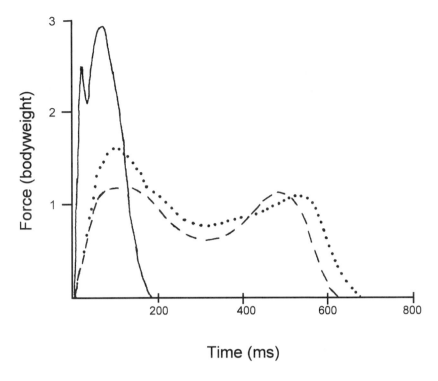

Figure 6.14. Typical vertical ground reaction forces (in units of bodyweight) for running (solid), walking (dashed), and step aerobic exercise (dotted).

FORCE-TIME PRINCIPLE

The applied manifestation of Newton's Second Law of Motion as the Impulse–Momentum Relationship is the **Force–Time Principle**. If a person can apply force over a longer period of time (large impulse), they will be able to achieve a greater speed (change in momentum) than if they used similar forces in a shorter time interval. Unfortunately, in many human movements there is not an unlimited amount of time to apply forces, and there are several muscle mechanical characteristics that complicate application of this principle. Recall from chapter 4 that maximizing the time force application is not always the best strategy for applying the **Force–Time Principle**. The movement of interest, muscle characteristics, and the mechanical strengths of tissues

all affect optimal application of forces to create motion.

There are a few movements that do allow movers to maximize the time of force application to safely slow down an object. In landing from a jump, the legs are extended at contact with the ground, so there is near maximal joint range of motion to flex the joints and absorb impact forces. A softball infielder is taught to lean forward and extend her glove hand to field a ground ball so that she can absorb the force of the ball over a longer time interval. Figure 6.15 illustrates two people catching balls: which athlete is using a technique that is correctly applying the Force–Time Principle? Young children often catch by trapping the object against the body and even turn their heads in fear. Even professional football players (6.15, below) occasionally rely on their

Figure 6.15. Catching the ball close to the body (the American football example) is a poor application of the Force–Time Principle because there is minimal time or range of motion to slow down the ball. The softball catcher has increased the time and range of motion that can be used to slow down the ball.

talent or sense of self-preservation more than coaching and use a similar catching technique. For how much time can forces be applied to slow the balls in these cases? The momentum of the ball in these situations is often so great that the force between the person's body and the ball builds up so fast that the ball bounces out of their grasp. If these people extended their arms and hands to the ball, the time the force is applied to slow down the ball could be more than ten times longer. Not only does this increase the chance of catching the ball, but it decreases the peak force and potential discomfort involved in catching.

Athletes taught to reach for the ground and "give" with ankle, hip, and knee flexion dramatically increase the time of force application in landing and decrease the peak ground reaction forces. Exactly how the muscles are positioned and pre-tensed prior to landing affects which muscle groups are used to cushion landing (DeVita & Skelly, 1992; Kovacs *et al.*, 1999; Zhang, Bates, & Dufek, 2000). How to teach this important skill has not been as well researched. The sound of an impact often tells an athlete about the severity of a collision, so this has been used as a teaching point in catching and landing. It has also been shown that focusing attention on decreasing the sound of landing is an effective strategy to decrease peak forces during landing (McNair, Prapavessis, & Callender, 2000). Increasing the "give" of the cushioning limbs increases the time of force appli-

Activity: Impulse–Momentum Relationship

Fill a few small balloons with water to roughly softball size. Throw the water balloon vertically and catch it. Throw the balloon several times trying to maximize the vertical height thrown. Imagine that the water balloon represents your body falling and the catching motions represent your leg actions in landing. What catching technique points modify the force and time of force application to the balloon to create a vertical impulse to reduce the momentum of the balloon to zero?

cation and decreases the tone and intensity of the sound created by the collision.

In some movements there are other biomechanical factors involved in the activity that limit the amount of time that force can be applied. In these activities, increasing time of force application would decrease performance, so the only way to increase the impulse is to rapidly create force during the limited time available. A good example of this is long jumping. Recall that in the kinematics chapter we learned that long jumpers have low takeoff angles (approximately 20°). The takeoff foot is usually on the board for only 100 ms, so there is little time to create vertical velocity. Skilled long jumpers train their neuromuscular system to strongly activate the leg muscles prior to foot strike. This allows the jumper to rapidly increase ground reaction forces so they can generate vertical velocity without losing too much horizontal velocity. Similar temporal limitations are at work in running or throwing. In many sports where players must throw the ball quickly to score or prevent an opponent scoring, the player may make a quicker throw than they would during maximal effort without time restrictions. A quick delivery may not use maximal throwing speed or the extra time it

takes to create that speed, but it meets the objective of that situation. Kinesiology professionals need to instruct movers as to when using more time of force application will result in safer and more effective movement, and when the use of longer force application is not the best movement strategy.

WORK–ENERGY RELATIONSHIP

The final approach to studying the kinetics of motion involves laws from a branch of physics dealing with the concepts of work and energy. Since much of the energy in the human body, machines, and on the earth are in the form of heat, these laws are used in thermodynamics to study the flow of heat energy. Biomechanics are interested in how mechanical energies are used to create movement.

Mechanical Energy

In mechanics, **energy** is the capacity to do work. In the movement of everyday objects, energy can be viewed as the mover of stuff (matter), even though at the atomic level matter and energy are more closely related. Energy is measured in Joules (J) and is a scalar quantity. Energy is a scalar because it represents an ability to do work that can be transferred in any direction. Energy can take many forms (for example, heat, chemical, nuclear, motion, or position). There are three mechanical energies that are due to an object's motion or position.

The energies of motion are linear and angular **kinetic energy**. Linear or translational kinetic energy can be calculated using the following formula: $KE_T = \frac{1}{2}m\mathbf{v}^2$. There are several important features of this formula. First, note that squaring velocity makes the energy of motion primarily dependent on the velocity of the object. The energy of motion varies with the square of the velocity, so doubling velocity increases the kinetic energy by a factor of 4 (2^2). Squaring velocity also eliminates the effect of the sign (+ or –) or vector nature of velocity. Angular or rotational kinetic energy can be calculated with a similar formula: $KE_R = \frac{1}{2}I\omega^2$. We will learn more about angular kinetics in chapter 7.

The mathematics of kinetic energy ($\frac{1}{2}mv^2$) looks surprisingly similar to momentum (mv). However, there are major differences in these two quantities. First, momentum is a vector quantity describing the quantity of motion in a particular direction. Second, kinetic energy is a scalar that describes how much work an object in motion could perform. The variable momentum is used to document the current state of motion, while kinetic energy describes the potential for future interactions. Let's consider a numerical example from American football. Imagine you are a small (80-kg) halfback spinning off a tackle with one yard to go for a touchdown. Who would you rather run into just before the goal line: a quickly moving defensive back or a very large lineman not moving as fast? Figure 6.16 illustrates the differences between kinetic energy and momentum in an inelastic collision.

Applying the impulse–momentum relationship is interesting because this will tell us about the state of motion or whether a touchdown will be scored. Notice that both defenders (small and big) have the same amount of momentum (–560 kg·m/s), but because the big defender has greater mass you will not fly backwards as fast as in the collision with the defensive back. The impulse–momentum relationship shows that you do not score either way (negative velocity after impact: V_2), but the defensive back collision looks very dramatic because you reverse directions with a faster negative velocity. The work–energy relationship tells us that the total mechanical energy of

the collision will be equal to the work the defender can do on you. Some of this energy is transferred into sound and heat, but most of it will be transferred into deformation of your pads and body! Note that the sum of the energies of the two athletes and the strong dependence of kinetic energy on velocity results in nearly twice (2240 versus 1280 J) as much energy in the collision with the defensive back. In short, the defensive back hurts the most because it is a very high-energy collision, potentially adding injury to the insult of not scoring.

There are two types of mechanical energy that objects have because of their position or shape. One is gravitational potential energy and the other is strain energy. Gravitational **potential energy** is the energy of the mass of an object by virtue of its position relative to the surface of the earth. Potential energy can be easily calculated with the formula: $PE = mgh$. Potential energy depends on the mass of the object, the acceleration due to gravity, and the height of the object. Raising an object with a mass of 35 kg a meter above the ground stores 343 J of energy in it ($PE = 35 \cdot 9.81 \cdot 1 = 343$). If this object were to be released, the potential energy would gradually be converted to kinetic energy as gravity accelerated the object toward the earth. This simple example of transfer of mechanical energies is an example of one of the most important laws of physics: the **Law of Conservation of Energy**.

The **Law of Conservation of Energy** states that energy cannot be created or destroyed; it is just transferred from one form to another. The kinetic energy of a tossed ball will be converted to potential energy or possibly strain energy in an object it may collide with. A tumbler taking off from a mat has kinetic energy in the vertical direction that is converted into potential energy on the way up, and back into kinetic energy on the way down. A bowler who increases the potential energy of the ball during the

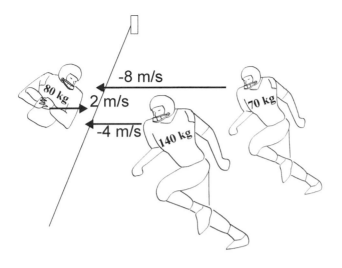

Kinetic Energy

$KE_{Back} = \frac{1}{2} \, 80(2)^2 = 160$ J

$KE_{Small} = \frac{1}{2} \, 70(-8)^2 = 2240$ J

$KE_{Big} = \frac{1}{2} \, 140(-4)^2 = 1120$ J

Total $_{Small}$ = 2400 J

Total $_{Big}$ = 1280 J

Conservation of Momentum

$m_1V_1 + m_2V_2 = m_1V_2 + m_2V_2$

Small $80(2) - 70(8) = (70 + 80) V_2$

$V_2 = -2.7$ m/s

Big $80(2) - 140(4) = (80 + 140) V_2$

$V_2 = -1.8$ m/s

Figure 6.16. Comparison of the kinetic energy (scalar) and momentum (vector) in a football collision. If you were the running back, you would not score a touchdown against either defender, but the work done on your body would be greater in colliding with the smaller defender because of their greater kinetic energy.

approach can convert this energy to kinetic energy prior to release (Figure 6.17). In a similar manner, in golf or tennis a forward swing can convert the potential energy from preparatory movement into kinetic energy. A major application area of conservation of energy is the study of heat or thermodynamics.

The First Law of Thermodynamics is the law of conservation of energy. This is the good news: when we energy is added into a machine, we get some other form of energy on the way out. Unlike these examples, examination of the next mechanical energy (strain energy) will illustrate the bad news of the Second Law of Thermodynamics: that it is impossible to create a machine that converts all input energy to some useful output energy. In other words, man-made devices will always lose

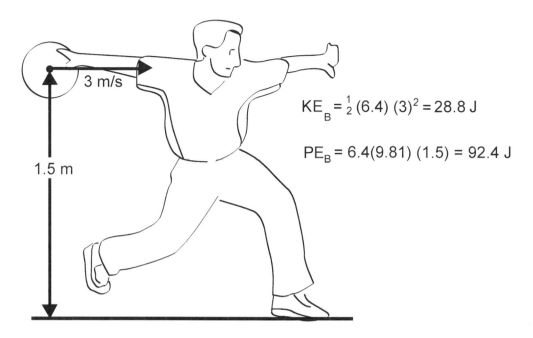

$$KE_B = \tfrac{1}{2}(6.4)(3)^2 = 28.8 \text{ J}$$

$$PE_B = 6.4(9.81)(1.5) = 92.4 \text{ J}$$

Figure 6.17. Raising a bowling ball in the approach stores more potential energy in the ball than the kinetic energy from the approach. The potential energy of the ball can be converted to kinetic energy in the downswing.

energy in some non-useful form and never achieve 100% efficiency. This is similar to the energy losses (hysteresis) in strain energy stored in deformed biological tissues studied in chapter 4.

Strain energy is the energy stored in an object when an external force deforms that object. Strain energy can be viewed as a form of potential energy. A pole vaulter stores strain energy in the pole when loading the pole by planting it in the box. Much of the kinetic energy stored in the vaulter's body during the run up is converted into strain energy and back into kinetic energy in the vertical direction. Unfortunately, again, not all the strain energy stored in objects is recovered as useful energy. Often large percentages of energy are converted to other kinds of energy that are not effective in terms of producing movement. Some strain energy stored in many objects is essentially lost because it is converted

into sound waves or heat. Some machines employ heat production to do work, but in human movement heat is a byproduct of many energy transformations that must be dissipated. Heat is often even more costly than the mechanical energy in human movement because the cardiovascular system must expend more chemical energy to dissipate the heat created by vigorous movement.

The mechanical properties of an object determine how much of any strain energy is recovered in restitution as useful work. Recall that many biomechanical tissues are viscoelastic and that the variable hysteresis (area between the loading and unloading force-displacement curves) determines the amount of energy lost to unproductive energies like heat. The elasticity of a material is defined as its stiffness. In many sports involving elastic collisions, a simpler variable can be used to get an estimate of the elasticity or energy losses of an object *relative to another object*. This variable is called the **coefficient of restitution** (COR and e are common abbreviations). The coefficient of restitution is a dimensionless number usually ranging from 0 (perfectly plastic collision: mud on your mother's kitchen floor) to near 1 (very elastic pairs of materials). The coefficient of restitution cannot be equal to or greater than 1 because of the second law of thermodynamics. High coefficients of restitution represent elastic collisions with little wasted energy, while lower coefficients of restitution do not recover useful work from the strain energy stored in an object.

The coefficient of restitution can be calculated as the relative velocity of separation divided by the relative velocity of approach of the two objects during a collision (Hatze, 1993). The most common use of the coefficient of restitution is in defining the relative elasticities of balls used in sports. Most sports have strict rules governing the dimensions, size, and specifications, including the ball and playing surfaces. Officials in basketball or tennis drop balls from a standard height and expect the ball to rebound to within a small specified range allowed by the rules. In these uniformly accelerated flight and impact conditions where the ground essentially doesn't move, e can be calculated with this formula: e = bounce/drop$^{1/2}$. If a tennis ball were dropped from a 1-meter height and it rebounded to 58 cm from a concrete surface, the coefficient of restitution would be $(58/100)^{1/2} = 0.76$. Dropping the same tennis ball on a short pile carpet might result in a 45-cm rebound, for an e = 0.67. The coefficient of restitution for a sport ball varies depending on the nature of the other object or surface it interacts with (Cross, 2000), the velocity of the collision, and other factors like temperature. Squash players know that it takes a few rallies to warm up the ball and increase its coefficient of restitution.

Also, putting softballs in a refrigerator will take some of the slugging percentage out of a strong hitting team.

Most research on the COR of sport balls has focused on the elasticity of a ball in the vertical direction, although there is a COR in the horizontal direction that affects friction and the change in horizontal ball velocity for oblique impacts (Cross, 2002). The horizontal COR strongly affects the spin created on the ball following impact. This is a complicated phenomenon because balls deform and can slide or rotate on a surface during impact. How spin, in general, affects the bounce of sport balls will be briefly discussed in the section on the spin principle in chapter 8.

Mechanical Work

All along we have been defining mechanical energies as the ability to do mechanical work. Now we must define **work** and understand that this mechanical variable is not exactly the same as most people's common perception of work as some kind of effort. The mechanical work done on an object is defined as the product of the force and displacement in the direction of the force ($\mathbf{W} = \mathbf{F} \cdot \mathbf{d}$). Joules are the units of work: one joule of work is equal to one Nm. In the English system, the units of work are usually written as foot-pounds (ft·lb) to avoid confusion with the angular kinetic variable torque, whose unit is the lb·ft. A patient performing rowing exercises (Figure 6.18) performs positive work ($\mathbf{W} = 70 \cdot 0.5 = +35$ Nm or Joules) on the weight. In essence, energy flows from the patient to the weights (increasing their potential energy) in the concentric phase of the exercise. In the eccentric phase of the exercise the work is negative, meaning that potential energy is being transferred from the load to the patient's body. Note that the algebraic formula assumes the force applied to the

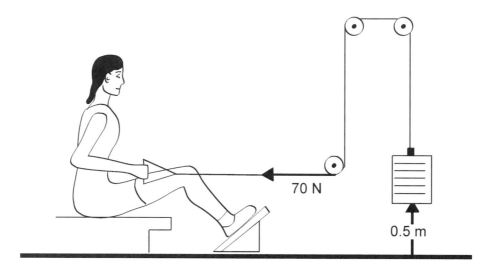

Figure 6.18. The mechanical work done on a weight in this rowing exercise is the product of the force and the displacement.

load is constant over the duration of the movement. Calculus is necessary to calculate the work of the true time-varying forces applied to weights in exercises. This example also assumes that the energy losses in the pulleys are negligible as they change the direction of the force created by the patient.

Note that mechanical work can only be done on an object when it is moved relative to the line of action of the force. A more complete algebraic definition of mechanical work in the horizontal (x) direction that takes into account the component of motion in the direction of the force on an object would be $W = (F \cos \theta) \cdot d_x$. For example, a person pulling a load horizontally on a dolly given the data in Figure 6.19 would do 435 Nm or Joules of work. Only the horizontal component of the force times the displacement of object determines the work done. Note also that the angle of pull in this example is like the muscle angle of pull analyzed earlier. The smaller the angle of pull, the greater the horizontal component of the force that does work to move the load.

The vertical component of pull does not do any mechanical work, although it may decrease the weight of the dolly or load and, thereby decrease the rolling friction to be overcome. What the best angle to pull in this situation load depends on many factors. Factoring in rolling friction and the strength (force) ability in various pulling postures might indicate that a higher angle of pull that doesn't maximize the horizontal force component may be "biomechanically" effective for this person. The inertia of the load, the friction under the person's feet, and the biomechanical factors of pulling from different postures all interact to determine the optimal angle for pulling an object. In fact, in some closed kinematic chain movements (like cycling) the optimal direction of force application does not always maximize the effectiveness or the component of force in the direction of motion (Doorenbosch *et al.*, 1997).

Mechanical work does not directly correspond to people's sense of muscular effort. Isometric actions, while taking considerable effort, do not perform mechanical

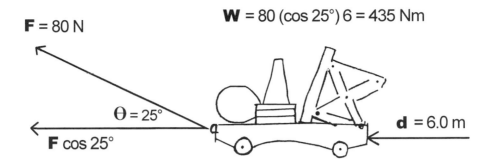

Figure 6.19. Mechanical work is calculated as displacement of the object in the direction of the force. This calculation is accurate if the 80-N force is constant during horizontal displacement of the dolly. If you were pulling this dolly, what angle of pull would you use?

work. This dependence on the object's displacement of mechanical work makes the work–energy relationship useful in biomechanical studies where the motion of an object may be of more interest than temporal factors.

This brings us to the **Work–Energy Relationship**, which states that the mechanical work done on an object is equal to the change in mechanical energy of that object. Biomechanical studies have used the work–energy relationship to study the kinetics of movements. One approach calculates the changes in mechanical energies of the segments to calculate work, while the other calculates mechanical power and integrates these data with respect to time to calculate work. The next section will discuss the concept of mechanical power.

Mechanical Power

Mechanical power is an important kinetic variable for analyzing many human movements because it incorporates time. Power is defined as the rate of doing work, so mechanical power is the time derivative of mechanical work or work divided by time ($P = W/t$). Note that a capital "P" is used because lower-case "p" is the symbol for momentum. Typical units of power are Watts (one J/s) and horsepower. One horsepower is equal to 746 W. Maximal mechanical power is achieved by the right combination of force and velocity that maximizes the mechanical work done on an object. This is clear from the other formula for calculating power: $P = F \cdot v$. Prove to yourself that the two equations for power are the same by substituting the formula for work **W** and do some rearranging that will allow you to substitute **v** for its mechanical definition.

If the concentric lift illustrated in Figure 6.18 was performed within 1.5 seconds, we could calculate the average power flow to the weights. The positive work done on the weights was equal to 35 J, so $P = W/t = 35/1.5 = 23.3$ W. Recall that these algebraic definitions of work and power calculate a mean value over a time interval for constant forces. The peak instantaneous power flow to the weight in Figure 6.18 would be higher than the average power calculated over the whole concentric phase of the lift. The Force–Motion Principle would say that the patient increased the vertical force on the resistance to more than the weight of the stack to positively accelerate it and would reduce this force to below the weight of the stack to gradually stop the weight at the end of the concentric phase. Instantaneous power flow to the weights also follows a complex pattern based on a combination of the force applied and the motion of the object.

What movements do you think require greater peak mechanical power delivered to a barbell: the lifts in the sport of Olympic weight lifting or power lifting? Don't let the names fool you. Since power is the rate of doing work, the movements with the greatest mechanical power must have high forces and high movement speeds. Olympic lifting has mechanical power outputs much higher than power lifting, and power lifting is clearly a misnomer given the true definition of power. The dead lift, squat, and bench press in power lifting are high-strength movements with large loads but very slow velocities. The faster movements of Olympic lifting, along with smaller weights, clearly create a greater power flow to the bar than power lifting. Peak power flows to the bar in power lifting are between 370 and 900 W (0.5–1.2 hp), while the peak power flow to a bar in Olympic lifts is often as great as 4000 W or 5.4 hp (see Garhammer, 1989). Olympic lifts are often used to train for "explosive" movements, and Olympic weight lifters can create significantly more whole-body mechanical power than other athletes (McBride, Triplett-McBride, Davie, & Newton, 1999).

Many people have been interested in the peak mechanical power output of whole-body and multi-segment movements. It is believed that higher power output is critical for quick, primarily anaerobic movements. In the coaching and kinesiology literature these movements have been described as "explosive." This terminology may communicate the point of high rates of force development and high levels of both speed and force, but a literal interpretation of this jargon is not too appealing! Remember that the mechanical power out-

put calculated for a human movement will strongly depend on the model used and the time interval used in the calculation. In addition, other biomechanical factors affect how much mechanical power is developed during movements.

The development of maximal power output in human movement depends on the direction of the movement, the number of segments used, and the inertia of the object. If we're talking about a simple movement with a large resistance, the right mix of force and velocity may be close to 30 to 45% of maximal isometric strength because of the Force–Velocity Relationship (Izquierdo *et al.*, 1999; Kaneko *et al.*, 1983; Wilson *et al.*, 1993). Figure 6.20 shows the *in vitro* concentric power output of skeletal muscle derived from the product of force and velocity in the Force–Velocity Relationship. In movements requiring multi-joint movements, specialized dynamometer measurements indicate that the best resistances for peak power production and training are likely to be higher than the 30–45% (Funato *et al.*, 1996; Newton *et al.*, 1996). The best conditioning for "explosive" movements

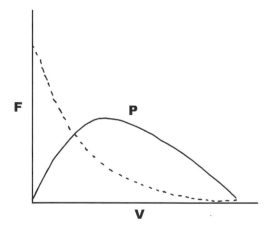

Figure 6.20. The *in vitro* mechanical power output of skeletal muscle. Note that peak power in concentric actions does not occur at either the extremes of force or velocity.

Interdisciplinary Issue: Efficiency

One area of great potential for interdisciplinary cooperation is in determining the efficiency of movement. This efficiency of human movement is conceptually different from the classical definition of efficiency in physics. Physics defines efficiency as the mechanical work output divided by the mechanical work input in a system, a calculation that helps engineers evaluate machines and engines. For endurance sports like distance running, adjusting a formula to find the ratio of mechanical energy created to metabolic cost appears to be an attractive way to study human movement (van Ingen Schenau & Cavanagh, 1999). Progress in this area has been hampered by the wide variability of individual performance and confusion about the various factors that contribute to this movement efficiency (Cavanagh & Kram, 1985). Cavanagh and Kram argued that the efficiency of running, for example, could be viewed as the sum of several efficiencies (e.g., biochemical, biomechanical, physiological, psychomotor) and other factors. Examples of the complexity of this area are the difficulty in defining baseline metabolic energy expenditure and calculating the true mechanical work because more work is done than is measured by ergometers. For instance, in cycle ergometry the mechanical work used to move the limbs is not measured. Biomechanists are also struggling to deal with the zero-work paradox in movements where there no net mechanical work is done, like in cyclic activities, co-contracting muscles, or forces applied to the pedal in an ineffective direction. Figure 6.21 illustrates the typical forces applied to a bicycle pedal at 90° (from vertical). The normal component of the pedal force does mechanical work in rotating the pedal (F_N), while the other component does no work that is transferred to the bike's flywheel. Movement efficiency is an area where cooperative and interdisciplinary research may be of interest to many scientists and may be an effective tool for improving human movement.

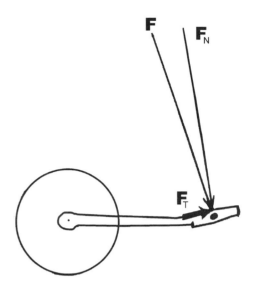

Figure 6.21. Only some of the force applied to a bicycle pedal creates work and mechanical power.

may be the use of moderate resistances (just less than strength levels that are usually >70% 1RM), which are moved as quickly as possible. Oftentimes these exercises use special equipment like the Plyometric Power System, which allows for the resistance to be thrown (Wilson *et al.*, 1993). The disadvantage of high-speed exercise is that it focuses training on the early concentric phase, leaving much of the range of motion submaximally trained. Even slow, heavy weight training exercises have large submaximal percentages (24–52%) of range of motion due to negative acceleration of the bar at the end of the concentric phase (Elliott, Wilson, & Kerr, 1989).

There are several field tests to estimate short-term explosive leg power, but the utility and accuracy of these tests are controversial. The Margaria test (Margaria, Aghemo, & Rovelli, 1966) estimates power from running up stairs, and various vertical jump equations (see Johnson & Bahamonde, 1996; Sayers, Harackiewicz, Harman, Frykman, & Rosenstein, 1999) have been proposed that are based on the original Sargent (1921) vertical jump test. Companies now sell mats that estimate the height and power of a vertical jump (from time and projectile equations). Although mechanical power output in such jumps is high, these tests and devices are limited because the resistance is limited to body mass, the many factors that affect jump height, and the assumptions used in the calculation. There has been a long history of criticism of the assumptions and logic of using vertical jump height to estimate muscular power (Adamson & Whitney, 1971; Barlow, 1971). Instantaneous measurements of power from force platforms or kinematic analysis are more accurate but are expensive and time-consuming. Future studies will help determine the role of mechanical power in various movements, how to train for these movements, and what field tests help coaches monitor athletes.

SEGMENTAL INTERACTION PRINCIPLE

Human movement can be performed in a wide variety of ways because of the many kinematic **degrees of freedom** our linked segments provide. In chapter 5 we saw that coordination of these kinematic chains ranges along a continuum from simultaneous to sequential. Kinetics provides several ways in which to examine the potential causes of these coordination patterns. The two expressions of Newton's second law and the work–energy relationship have been employed in the study of the coordination of movement. This section proposes a Principle of Segmental Interaction that can be used to understand the origins of movement so that professionals can modify movement to improve performance and reduce risk of injury.

The **Segmental Interaction Principle** says that forces acting between the segments of a body can transfer energy between segments. The biomechanics literature has referred to this phenomenon in several ways (Putnam, 1993). The contribution of body segments to movement has been called coordination of temporal impulses (Hochmuth & Marhold, 1978), the kinetic link principle (Kreighbaum & Barthels, 1996), summation of speed (Bunn, 1972), summation or continuity of joint torques (Norman, 1975), the sequential or proximal-to-distal sequencing of movement (Marshall & Elliott, 2000), and the transfer of energy or transfer of momentum (Lees & Barton, 1996; Miller, 1980). The many names for this phenomenon and the three ways to document kinetics are a good

---

Application: Strength vs. Power

The force–velocity relationship and domains of strength discussed in chapter 4, as well as this chapter's discussion of mechanical power should make it clear that muscular strength and power are not the same thing. Like the previous discussion on power lifting, the common use of the term *power* is often inappropriate. Muscular strength is the expression of maximal tension in isometric or slow velocities of shortening. We have seen that peak power is the right combination of force and velocity that maximizes mechanical work. In cycling, the gears are adjusted to find this peak power point. If cadence (pedal cycles and, consequently, muscle velocity of shortening) is too high, muscular forces are low and peak power is not achieved. Similarly, power output can be submaximal if cadence is too slow and muscle forces high. The right mix of force and velocity seems to be between 30 and 70% of maximal isometric force and depends on the movement. Kinesiology professionals need to keep up with the growing research on the biomechanics of conditioning and sport movements. Future research will help refine our understanding of the nature of specific movements and the most appropriate exercise resistances and training programs.

indication of the difficulty of the problem and the controversial nature of the causes of human motion.

Currently it is not possible to have definitive answers on the linear and angular kinetic causes for various coordination strategies. This text has chosen to emphasize the forces transferred between segments as the primary kinetic mechanism for coordination of movement. Most electromyographic (EMG) research has shown that in sequential movements muscles are activated in short bursts that are timed to take advantage of the forces and geometry between adjacent segments (Feldman *et al.*, 1998; Roberts, 1991). This coordination of muscular kinetics to take advantage of "passive dynamics" (gravitational, inertial forces) has been observed in the swing limb during walking (Mena, Mansour, & Simon, 1981), running (Phillips, Roberts, & Huang, 1983), kicking (Roberts, 1991), throwing (Feltner, 1989; Hirashima, Kadota, Sakurai, Kudo, & Ohtsuki, 2002), and limb adjustments to unexpected obstacles (Eng, Winter, & Patla, 1997).

Some biomechanists have theorized that the segmental interaction that drives the sequential strategy is a transfer of energy from the proximal segment to the distal segment. This theory originated from observations of the close association between the negative acceleration of the proximal segment (see the activity on Segmental Interaction below) with the positive acceleration of the distal segment (Plagenhoef, 1971; Roberts, 1991). This mechanism is logically appealing because the energy of large muscle groups can be transferred distally and is consistent with the large forces and accelerations of small segments late in baseball pitching (Feltner & Dapena, 1986; Fleisig, Andrews, Dillman, & Escamilla, 1995; Roberts, 1991). Figure 6.22 illustrates a schematic of throwing where the negative angular acceleration of the arm (α_A) creates a backward elbow joint force (F_E) that accel-

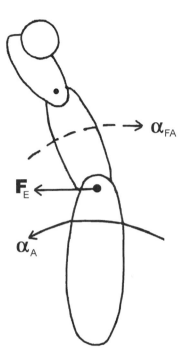

Figure 6.22. Simple sagittal plane model of throwing illustrates the Segmental Interaction Principle. Joint forces (F_E) from a slowing proximal segment create a segmental interaction to angularly accelerate the more distal segments (α_{FA}).

erates the forearm (α_{FA}). This view of the Segment Interaction Principle states that slowing the larger proximal segment will transfer energy to the distal segment. It is clear that this movement strategy is highly effective in creating high-speed movements of distal segments, but the exact mechanism of the segmental interaction principle is not clear.

When you get down to this level of kinetics, you often end up with a chicken-or-egg dilemma. In other words, which force/torque was created first and which is the reaction force/torque (Newton's third law)? There are some scholars who have derived equations that support the proximal-to-distal transfer of energy (Hong, Cheung, & Roberts, 2000; Roberts, 1991), while others show that the acceleration of the distal segment causes slowing of the proximal segment (Putnam, 1991, 1993; Sorensen, Zacho, Simonsen, Dyhre-Poulsen, & Klausen, 1996). Whatever the underlying mechanism or direction of transfer, fast human movements utilize a sequential (proximal-to-distal) coordination that relies on the transfer of forces/energy between segments. We are truly fortunate to have so many muscles and degrees of freedom to create a wide variety and speeds of motion.

A good example of the controversy related to the Segmental Interaction Principle is the role of the hand and wrists in the golf swing. Skilled golf shots can be accurately modeled as a two-segment (arm and club) system with motion occurring in a diagonal plane. Golf pros call this the swing plane. Some pros say the golfer should actively drive the club with wrist action, while others teach a relaxed or more passive wrist release. A recent simulation study found that correctly timed wrist torques could increase club head speed by 9% (Sprigings & Neal, 2000), but the small percentage and timing of these active contributions suggests that proximal joint forces are the primary accelerator of the club. Jorgensen (1994) has provided simple qualitative demonstrations and convincing kinetic data that support the more relaxed use of wrist action and explain how weight shifts can be timed to accelerate the golf club.

It is clear that forces are transferred between segments to contribute to the motion of the kinematic chain (Zajac & Gordon, 1989). The exact nature of that segmental interaction remains elusive, so kinesiology professionals can expect performers to have a variety (sequential to simultaneous) of combinations of joint motion. It would be unwise to speculate too much on the muscular origins of that transfer. This view is consistent with the EMG and biomechanical modeling research reviewed in chapter 3. So how can kinesiology professionals prescribe conditioning exercises and learning progressions so as to maximize the segmental interaction effect? Currently, there are few answers, but we can make a few tentative generalizations about conditioning and learning motor skills.

Physical conditioning for any human movement should clearly follow the training principle of *specificity*. Biomechanically, this means that the muscular actions and movements should emulate the movement as much as possible. Since the exact kinetic mechanism of segmental interaction is not clear, kinesiology professionals should select exercises that train all the muscles involved in a movement. In soccer kicking, it is not clear whether it is the activity of the quadriceps or hip flexors that predominantly contribute to acceleration of the lower leg. Selecting exercises that train both

Activity: Segmental Interaction

Segmental interaction or the transfer of energy from a proximal to a distal segment can be easily simulated using a two-segment model. Suspend a rigid stick (ruler, yardstick, racket) between the tips of your index finger and thumb. Using your hand/forearm as the proximal segment and the stick as the distal segment, simulate a kick. You can make the stick extend or kick without any extensor muscles by using intersegmental reaction forces. Accelerate your arm in the direction of the kick (positive). When you reach peak speed, rapidly slow (negatively accelerate) your arm and observe the positive acceleration of the stick. Positive acceleration of your arm creates an inertial lag in the stick, while negative acceleration of your arm creates a backward force at the joint, which creates a torque that positively accelerates the stick.

muscles is clearly indicated. More recent trends in rehabilitation and conditioning have focused on training with "functional" movements that emulate the movement,

Interdisciplinary Issue: Kinematic Chain

A *kinematic chain* is an engineering term that refers to a series of linked rigid bodies. The concept of kinematic chains was developed to simplify the mathematics of the kinematics and kinetics of linked mechanical systems. A classic biomechanics textbook (Steindler, 1955) adapted this terminology to refer to the linked segments of the human body as a "kinetic chain" and to classify movements as primarily "open" or "closed" kinetic chains. A closed kinetic chain is a movement where the motion of the distal segment is restrained by "considerable external resistance." Over the years, the rehabilitation and conditioning professions have adopted this terminology, referring to open kinetic chain exercises (knee extension) and closed kinetic chain exercises (leg press or squat). Considerable research has focused on the forces and muscle activation involved in various exercises classified as open or closed kinetic chains. This research has shown both similarities and differences in muscular function between similar open and closed kinetic chain movements. There are, however, problems in uniquely defining a closed chain or what constitutes "considerable resistance." The vague nature of the classification of many exercises has prompted calls to avoid this terminology (Blackard et al., 1999; di Fabio, 1999; Dillman et al., 1994).

rather than isolating specific muscle groups. The resistance, body motion, speed, and balance aspects of "functional" exercises may be more specific forms of training; unfortunately, there has been limited research on this topic.

Learning the sequential coordination of a large kinematic chain is a most difficult task. Unfortunately, there have been relatively few studies on changes in joint kinetics accompanied by learning. Assuming that the energy was transferred distally in a sequential movement (like our immature volleyball spike in the previous chapter), it would not be desirable to practice the skill in parts because there would be no energy to learn to transfer. Recent studies have reinforced the idea that sequential skills should be learned in whole at submaximal speed, rather than in disconnected parts (see Sorensen, Zacho, Simonsen, Dyhre-Poulsen, & Klausen, 2000). Most modeling and EMG studies of the vertical jump have also shown the interaction of muscle activation and coordination (Bobbert & van Zandwijk, 1999; Bobbert & van Soest, 1994; van Zandwijk, Bobbert, Munneke, & Pas, 2000), while some other studies have shown that strength parameters do not affect coordination (Tomioka, Owings, & Grabiner, 2001). Improvements in computers, software, and biomechanical models may allow more extensive studies of the changes in kinetics as skills are learned. Currently, application of the **Segmental Interaction Principle** involves corrections in body positioning and timing. Practice should focus on complete repetitions of the whole skill performed at submaximal speeds. Improvement should occur with many practice repetitions, while gradually increasing speed. This perspective is consistent with more recent motor learning interest in a dynamical systems theory understanding of coordination, rather than centralized motor program (Schmidt & Wrisberg, 2000).

Application: Arm Swing Transfer of Energy

Many movements incorporate an arm swing that is believed to contribute to performance. How much does arm swing contribute to vertical jump performance? Several studies have shown that the height of a jump increases by about 10% with compared to those without arm swing (see Feltner, Fraschetti, & Crisp, 1999). The possible mechanism is transfer of energy or momentum from the arms to the rest of the body. Logically, vigorous positive (upward) acceleration of the arms creates a downward reaction force on the body that increases the vertical ground reaction force. It has also been hypothesized that this downward force creates a pre-loading effect on the lower extremities that limits the speed of knee extension, allowing greater quadriceps forces because of the Force–Velocity Relationship. A detailed kinetic study (Feltner et al., 1999) found that augmenting knee torques early in a jump with arm swings combined with slowing of trunk extension late in the jump may be the mechanisms involved in a good arm swing during a vertical jump. Late in the jump, the arms are negatively accelerated, creating a downward force at the shoulder that slows trunk extension and shortening of the hip extensors. While the arms do not weigh a lot, the vigor of these movements does create large forces, which can be easily seen by performing this arm swing pattern standing on a force platform.

What segmental interactions create and transfer this energy? This answer is less clear and depends on the model and kinetic variable used during analysis. The muscular and segmental contributions to a vertical jump have been analyzed using force platforms (Luthanen & Komi, 1978a,b), computer modeling (Bobbert & van Soest, 1994; Pandy, Zajac, Sim, & Levine, 1990), joint mechanical power calculations (Fukashiro & Komi, 1987; Hubley & Wells, 1983; Nagano, Ishige, & Fukashiro, 1998), angular momentum (Lees & Barton, 1996), and net joint torque contributions to vertical motion (Feltner et al., 1999; Hay, Vaughan, & Woodworth, 1981). While the jumping technique may look quite similar, there is considerable between-subject variation in the kinetics of the vertical jump (Hubley & Wells, 1983). The problems involved in partitioning contributions include defining energy transfer, energy transfer of biarticular muscles, muscle co-activation, and bilateral differences between limbs. While there is much yet to learn, it appears that the hip extensors contribute the most energy, closely followed by the knee extensors, with smaller contributions by the ankle plantar flexors. Conditioning for vertical jumping should utilize a variety of jumps and jump-like exercises. If specific muscle groups are going to be isolated for extra training, the hip and knee extensors appear to be the groups with the greatest contribution to the movement.

SUMMARY

Linear kinetics is the study of the causes of linear motion. There are several laws of mechanics that can be applied to a study of the causes of linear motion: Newton's laws, the impulse–momentum relationship, and the work–energy relationship. The most common approach involves Newton's Laws of Motion, called the laws of *Inertia*, *Momentum/Acceleration*, and *Reaction*. Inertia is the tendency of all objects to resist changes in their state of motion. The Inertia Principle suggests that reducing mass will make objects easier to accelerate, while increasing mass will make objects more stable and harder to accelerate. Applying the Inertia Principle might also mean using more mass in activities where there is time to overcome the inertia, so that it can be used later in the movement. When two objects are in contact, the forces of interaction between the bodies are resolved into right-angle directions: normal reaction and friction. The Impulse–Momentum Relationship says that the change in momentum of an object is equal to the impulse of the resultant forces acting on the object. This is Newton's second law when applied over a time interval. The real-world application of this relationship is the Force–Time Principle. Energy is the capacity to do mechanical work; mechanical energies include strain, potential, and kinetic energy. The Work–Energy Relationship says that mechanical work equals the change in mechanical energy. Mechanical power is the rate of doing work, and can also be calculated by the product of force and velocity. The Segmental Interaction Principle says that energy can be transferred between segments. While the exact nature of these transfers has been difficult to determine, both simultaneous and sequentially coordinated movements take advantage of the energy transferred through the linked segment system of the body.

REVIEW QUESTIONS

1. Which has more inertia, a 6-kg bowling ball sitting on the floor or one rolling down the lane? Why?

2. What are the two ways to express Newton's second law?

3. When might it be advantageous for a person to increase the inertia used in a movement?

4. Do smaller or larger muscle angles of pull on a distal segment tend to create more joint rotation? Why?

5. What are strategies to increase the friction between a subject's feet and the floor?

6. What two things can be changed to increase the impulse applied to an object? What kinds of human movement favor one over the other?

7. If the force from the tibia on the femur illustrated below was 1000 N acting at 30° to the femur, what are the longitudinal (causing knee compression) and normal (knee shear) components of this force? Hint: move one component to form a right triangle and solve.

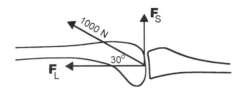

8. Give human movement examples of the three mechanical energies.

9. Compare and contrast muscular strength and muscular power.

10. How is momentum different from kinetic energy?

11. A rock climber weighing 800 N has fallen and is about to be belayed (caught with a safety rope) by a 1500-N vertical force. Ignoring the weight of the rope and safety harness, what is the vertical acceleration of the climber? Hint: remember to sum forces with correct signs (related to direction).

12. Draw a free-body diagram of a proximal segment of the body showing all forces from adjacent segments. Draw a free body diagram of an adjacent segment using Newton's third law to determine the size and direction at the joint.

13. What are the potential kinetic mechanisms that make a sequential motion of segments in high-speed movements the optimal coordination?

14. Do the angles of pull (relative to the body) of free weights change during an exercises? Why?

15. An Olympic lifter exerts a 4000-N upward (vertical) force to a 30-kg barbell. What direction will the bar tend to move, and what is its vertical acceleration?

KEY TERMS

conservation of energy (Law of
  Conservation of Energy)
degrees of freedom
direct dynamics
energy
force platform
force–time principle
friction
impulse
impulse–momentum relationship
inverse dynamics
kinetic energy
Law of Acceleration
Law of Inertia
Law of Reaction
momentum
normal reaction
potential energy
power (mechanical)
strain energy
work (mechanical)
work–energy relationship

SUGGESTED READING

Abernethy, P., Wilson, G., & Logan, P. (1995). Strength and power assessment: Issues, controversies and challenges. *Sports Medicine*, **19**, 401–417.

Cavanagh, P. R., & LaFortune, M. A. (1980). Ground reaction forces in distance running. *Journal of Biomechanics*, **15**, 397–406.

Dowling, J. J., & Vamos, L. (1993). Identification of kinetic and temporal factors related to vertical jump performance. *Journal of Applied Biomechanics*, **9**, 95–110.

Jorgensen, T. P. (1994). *The physics of golf*. New York: American Institute of Physics.

Lees, A., & Barton, G. (1996). The interpretation of relative momentum data to assess the contribution of the free limbs to the generation of vertical velocity in sports activities. *Journal of Sports Sciences*, **14**, 503–511.

McPoil, T. G., Cornwall, M. W., & Yamada, W. (1995). A comparison of two in-shoe plantar pressure measurement systems. *The Lower Extremity*, **2**, 95–103.

Schieb, D. A. (1987, January). The biomechanics piezoelectric force plate. *Soma*, 35–40.

Zatsiorsky, V. M. (2002). *Kinetics of human motion*. Champaign, IL: Human Kinetics.

Zajac, F. E. (2002). Understanding muscle coordination of the human leg with dynamical simulations. *Journal of Biomechanics*, **35**, 1011–1018.

Zajac, F. E., & Gordon, M. E. (1989). Determining muscle's force and action in multi-articular movement. *Exercise and Sport Sciences Reviews*, **17**, 187–230.

WEB LINKS

Linear Kinetics—Page on the kinetics of winter olympic sports by Debra King and colleagues from Montana State University.
    http://btc.montana.edu/olympics/physbio/physics/dyn-intro.html

Kwon3D—Theoretical background information for kinematic and kinetic software developed by Young-Hoo Kwon of Texas Woman's University.
    http://kwon3d.com/theories.html

Kinetics simulations of collisions and an air track—part of mechanics simulations at explorescience.com.
    http://www.explorescience.com/activities/activity_list.cfm?categoryID=10

CHAPTER 7

Angular Kinetics

Angular kinetics explains the causes of rotary motion and employs many variables similar to the ones discussed in the previous chapter on linear kinetics. In fact, Newton's laws have angular analogues that explain how torques create rotation. The net torque acting on an object creates an angular acceleration inversely proportional to the angular inertia called the moment of inertia. Angular kinetics is quite useful because it explains the causes of joint rotations and provides a quantitative way to determine the center of gravity of the human body. The application of angular kinetics is illustrated with the principles of **Inertia** and **Balance**.

TORQUE

The rotating effect of a force is called a **torque** or **moment of force**. Recall that a moment of force or torque is a vector quantity, and the usual two-dimensional convention is that counterclockwise rotations are positive. Torque is calculated as the product of force (**F**) and the **moment arm**. The moment arm or leverage is the perpendicular displacement ($d_\perp$) from the line of action of the force and the axis of rotation (Figure 7.1). The biceps femoris pictured in Figure 7.1 has moment arms that create hip extension and knee flexion torques. An important point is that the moment arm is always the shortest displacement between the force line of action and axis of rotation. This text will use the term torque synony-

mously with moment of force, even though there is a more specific mechanics-of-materials meaning for torque.

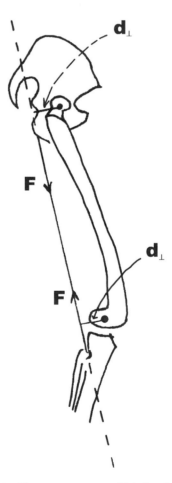

Figure 7.1. The moment arms ($d_\perp$) for the biceps femoris muscle. A moment arm is the right-angle distance from the line of action of the force to the axis of rotation.

In algebraic terms, the formula for the torque is $T = F \cdot d_\perp$, so that typical units of torque are N·m and lb·ft. Like angular kinematics, the usual convention is to call counterclockwise (ccw) torques positive and clockwise ones negative. Note that the size of the force and the moment arm are *equally* important in determining the size of the torque created. This has important implications for maximizing performance in many activities. A person wanting to create more torque can increase the applied force or increase their effective moment arm. Increasing the moment arm is often easier and faster than months of conditioning! Figure 7.2 illustrates two positions where a therapist can provide resistance with a hand dynamometer to manually test the isometric strength of the elbow extensors. By positioning their arm more distal (position 2), the therapist increases the moment arm and decreases the force they must create to balance the torque created by the patient and gravity (T_p).

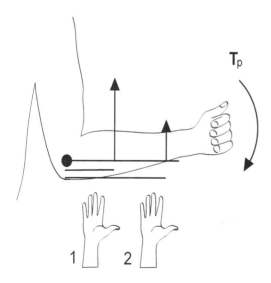

T_p

Figure 7.2. Increasing the moment arm for the therapist's (position 2) manual resistance makes it easier to perform a manual muscle test that balances the extensor torque created by the patient (T_p).

Activity: Torque and Levers

Take a desk ruler ($\approx$12-inch) and balance it on a sturdy small cylinder like a highlighter. Place a dime at the 11-inch position and note the behavior of the ruler. Tap the 1-inch position on the ruler with your index finger and note the motion of the dime. Which torque was larger: the torque created by the dime or your finger? Why? Tap the ruler with the same effort on different positions on the ruler with the dime at 11 inches and note the motion of the dime. Modify the position (axis of rotation) of the highlighter to maximize the moment arm for the dime and note how much force your finger must exert to balance the lever in a horizontal position. How much motion in the dime can you create if you tap the ruler? In these activities you have built a simple machine called a *lever*. A lever is a nearly rigid object rotated about an axis. Levers can be built to magnify speed or force. Most human body segment levers magnify speed because the moment arm for the effort is less than the moment arm for the resistance being moved. A biceps brachii must make a large force to make a torque larger than the torque created by a dumbbell, but a small amount of shortening of the muscle creates greater rotation and speed at the hand. Early biomechanical research was interested in using anatomical leverage principles for a theory of high-speed movements, but this turned out to be a dead end because of the discovery of sequential coordination of these movements (Roberts, 1991).

Let's look at another example of applying forces in an optimal direction to maximize torque output. A biomechanics student takes a break from her studies to bring a niece to the playground. Let's calculate the torque the student creates on the merry-go-round by the force F_1 illustrated in Figure 7.3. Thirty pounds of force times the moment arm of 4 feet is equal to 120 lb·ft of torque. This torque can be considered positive because it acts counterclockwise. If on the second spin the student pushes with the same magnitude of force (F_2) in a different direction, the torque and angular motion created would be smaller because of the smaller moment arm (d_B). Use the conversion factor in Appendix B to see how many N·m are equal to 120 lb·ft of torque.

Good examples of torque measurements in exercise science are the joint torques measured by isokinetic dynamometers. The typical maximum isometric torques of several muscle groups for males are listed in Table 7.1. These torques should give you a good idea of some "ballpark" maximal values for many major joints. Peak

TABLE 7.1		
Typical Isometric Joint Torques Measured by Isokinetic Dynamometers		
	Peak torques	
	N·m	lb·ft
Trunk extension	258	190
Trunk flexion	177	130
Knee extension	204	150
Knee flexion	109	80
Hip extension	150	204
Ankle plantar flexion	74	102
Elbow flexion	20	44.6
Wrist flexion	8	11
Wrist extension	4	5

torques from inverse dynamics in sporting movements can be larger than those seen in isokinetic testing because of antagonist activity in isokinetics testing, segment interaction in dynamic movements, the stretch-shortening cycle, and eccentric muscle actions. Most isokinetic norms are normalized to bodyweight (e.g., lb·ft/lb) and categorized by gender and age. Recall that

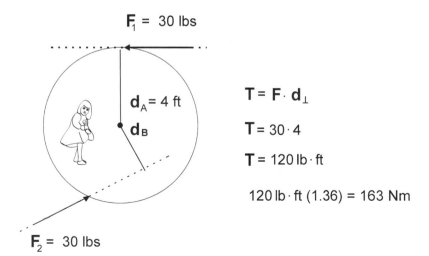

$$T = F \cdot d_\perp$$

$$T = 30 \cdot 4$$

$$T = 120 \, lb \cdot ft$$

$$120 \, lb \cdot ft \, (1.36) = 163 \, Nm$$

Figure 7.3. Calculating the torque created by a person pushing on a merry-go-round involves multiplying the force times its moment arm. This torque can be converted to other units of torque with conversion factors (Appendix B).

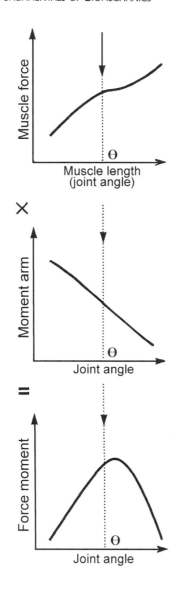

Figure 7.4. Joint torque–angle diagrams represent the strength curves of muscle groups. The shapes of joints vary based primarily upon the combined effect of changes in muscle length properties and muscle moment arms. Reprinted by permission from Zatsiorsky (1995).

the shape of the torque-angle graphs from isokinetic testing reflects the integration of many muscle mechanical variables. The angle of the joints affects the torque that the muscle group is capable of producing be-

cause of variations in moment arm, muscle angle of pull, and the force–length relationship of the muscle. There are several shapes of torque-angle diagrams, but they most often look like an inverted "U" because of the combined effect of changes in muscle moment arm and force–length relationship (Figure 7.4).

Torque is a good variable to use for expressing muscular strength because it is not dependent on the point of application of force on the limb. The torque an isokinetic machine (T) measures will be the same for either of the two resistance pad locations illustrated in Figure 7.5 if the subject's effort is the same. Sliding the pad toward the subject's knee will decrease the moment arm for the force applied by the subject, increasing the force on the leg (F_2) at that point to create the same torque. Using torque instead of force created by the subject allows for easier comparison of measurements between different dynamometers.

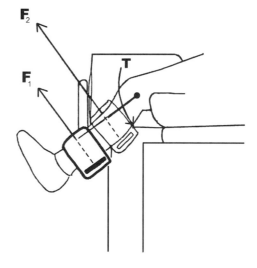

Figure 7.5. Isokinetic dynamomters usually measure torque because torque does not vary with variation in pad placement. Positioning the pad distally decreases the force the leg applies to the pad for a given torque because the moment arm for the leg is larger.

Application: Muscle-Balance and Strength Curves

Recall that testing with an isokinetic dynamometer documents the strength curves (torque–joint angle graphs) of muscle groups. Normative torques from isokinetic testing also provide valuable information on the ratio of strength between opposing muscle groups. Many dynamometers have computerized reports that list test data normalized to bodyweight and expressed as a ratio of the peak torque of opposing muscle groups. For example, peak torques created by the hip flexors tend to be 60 to 75% of peak hip extensor torques (Perrin, 1993). Another common strength ratio of interest is the ratio of the quadriceps to the hamstrings. This ratio depends on the speed and muscle action tested, but peak concentric hamstring torque is typically between 40 and 50% of peak concentric quadriceps torque (Perrin, 1993), which is close to the physiological cross-sectional area difference between these muscle groups. Greater emphasis has more recently been placed on more functional ratios (see Aagaard, Simonsen, Magnusson, Larsson, & Dyhre-Poulsen, 1998), like hamstring eccentric to quadriceps concentric strength ($H_{ecc}:Q_{con}$), because hamstrings are often injured ("pulled" in common parlance) when they slow the vigorous knee extension and hip flexion before foot strike in sprinting. In conditioning and rehabilitation, opposing muscle group strength ratios are often referred to as muscle balance. Isokinetic (see Perrin, 1993) and hand dynamometer (see Phillips, Lo, & Mastaglia, 2000) testing are the usual clinical measures of strength, while strength and conditioning professionals usually use one-repetition maxima (1RM) for various lifts. These forms of strength testing to evaluate muscle balance are believed to provide important sources of information on the training status, performance, and potential for injury of athletes. In rehabilitation and conditioning settings, isokinetic and other forms of strength testing are useful in monitoring progress during recovery. Athletes are cleared to return to practice when measurements return to some criterion/standard, a percentage of pre-injury levels, or a percentage of the uninvolved side of their body. It is important for kinesiology professionals to remember that the strength (torque capability) of a muscle group is strongly dependent on many factors: testing equipment, protocol, and body position, among others, affect the results of strength testing. If standards in testing are being used to qualify people for jobs or athletic participation, there needs to be clear evidence correlating the criterion test and standard with safe job performance.

SUMMING TORQUES

The state of an object's rotation depends on the balance of torques created by the forces acting on the object. Remember that summing or adding torques acting on an object must take into account the vector nature of torques. All the muscles of a muscle group sum together to create a joint torque in a particular direction. These muscle group torques must also be summed with torques from antagonist muscles, ligaments, and external forces to determine the net torque at a joint. Figure 7.6 illustrates the forces of the anterior deltoid and long head of the biceps in flexing the shoulder in the sagittal plane. If ccw torques are positive, the torques created by these muscles would be positive. The net torque of these two muscles is the sum of their individual torques, or 6.3 N·m (60 · 0.06 + 90 · 0.03 = 6.3 N·m). If the weight of this person's arm multiplied by its moment arm created a gravitational torque of –16 N·m, what is the net torque acting at the shoulder? Assuming there are no other shoulder flexors or exten-

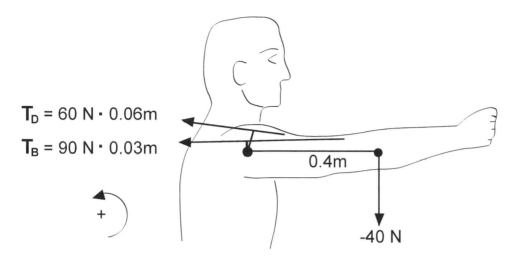

T_D = 60 N · 0.06m

T_B = 90 N · 0.03m

0.4m

+

-40 N

Figure 7.6. The shoulder flexion torques of anterior deltoid and long head of the biceps can be summed to obtain the resultant flexion torque acting to oppose the gravitational torque from the weight of the arm.

sors active to make forces, we can sum the gravitational torque (–16 N·m) and the net muscle torque (6.3 N·m) to find the resultant torque of –9.7 N·m. This means that there is a resultant turning effect acting at the shoulder that is an extension torque, where the shoulder flexors are acting eccentrically to lower the arm. Torques can be summed about any axis, but be sure to multiply the force by the moment arm and then assign the correct sign to represent the direction of rotation before they are summed.

Recall the isometric joint torques reported in Table 7.1. Peak joint torques during vigorous movement calculated from inverse dynamics are often larger than those measured on isokinetic dynamometers (Veloso & Abrantes, 2000). There are several reasons for this phenomenon, including transfer of energy from biarticular muscles, differences in muscle action, and coactivation. Coactivation of antagonist muscles is a good example of summing opposing torques. EMG research has shown that isokinetic joint torques underestimate net agonist muscle torque because of coactivation of antagonist muscles (Aagaard, Simonsen,

Andersen, Magnusson, Bojsen-Moller, & Dyhre-Poulsen, 2000: Kellis & Baltzopoulos, 1997, 1998).

ANGULAR INERTIA (MOMENT OF INERTIA)

A *moment of force* or *torque* is the mechanical effect that creates rotation, but what is the resistance to angular motion? In linear kinetics we learned that mass was the mechanical measure of inertia. In angular kinetics, inertia is measured by the **moment of inertia**, a term pretty easy to remember because it uses the terms *inertia* and *moment* from moment of force. Like the mass (linear inertia), **moment of inertia** is the resistance to angular acceleration. While an object's mass is constant, the object has an infinite number of moments of inertia! This is because the object can be rotated about an infinite number of axes. We will see that rotating the human body is even more interesting because the links allow the configuration of the body to change along with the axes of rotation.

The symbol for the moment of inertia is I. Subscripts are often used to denote the axis of rotation associated with a moment of inertia. The smallest moment of inertia of an object in a particular plane of motion is about its center of gravity (I_0). Biomechanical studies also use moments of inertia about the proximal (I_P) and distal (I_D) ends of body segments. The formula for a rigid-body moment of inertia about an axis (A) is $I_A = \Sigma mr^2$. To determine the moment of inertia of a ski in the transverse plane about an anatomically longitudinal axis (Figure 7.7), the ski is cut into eight small masses (m) of know radial distances (r) from the axis. The sum of the product of these masses and the squared radius is the moment of inertia of the ski about that axis. Note that the SI units of moment of inertia are kg·m².

The formula for moment of inertia shows that an object's resistance to rotation depends more on distribution of mass (r^2)

Activity: Moment of Inertia

Take a long object like a baseball bat, tennis racket, or golf club and hold it in one hand. Slowly swing the object back and forth In a horizontal plane to eliminate gravitational torque from the plane of motion. Try to sense how difficult it is to initiate or reverse the object's rotation. You are trying to subjectively evaluate the moment of inertia of the object. Grab the object in several locations and note how the moment of inertia changes. Add mass to the object (e.g., put a small book in the racket cover) at several locations. Does the moment of inertia of the object seem to be more related to mass or the location of the mass?

than mass (m). This large increase in moment of inertia from changes in location of mass relative to the axis of rotation (because **r** is squared) is very important in human movement. Modifications in the mo-

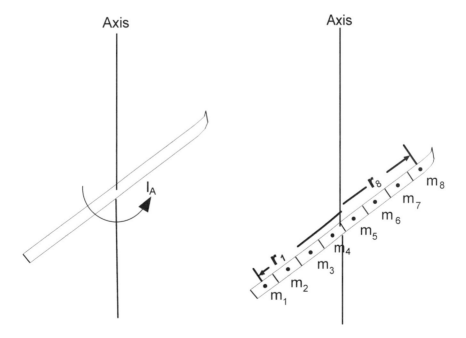

Figure 7.7. The moment of inertia of a ski about a specific axis can be calculated by summing the products of the masses of small elements (m) and the square of the distance from the axis (r).

ment of inertia of body segments can help or hinder movement, and the moment of inertia of implements or tools can dramatically affect their effectiveness.

Most all persons go through adolescence with some short-term clumsiness. Much of this phenomenon is related to motor control problems from large changes in limb moment of inertia. Imagine the balance and motor control problems from a major shift in leg moment of inertia if a young person grows two shoe sizes and 4 inches in a 3-month period. How much larger is the moment of inertia of this teenager's leg about the hip in the sagittal plane if this growth (dimension and mass) was about 8%? Would the increase in the moment of inertia of the leg be 8% or larger? Why?

When we want to rotate our bodies we can skillfully manipulate the moment of in-ertia by changing the configuration of our body segments relative to the axis of rotation. Bending the joints of the upper and lower extremities brings segmental masses close to an axis of rotation, dramatically decreasing the limb's moment of inertia. This bending allows for easier angular acceleration and motion. For example, the faster a person runs the greater the knee flexion in the swing limb, which makes the leg easy to rotate and to get into position for another footstrike. Diving and skilled gymnastic tumbling both rely on decreasing the moment of inertia of the human body to allow for more rotations, or increasing the length of the body to slow rotation down. Figure 7.8 shows the dramatic differences in the moment of inertia for a human body in the sagittal plane for different body segment configurations relative to the axis of rotation.

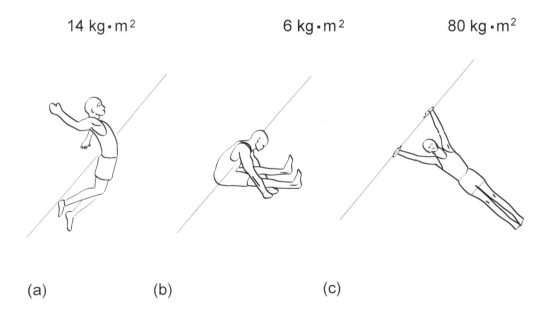

$14 \text{ kg} \cdot \text{m}^2$ $6 \text{ kg} \cdot \text{m}^2$ $80 \text{ kg} \cdot \text{m}^2$

(a)   (b)   (c)

Figure 7.8. The movement of body segments relative to the axis of rotation makes for large variations in the moment of inertia of the body. Typical sagittal plane moments of inertia and axes of rotation for a typical athlete are illustrated for long jump (a,b) and high bar (c) body positions.

Variations in the moment of inertia of external objects or tools are also very important to performance. Imagine you are designing a new unicycle wheel. You design two prototypes with the same mass, but with different distributions of mass. Which wheel design (see Figure 7.9) do you think would help a cyclist maintain balance: wheel A or wheel B? Think about the movement of the wheel when a person balances on a unicycle. Does agility (low inertia) or consistency of rotation (high inertia) benefit the cyclist? If, on the other hand, you are developing an exercise bike that would provide slow and smooth changes in resistance, which wheel would you use? A heavy ski boot and ski dramatically affect the moments of inertia of your legs about the hip joint. Which joint axis do you think is most affected?

The moment of inertia of many sport implements (golf clubs and tennis rackets) is commonly called the "swing weight." A longer implement can have a similar swing weight to a shorter implement by keeping mass proximal and making sure the added length has low mass. It is important to realize that the three-dimensional nature of sports equipment means that there are moments of inertia about the three principal or dimensional axes of the equipment. Tennis players often add lead tape to their rackets so as to increase shot speed and racket stability. Tape is often added to the perimeter of the frame for stability (by increasing the polar moment of inertia) against off-center impacts in the lateral directions. Weight at the top of the frame would not affect this lateral stability, but would increase the moments of inertia for swinging the racket forward and upward. The large radius of this mass (from his grip to the tip of the racket), however, would make the racket more difficult to swing. Recent baseball/softball bat designs allow for variations in where bat mass is located, making for wide variation in the moment of inertia for a swing. It turns out that an individual

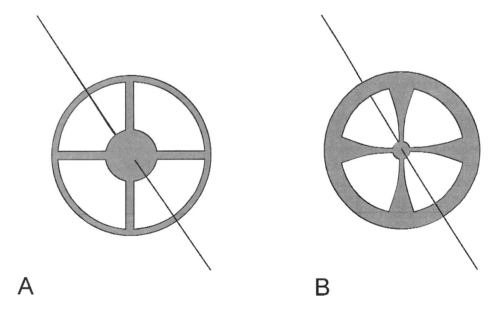

A   **B**

Figure 7.9. The distribution of mass most strongly affects moment of inertia, so wheel A with mass close to the axle would have much less resistance to rotation than wheel B. Wheel A would make it easier for a cyclist to make quick adjustments of the wheel back and forth to balance a unicycle.

batting style affects optimal bat mass (Bahill & Freitas, 1995) and moment of inertia (Watts & Bahill, 2000) for a particular batter.

You can now see that the principle of inertia can be extended to angular motion of biomechanical systems. This application of the concepts related to moment of inertia are a bit more complex than mass in linear kinetics. For example, a person putting on snowshoes will experience a dramatic increase (larger than the small mass of the shoes implies) in the moment of inertia of the leg about the hip in the sagittal plane because of the long radius for this extra mass. A tennis player adding lead tape to the head of their racket will more quickly modify the angular inertia of the racket than its linear inertia. Angular inertia is most strongly related to the distribution of mass, so an effective strategy to decrease this inertia is to bring segment masses close to the axis of rotation. Coaches can get players to "compact" their extremities or body to make it easier to initiate rotation.

NEWTON'S ANGULAR ANALOGUES

Newton's laws of motion also apply to angular motion, so each may be rephrased using angular variables. The angular analogue of Newton's third law says that for every torque there is an equal and opposite torque. The angular acceleration of an object is proportional to the resultant torque, is in the same direction, and is inversely proportional to the moment of inertia. This is the angular expression of Newton's second law. Likewise, Newton's first law demonstrates that objects tend to stay in their state of angular motion unless acted upon by an unbalanced torque. Biomechanists often use rigid body models of the human body and apply Newton's laws to calculate the net forces and torques acting on body segments.

This working backward from video measurements of acceleration (second derivatives) using both the linear and angular versions of Newton's second law is called **inverse dynamics**. Such analyses to understand the resultant forces and torques that create movement were first done using laborious hand calculations and graphing (Bressler & Frankel, 1950; Elftman, 1939), but they are now done with the assistance of powerful computers and mathematical computation programs. The resultant or net joint torques calculated by inverse dynamics do not account for co-contraction of muscle groups and represent the sum of many muscles, ligaments, joint contact, and other anatomical forces (Winter, 1990).

Despite the imperfect nature of these net torques (see Hatze, 2000; Winter 1990), inverse dynamics provides good estimates of the net motor control signals to create human movement. Figure 7.10 illustrates the net joint torques at the hip and knee in a soccer toe kick. These torques are similar

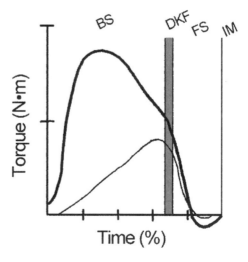

Figure 7.10. The net joint hip and knee joint torques in a soccer kick calculated from inverse dynamics. The backswing (BS), range of deepest knee flexion (DKF), forward swing (FS), and impact (IMP) are illustrated. Adapted with permission from Zernicke and Roberts (1976).

to the torques recently reported in a three-dimensional study of soccer kicks (Nunome *et al.*, 2002). The kick is initiated by a large hip flexor torque that rapidly decreases before impact with the soccer ball. The knee extensor torque follows the hip flexor torque and also decreases to near zero at impact. This near-zero knee extensor torque could be expected because the foot would be near peak speed at impact, with the body protecting the knee from hyperextension. If the movement were a punt, there would usually be another rise and peak in hip flexor torque following the decline in knee torque (Putnam, 1983). It is pretty clear from this planar (2D) example of inverse dynamics that the hip flexor musculature may make a larger contribution to kicking than the knee extensors. It is not as easy to calculate or interpret 3D kinetics since a large joint torque might have a very small resistance arm and not make a large contribution to a desired motion, or a torque might be critical to positioning a segment for another torque to be able to accelerate the segment (Sprigings, Marshall, Elliott, & Jennings, 1994).

The resultant joint torques calculated in inverse dynamics are often multiplied by the joint angular velocity to derive net *joint powers*. When the product of a net joint torque and joint angular velocity are positive (in the same direction), the muscle action is hypothesized to be primarily concentric and generating positive work. Negative joint powers are hypothesized to represent eccentric actions of muscle groups slowing down an adjacent segment. These joint powers can be integrated with respect to time to calculate the net work done at joints. Other studies first calculated mechanical energies (kinetic and potential energies), and summed them to estimate work and eventually calculate power. Unfortunately, these summing of mechanical energy analyses do not agree well with direct calculation of joint power from

torques because of difficulties in modeling the transfer of mechanical energies between external forces and body segments (Aleshinsky, 1986a,b; Wells, 1988) and coactivation of muscles (Neptune & van den Bogert, 1998).

EQUILIBRIUM

An important concept that grows out of Newton's first and second laws is *equilibrium*. **Mechanical equilibrium** occurs when the forces and torques acting on an object sum to zero. Newton's first law refers to the two conditions of **static equilibrium** ($\Sigma F = 0$, $\Sigma T = 0$), where an object is motionless or moving at a constant velocity. **Dynamic equilibrium** is used to refer to the kinetics of accelerated bodies using Newton's second law ($\Sigma F = m \cdot a$, $\Sigma T = I \cdot \alpha$). In a sense, dynamic equilibrium fits the definition of equilibrium if you rearrange the equations (i.e., $\Sigma F - m \cdot a = 0$). The $m \cdot a$ term in the previous equation is often referred to as the **inertial force**. This inertial force is not a real force and can cause confusion in understanding the kinetics of motion.

This text will focus on static equilibrium examples because of their simplicity and because summation of forces and torques is identical to dynamic equilibrium. Biomechanics studies often use static or quasi-static analyses (and thus employ static equilibrium equations and avoid difficulties in calculating accurate accelerations) in order to study slow movements with small accelerations. The occupational lifting standards set by the National Institute for Occupational Safety and Health (NIOSH) were based in large part on static biomechanical models and analyses of lifting. Static equilibrium will also be used in the following section to calculate the center of gravity of the human body.

Equilibrium and angular kinetics are the mechanical tools most often used in the study of balance. We will see in the next two sections that the center of gravity of the human body can be calculated by summing moments in a static equilibrium form, and these kinds of data are useful in examining the state of mobility and stability of the body. This control of stability and ability to move is commonly called *balance*. What mechanics tells us about balance is summarized in the Principle of Balance.

CENTER OF GRAVITY

A natural application of angular kinetics and anthropometrics is the determination of the center of gravity of the body. The **center of gravity** is the location in space where the weight (gravitational force) of an object can be considered to act. The center of small rigid objects (pencil, pen, bat) can be easily found by trying to balance the object on your finger. The point where the object balances is in fact the center of gravity, which is the theoretical point in space where you could replace the weight of the whole object with one downward force. There is no requirement for this location to be in a high-mass area, or even within or on the object itself. Think about where the center of gravity of a basketball would be.

The center of gravity of the human body can move around, because joints allow the masses of body segments to move. In the anatomical position, the typical location of a body's center of gravity in the sagittal plane is at a point equivalent to 57 and 55% of the height for males and fe-

Interdisciplinary Issue: The Spine and Low-Back Pain

One of the most common complaints is low-back pain. The medical literature would say that the etiology (origin) of these problems is most often idiopatic (of unknown origin). The diagnostic accuracy of advanced imaging techniques like magnetic resonance imaging (MRI) for identifying spinal abnormalities (e.g., disk herniation) that correlate with function and symptoms of low-back pain is poor (Beattie & Meyers, 1998). The causes of low-back pain are complicated and elusive. Biomechanics can contribute clues that may help solve this mystery. Mechanically, the spine is like a stack of blocks separated by small cushions (McGill, 2001). Stability of the spine is primarily a function of the ligaments and muscles, which act like the guy wires that stabilize a tower or the mast of a boat. These muscles are short and long and often must simultaneously stabilize and move the spine. Total spine motion is a summation of the small motions at each intervertebral level (Ashton-Miller & Schultz, 1988). Biomechanical studies of animal and cadaver spines usually examine loading and rotation between two spinal levels in what is called a motion segment. Individuals even exhibit different strategies for rotation of motion segments in simple trunk flexion movements (Gatton & Pearcy, 1999; Nussbaum & Chaffin, 1997), so that neuromuscular control likely plays a role in injury and rehabilitation (Ebenbichler, Oddsson, Kollmitzer, & Erim, 2001). Occasionally a subject is unfortunate and gets injured in a biomechanical study. Cholewicki and McGill (1992) reported x-ray measurements of the "buckling" of a single spinal segment that occurred during a heavy deadlift. Biomechanics research using computer models and EMG are trying to understand how muscles and loads affect the spine. This information must be combined with occupational, epidemiological, neurologic, and rehabilitative research to understand the development and treatment of low-back pain.

males, respectively (Hay & Reid, 1982). Can you name some structural and weight distribution differences between the genders that account for this general difference? Knowing where the force of gravity acts in various postures of the human body allows biomechanists to study the kinetics and stability of these body positions.

There are two main methods used to calculate the center of gravity of the human body, and both methods employ the equations of static equilibrium. One lab method, which requires a person to hold a certain body position, is called the **reaction change** or *reaction board* method. The other method used in research is called the *segmental method*. The **segmental method** uses anthropometric data and mathematically breaks up the body into segments to calculate the center of gravity.

The reaction board method requires a rigid board with special feet and a scale (2D) or scales (3D) to measure the ground reaction force under the feet of the board. The "feet" of a reaction board are knife-like edges or small points similar to the point of a nail. A 2D reaction board, a free-body diagram, and static equilibrium equations to calculate the center of gravity in the sagittal plane are illustrated in Figure 7.11. Note that the weight force of the board itself is not included. This force can be easily added to the computation, but an efficient biomechanist zeros the scale with the board in place to exclude extra terms from the calculations. The subject in Figure 7.11 weighs 185 pounds, the distance between the edges is 7 feet, and the scale reading is 72.7 pounds. With only three forces acting on this system and everything known but the location of the center of gravity, it is rather simple to apply the static equilibrium equation for torque and solve for the center of gravity ($d_\perp$). Note how the sign of the torque created by the subject's body is negative according to convention, so a negative

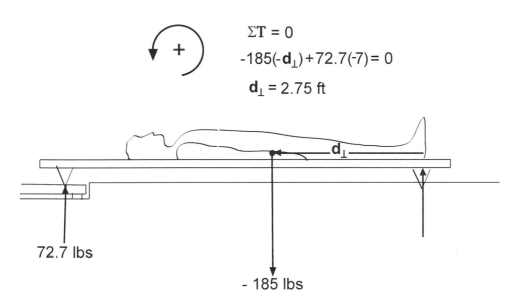

$$\Sigma T = 0$$
$$-185(-d_\perp) + 72.7(-7) = 0$$
$$d_\perp = 2.75 \text{ ft}$$

72.7 lbs

- 185 lbs

Figure 7.11. Application of static equilibrium and a reaction board to calculate whole body center of gravity. Summing torques about the reaction board edge at the feet and solving for the moment arm ($d_\perp$) for gravity locates the center of gravity.

$d_\perp$ (to the left) of the reaction board edge fits this standard, and horizontal displacement to the left is negative. In this case, the subject's center of gravity is 2.75 feet up from the edge of the reaction board. If the subject were 5.8 feet in height, his center of gravity in this position would be 47% of his or her height.

In the segmental method, the body is mathematically broken up into segments. The weight of each segment is then estimated from mean anthropometric data. For example, according to Plagenhoef, Evans, & Abdelnour (1983), the weight of the forearm and hand is 2.52 and 2.07% for a man and a woman, respectively. Mean anthropometric data are also used to locate the segmental centers of gravity (percentages of segment length) from either the proximal or distal point of the segment. Figure 7.12 depicts calculation of the center of gravity of a high jumper clearing the bar using a three-segment biomechanical model. This simple model (head+arms+trunk, thighs, legs+feet) illustrates the segmental method of calculating the center of gravity of a

Segment Centers of Gravity (X,Y)
Shank/feet (3.2,6.6)
Thighs (5.0,10.5)
Head/Arms/Trunk (11.0,9.0)

$$\Sigma T_o = 0$$

$$182\,(d_\perp) - 22(3.2) - 36(5) - 124(11) = 0$$

$$d_\perp = 8.9 \text{ in}$$

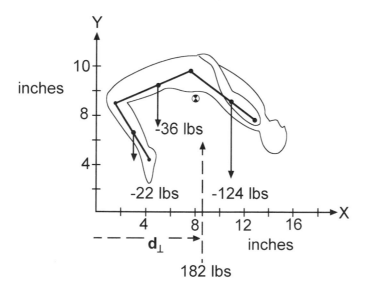

Figure 7.12. Calculating the horizontal position of the whole body center of gravity of a high jumper using the segmental method and a three-segment model of the body. Most sport biomechanical models use more segments, but the principle for calculating the center of gravity is the same.

linked biomechanical system. Points on the feet, knee, hip, and shoulder are located and combined with anthropometric data to calculate the positions of the centers of gravity of the various segments of the model. Most biomechanical studies use rigid-body models with more segments to more accurately calculate the whole-body center of gravity and other biomechanical variables. If a biomechanist were studying a high jump with high-speed video (120 Hz), a center of gravity calculation much like this would be made for every image (video snapshot) of the movement.

The segmental method is also based on static equilibrium. The size and location (moment arm) of the segmental forces are used to calculate and sum the torques created by each segment. If this body posture in the snapshot were to be balanced by a torque in the opposite direction (product of the whole bodyweight acting in the opposite direction times the center of gravity location: $182 \cdot \mathbf{d}_\perp$), the total torque would be zero. By applying the law of statics and summing torques about the origin of our frame of reference, we calculate that the person's bodyweight acts 8.9 inches from the origin. These distances are small because the numbers represent measurements on an image. In a 2D biomechanical analysis, the image-size measurements are scaled to real-life size by careful set-up procedures and imaging a control object of known dimensions.

Finding the height of the center of gravity is identical, except that the y coordinates of the segmental centers of gravity are used as the moment arms. Students can then imagine the segment weight forces acting to the left, and the height of the center of gravity is the y coordinate that, multiplied by the whole bodyweight acting to the right, would cancel out the segmental torques toward the left. Based on the subject's body position and the weights of the three segments, guess the height in centimeters of the center of gravity. Did the center of gravity pass over the bar? Finish the calculation in Figure 7.12 to check your guess. The segmental method can be applied using any number of segments, and in all three dimensions during 3D kinematic analysis. There are errors associated with the segmental method, and more complex calculations are done in situations where errors (e.g., trunk flexion/extension, abdominal obesity) are likely (Kingma, Toussaint, Commissaris, Hoozemans, & Ober, 1995).

Activity: Center of Gravity and Moment of Inertia

Take a 12-inch ruler and balance it on your finger to locate the center of gravity. Lightly pinch the ruler between your index finger and thumb at the 1-inch point, and allow the ruler to hang vertically below your hand. Swing the ruler in a vertical plane and sense the resistance of the ruler to rotation. Tape a quarter to various positions on the ruler and note how the center of gravity shifts and how the resistance to rotation changes. Which changes more: center of gravity or moment of inertia? Why? What factors make it difficult to sense changes in ruler moment of inertia?

PRINCIPLE OF BALANCE

We have seen than angular kinetics provides mathematical tools for understanding rotation, center of gravity, and rotational equilibrium. The movement concept of balance is closely related to these angular kinetic variables. **Balance** is a person's ability to control their body position relative to some base of support (Figure 7.13). This ability is needed in both static equilibrium

Figure 7.13. Balance is the degree of control a person has over their body. Balance is expressed in static (track start) and dynamic conditions (basketball player boxing out an opponent). Track image used with permission from Getty Images.

conditions (e.g., handstand on a balance beam) and during dynamic movement (e.g., shifting the center of gravity from the rear foot to the forward foot). Balance can be enhanced by improving body segment positioning or posture. These adjustments should be based on mechanical principles. There are also many sensory organs and cognitive processes involved in the control of movement (balance), but this section focuses on the mechanical or technique factors affecting balance and outlines application of the Principle of Balance.

Before we apply this principle to several human movements, it is important to examine the mechanical paradox of stability and mobility. It turns out that optimal posture depends on the right mix of stability and mobility for the movement of interest. This is not always an easy task, because stability and mobility are inversely related. Highly stable postures allow a person to re-

sist changes in position, while the initiation of movement (mobility) is facilitated by the adoption of a less stable posture. The skilled mover learns to control the position of their body for the right mix of stability and mobility for a task.

The biomechanical factors that can be changed to modify stability/mobility are the base of support and the position of the center of gravity relative to the base of support. The base of support is the two-dimensional area formed by the supporting segments or areas of the body (Figure 7.14). A large base of support provides greater stability because there is greater area over which to keep the bodyweight. Much of the difficulty in many gymnastic balancing skills (e.g., handstand or scale) comes from the small base of support on which to center bodyweight.

The posture of the body in stance or during motion determines the position of

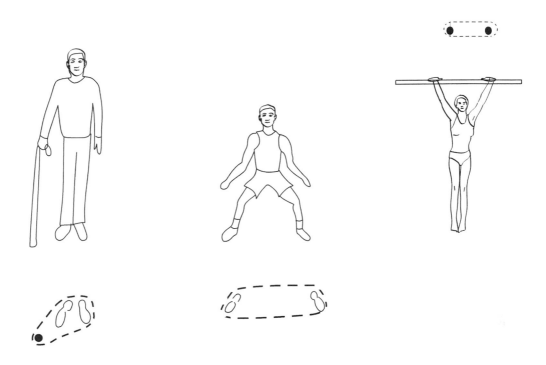

Figure 7.14. The base of support is the two-dimensional area within all supporting or suspending points of the biomechanical system.

the center of gravity relative to the base of support. Since gravity is the major external force our body moves against, the horizontal and vertical positions of the center of gravity relative to the base of support are crucial in determining the stability/mobility of that posture. The horizontal distance from the edge of the base of support to the center of gravity (line of action of gravity) determines how far the weight must be shifted to destabilize a person (Figure 7.15a). If the line of gravity falls outside the base of support, the gravitational torque tends to tip the body over the edge of the base of support. The vertical distance or height of the center of gravity affects the geometric stability of the body. When the position of the center of gravity is higher, it is easier to move beyond the base of support than in postures with a lower center of gravity. Positioning the line of gravity outside the base of support can facilitate the

rotation of the body by the force of gravity (Figure 7.15b).

Recall that the inertia (mass and moment of inertia), and other external forces like friction between the base and supporting surface all affect the equilibrium of an object. There are also biomechanical factors (muscle mechanics, muscle moment arms, angles of pull, and so on) that affect the forces and torques a person can create to resist forces that would tend to disrupt their balance. The general base of support and body posture technique guidelines in many sports and exercises must be based on integration of the biological and mechanical bases of movement. For example, many sports use the "shoulder width apart" cue for the width of stances because this base of support is a good compromise between stability and mobility. Wider bases of support would increase potential stability but put the limbs in a poor position to create

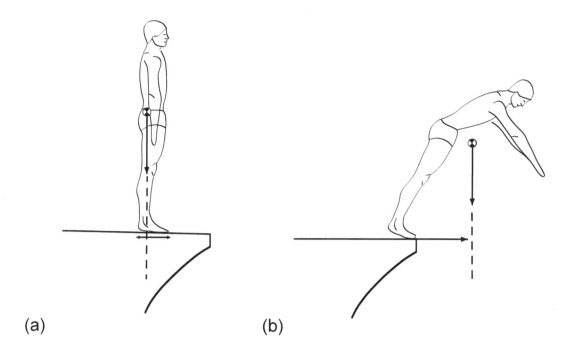

(a)   (b)

Figure 7.15. The position of the line of gravity relative to the limits of the base of support determines how far the weight must be shifted for gravity to tend to topple the body (a) or the size of the gravitational torque helps create desired rotation (b).

torques and expend energy, creating opposing friction forces to maintain the base of support.

The Principle of Balance is based on the mechanical tradeoff between stability and mobility. The Principle of Balance is similar to the Coordination Continuum because the support technique can be envisioned as a continuum between high stability and high mobility. The most appropriate technique for controlling your body depends on where the goal of the movement falls on the stability–mobility continuum. Coaches, therapists, and teachers can easily improve the ease of maintaining stability or initiating movement (mobility) in many movements by modifying the base of support and the positions of the segments of the body. It is important to note that good mechanical posture is not always required for good balance. High levels of skill and mus-

cular properties allow some people to have excellent balance in adverse situations. A skater gliding on one skate and a basketball player caroming off defenders and still making a lay-up are examples of good balance in less than ideal conditions.

Imagine that a physical therapist is helping a patient recover from hip joint replacement surgery. The patient has regained enough strength to stand for short lengths of time, but must overcome some discomfort and instability when transitioning to walking. The patient can walk safely between parallel bars in the clinic, so the therapist has the patient use a cane. This effectively increases the base of support, because the therapist thinks increasing stability (and safety) is more important. If we combine angular kinetics with the Principle of Balance, it is possible to determine on what side of the body the cane should be held. If the cane were held on the same (affected) side, the base of support would be larger, but there would be little reduction in the pain of the hip implant because the gravitational torque of the upper body about the stance hip would not be reduced. If the patient held the cane in the hand on the opposite (unaffected) side, the base of support would also be larger, and the arm could now support the weight of the upper body, which would reduce the need for hip abductor activity by the recovering hip. Diagram the increase in area of the base of support from a single-leg stance in walking to a single-leg stance with a cane in each hand. Estimate the percentage increase in base of support area using the cane in each hand.

Classic examples of postures that would maximize mobility are the starting positions during a (track or swimming) race where the direction of motion is known. The track athlete in Figure 7.16 has elongated his stance in the direction of his start, and in the "set" position moves his center of gravity near the edge of his base of support. The blocks are not extended too far backwards because this interacts with the athlete's ability to shift weight forward and generate forces against the ground. For a summary of the research on the effect of various start postures on sprint time, see Hay (1993). Hay also provides a good summary of early research on basic footwork and movement technique factors in many sports.

In many sports, athletes must take on defensive roles that require quick movement in many directions. The Principle of Balance suggests that postures that foster mobility over stability have smaller bases of support, with the center of gravity of the

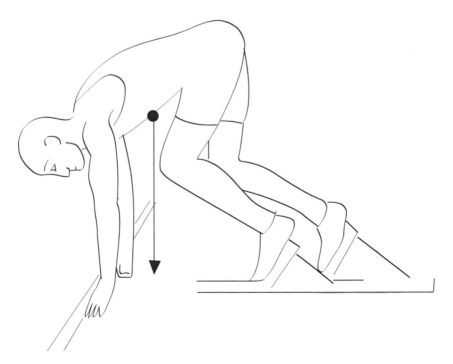

Figure 7.16. The starting position of a sprinter in the blocks shifts the line of gravity toward the front of the stance and the intended direction of motion. This stance favors mobility forward over stability.

body not too close to the base of support. When athletes have to be ready to move in all directions, most coaches recommend a slightly staggered (one foot slightly forward) stance with feet about shoulder width apart. Compare the stance and posture of the volleyball and basketball players in Figure 7.17. Compare the size of the base of support and estimate the location of the center of gravity in both body positions. What posture differences are apparent, and are these related to the predominant motion required in that sport? Bases of support need only be enlarged in directions where stability is needed or the direction of motion is known.

There are movement exceptions to strict application of the Principle of Balance because of high skill levels or the interaction of other biomechanical factors. In well-learned skills like walking, balance is easily maintained without conscious attention over a very narrow base of support. Gymnasts can maintain balance on very small bases of support as the result of con-

siderable skill and training. A platform diver doing a handstand prior to a dive keeps their base of support smaller than one shoulder width because extra side-to-side stability is not needed and the greater shoulder muscle activity that would be required if the arms were not directly underneath the body. Another example might be the jump shot in basketball. Many coaches encourage shooters to "square up" or face the basket with the body when shooting. Ironically, the stance most basketball players spontaneously adopt is staggered, with the shooting side foot slightly forward. This added base of support in the forward–backward direction allows the player to transition from pre-shot motion to the primarily vertical motion of the jump. It has also been hypothesized that this stagger in the stance and trunk (not squaring up) helps the player keep the shooting arm aligned with the eyes and basket, facilitating side-to-side accuracy (Knudson, 1993).

Balance is a key component of most motor skills. While there are many factors

Figure 7.17. Comparison of the ready positions of a basketball player and a volleyball player. How are the mechanical features of their stance adapted to the movement they are preparing for?

that affect the ability to control body mobility and stability, biomechanics focuses on the base of support and position of the center of gravity. Mechanically, stability and mobility are inversely related. Coaches can apply the Principle of Balance to select the base of support and postures that will provide just the right mix of stability/mobility for a particular movement. Angular kinetics is the ideal quantitative tool for calculating center of gravity, and for examining the torques created by gravity that the neuromuscular system must balance.

Interdisciplinary Issue: Gender Differences

It is generally considered that the lower center of gravity in women gives them better balance than men. What is the biomechanical significance of the structural and physiological differences between men and women? While there is substantial research on the physiological differences between the genders, there is less comparative research on the biomechanical differences. Motor control and ergonomic studies have observed significant differences in joint angles during reaching (Thomas, Corcos, & Hasan, 1998) and lifting (Lindbeck & Kjellberg, 2000). Greater interest in gender differences seems to focus on issues related to risk of injury, for example, to like the anterior collateral ligament (ACL) (Charlton, St. John, Ciccotti, Harrison, & Schweitzer, 2002; Malinzak, Colby, Kirkendall, Yu, & Garrett, 2001).

SUMMARY

The key mechanical variable in understanding the causes of rotary motion is the moment of force or torque. The size of the torque that would rotate an object is equal to the force times its moment arm. The moment of inertia is a variable expressing the angular inertia of an object about a specific axis of rotation. The moment of inertia most strongly depends on the distribution of mass relative to the axis of rotation of interest. When all the torques acting on an object sum to zero, the object is said to be in static equilibrium. The equations of static equilibrium are often used to calculate the center of gravity of objects. Biomechanics most often uses the reaction change and segmental methods to calculate the center of gravity of the human body. Balance is the ability of a person to control their body position relative to some base of support. The Balance Principle deals with the mechanical factors that affect balance, and the tradeoff between stability and mobility in various body postures.

REVIEW QUESTIONS

1. What are the two most important parameters that determine the size of a torque or moment of force?

2. What is the inertial resistance to angular acceleration object about an axis, and what factors affect its size?

3. Give examples of how the human body can position itself to increase or decrease its inertial resistance to rotation.

4. Calculate the shoulder flexion torque required to hold an 80-lb barbell just above your chest in a bench press. The horizontal distance from your shoulder axis to the barbell is 0.9 feet.

5. Restate Newton's three laws of motion in angular kinetic terms.

6. Explain how static equilibrium can be used to calculate the center of gravity of the human body.

7. Draw or trace a few freeze-frame images of the human body in several positions from sport or other human movements. Estimate the location of the center of gravity.

8. A mischievous little brother runs ahead of his sister and through a revolving door at a hotel. The little brother pushes in the opposite direction of his sister trying to exit. If the brother pushes with a maximum horizontal force of 40 pounds acting at a right angle and 1.5 feet from the axis of the revolving door, how much force will the sister need to create acting at 2.0 feet from the axis of rotation to spoil his fun?

9. What mechanical factors can be used to maximize stability? What does this do to a person's mobility?

10. What movement factors can a kinesiology professional qualitatively judge that show a person's balance in dynamic movements?

11. Say the force F_2 applied by the student in Figure 7.3 acted 55° in from the tangent to the merry-go-round. Calculate the torque created by the student.

12. Draw a free-body diagram of a person standing on a reaction board (hint: the system is the body plus the board). Estimate the length of the board and the horizontal distance to the person's center of gravity. Calculate the reaction force on the board if you were the person on it.

13. If the rotary component of a brachialis force is 70 N and the muscle attaches 0.4 m from the axis of rotation, what is the flexor torque created by the muscle? What other information do you need in order to calculate the resultant force created by the brachialis?

KEY TERMS

Balance Principle
center of gravity
inertial force
moment arm
moment or moment of force
moment of inertia
reaction change
segmental method
static equilibrium
torque

SUGGESTED READING

Brown, L. E. (Ed.) (2000). *Isokinetics in human performance*. Champaign, IL: Human Kinetics.

Chaffin, B. D., Andersson, G. B. J., & Martin, B. J. (1999). *Occupational biomechanics* (3rd ed.). New York: Wiley.

Huxham, F. E., Goldie, P. A., & Patla, A. E. (2001). Theoretical considerations in balance assessment. *Australian Journal of Physiotherapy*, **47**, 89–100.

Mann, R. V. (1981). A kinetic analysis of sprinting. *Medicine and Science in Sports and Exercise*, **13**, 325–328.

McGill, S. M., & Norman, R. W. (1985). Dynamically and statically determined low back moments during lifting. *Journal of Biomechanics*, **18**, 877–886.

Murray, M. P., Seireg, A., & Scholz, R. C. (1967). Center of gravity, center of pressure, and supportive forces during human activities. *Journal of Applied Physiology*, **23**, 831–838.

Winter, D. A. (1984). Kinematic and kinetic patterns of human gait: Variability and compensating effects. *Human Movement Science*, **3**, 51–76.

Winter, D. A. (1995). Human balance and posture control during standing and walking. *Gait and Posture*, **3**, 193–214.

Winters, J. M., & Woo, S. L.-Y. (Eds.) (1990). *Multiple muscle systems*. New York: Springer.

Zatsiorsky, V. M. (2002). *Kinetics of human motion*. Champaign, IL: Human Kinetics.

Zernicke, R. F., & Roberts, E. M. (1976). Human lower extremity kinetic relationships during systematic variations in resultant limb velocity. In P. V. Komi (Ed.), *Biomechanics V–B* (pp. 41-50). Baltimore: University Park Press.

WEB LINKS

Torque tutorial—part of the physics tutorials at University of Guelph.
        http://eta.physics.uoguelph.ca/tutorials/torque/Q.torque.intro.html

Support moment—torques in leg joints in walking are examined in this teach-in exercise from the Clinical Gait Analysis website.
        http://guardian.curtin.edu.au/cga/teach-in/support/

Torque, center of mass, and moment of inertia simulations—part of mechanics simulations at explorescience.com.
        http://www.explorescience.com/activities/activity_list.cfm?categoryID=10

CHAPTER 8

Fluid Mechanics

External forces that have a major effect on most human movements are related to immersion in or flow of fluids past a body. This chapter reviews the mechanical effect of moving through air and water, the two most common fluids encountered in human movement. Fluid forces usually result in considerable resistance to high-velocity movements through fluids, so many sport techniques and pieces of equipment are designed to minimize fluid resistance. Fluid forces, however, can also be used to create movement, like in the skillful application of spin to projectiles. This chapter concludes with application of this use of fluid forces in the **Principle of Spin**.

FLUIDS

You may have studied the various states of matter in physics or noticed that many substances are not easily classified as totally solid or liquid. Mechanically, fluids are defined as substances that *flow or continuously deform when acted upon by shear forces*. A thorough review of all the nuances of fluid mechanics is not possible, so key concepts related to the supporting force of immersion in fluids and the forces that arise from moving through fluids will be reviewed. Several references are cited to guide students interested in digging deeper into the nuances of fluid mechanics.

FLUID FORCES

For the purposes of this chapter, the major fluid forces that affect human motion are classified according to an object's position or velocity within a fluid. When an object is placed in a fluid there is a resultant upward force or supporting fluid force called **buoyancy**. The fluid force related to how the fluid flows past the object is resolved into right-angle components called **lift** and **drag**. In most movement, people have considerable control over factors that affect these forces. Let's see how these fluid forces affect human movement.

Buoyancy

The vertical, supporting force of a fluid is called buoyancy. When an inanimate object is put in a fluid (like water), the vector sum of gravity and the buoyant force determines whether or not the object will float (Figure 8.1). The **Archimedes Principle** states that the size of the buoyant force is equal to the weight of the fluid displaced by the object. Folklore says that the famous

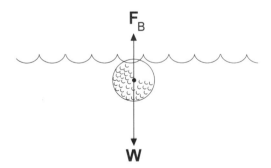

Figure 8.1. The resultant vector of gravity (**W**) and buoyancy (**F**$_B$) will determine if an inanimate object floats. This golf ball will sink to the bottom of the water hazard.

Greek physicist/mathematician realized this important principle when noticing water level changes while taking a bath.

A sailboat floats at a level where the weight of the boat and contents are equal in size to the weight of the volume of water displaced. Flotation devices used for water exercise and safety increase the buoyancy of a person in two ways: having a lower density (mass/volume) than water and having a hollow construction. These flotation devices displace water that weighs more than the device, increasing the buoyancy of the person.

In a gravitational field the mass of a fluid is attracted in a particular direction. The weight of water is typically 9800 N per cubic meter, but this figure gradually increases for water at greater depths. The deeper a scuba diver descends, the greater the fluid pressure around them (because of the greater mass of water essentially "on top" of them). This increased pressure in a particular volume of fluid means that the volume of water weighs more than a similar volume of water at the surface, so the buoyant force on objects tends to slightly increase as depth increases. A similar phenomenon occurs as we descend from a mountain, where the fluid pressure of the atmosphere on us increases. The buoyant force on the human body from the "sea" of atmospheric gases also depends on our depth (opposite of elevation), but is usually a fraction of a pound and can be ignored in vertical kinetic calculations of human movement.

The density of the human body is very close to that of water, largely due to the high water content of all tissue. Lean tissue (muscle and bone) have densities greater than water, while body fat tends to be less dense than water. The buoyant force on a swimmer varies with changes in body composition and when the person inhales or exhales. Taking a deep breath expands the chest, which increases the volume of the

Activity

The next time you are at a pool, see if you can detect an increase in buoyant force with increasing depth. Hold a large sport ball (water polo, soccer, football) in one hand and gradually submerge it. Note the downward vertical force you exert to balance the buoyant force of the ball as it descends. Also note the horizontal forces you must exert to keep your hand forces balanced with the buoyant force and gravity! Another simple activity is to mark the water line on a floating ping pong ball. Tape dimes to the ball and find the maximum buoyant force of a ping pong ball. Does a forcibly submerged ball have potential energy?

body and increases the buoyant force. It you have ever taught a swimming class you know that people typically fall into three groups based on their somotype and body composition: floaters, conditional floaters, and sinkers. The majority of your swim class can easily float when holding their breath (conditional floaters). There will be a few folks who easily float (floaters) or cannot float (sinkers) without some form of propulsion or flotation device.

The buoyant force in water acts upward at the center of buoyancy. The **center of buoyancy** is essentially the centroid of the volume of water displaced by an object. In the human body, the trunk makes up most of the volume, so the center of buoyancy is located 1–2 cm superior (McLean & Hinrichs, 2000a) to the center of gravity (Figure 8.2). Since so much body volume is in the upper trunk, moving the rest of the body makes smaller changes in the center of buoyancy than in the center of gravity. Note that the weight force and buoyant force create a force couple that will tend to rotate the swimmer's legs down until the

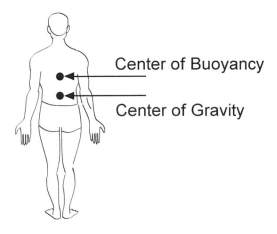

Figure 8.2. The center of buoyancy of the human body is superior to the center of gravity because of the large volume of the upper body.

We have seen that objects in a fluid experience a supporting force related to the position of the object in the fluid and the density of the object. The next section will deal with the interaction forces between an object and the fluid when there is relative motion between the two. These fluid motion forces can be quite large. The fluid forces between the air and your body are nearly identical if you are falling at 120 km/hr while skydiving in a specific body position or if you are apparently still on top of a column of 120 km/hr airflow in a simulator. In both these situations the drag forces on the body are equal to your body weight. In the first case the body is falling through essentially still air while in the second case the body is essentially stationary with air flowing over it.

Drag

The fluid force resisting motion between an object and a fluid is called **drag**. Drag acts in the same direction (parallel) as the relative flow of the fluid past an object and in the opposite direction of the object's motion in the fluid. Drag forces act on the fisherman (creek) and the fly (air) due to the relative motion of the fluid past the objects (see Figure 8.3). If there are no propulsive forces acting on the object, like a projectile (see chapter 5, p. 113), the drag force tends to slow down the motion of the projectile through the fluid. Since the drag force acts parallel to the relative flow of the fluid, it is much like the contact force of friction studied in chapter 6.

Research has shown that the size of the drag force (F_D) that must be overcome in a fluid can be calculated using the following formula: $F_D = \frac{1}{2}C_D \, \rho \, A_P \, V^2$. The coefficient of drag (C_D) is a dimensionless number much like the coefficient of friction or restitution. We will see later that C_D depends on many object and fluid flow factors. Drag

weight and buoyant force are nearly colinear. Swimmers still scared of the water have great difficulty floating on their back because they tend to pike and lift the head/upper trunk out of the water. The resulting loss of buoyant force (from less water displacement) tends to dip the swimmer's head deeper into the water. If you are having difficulty getting a swimmer to relax and do a back float, how can you shift their limbs to shift the center of gravity and maintain a large buoyant force?

Application: Hydrotherapy

Therapeutic exercises in water utilize its buoyant force to unload the lower extremity. The amount of unloading of the body can be easily manipulated by the extent of submersion. This exercise modality differs from suspension systems that unload the body by pulleys lifting up the trunk because of other fluid forces. The flow of water also creates lift and drag forces that have been shown to create differences in muscle activation in exercise (Poyhonen, Kryolainen, Keskien, Hautala, Savolainen, & Malkia, 2001). Therapy pools that create currents for exercise likely exaggerate the neuromuscular differences between these movements and dry land movement.

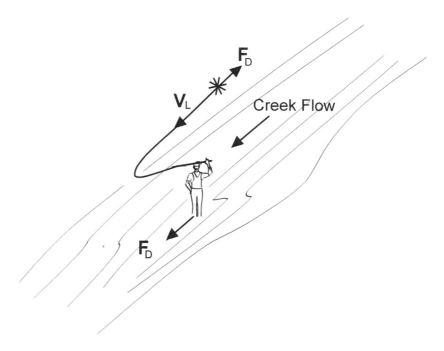

Figure 8.3. The fluid force of drag ($\mathbf{F}_D$) acts in a direction opposing the relative flow of fluid past the object.

also depends on fluid density (ρ) and the projected frontal area (A_P) in the path of the fluid flow. The most important factor affecting drag is the relative velocity ($\mathbf{V}^2$) of the fluid past the object.

Like the velocity term in kinetic energy, the force of drag varies with the square of the relative fluid velocity. This means that, all other things being equal, a cyclist that doubles and then triples his pace increases drag by 4 and 9 times compared to his initial speed! This explains why running faster or into a strong breeze feels much more difficult. The importance of the adjective "relative" can be easily appreciated by noting that it is easier to run with a strong breeze *behind* you. The dramatic effect of drag on sprint performance has forced the International Amateur Athletic Federation to not ratify sprint records if the wind assisting a runner exceeds 2.0 m/s. World records are always a controversial issue, but current weighting of records in many events does not taken into account the effect of altitude (Mureika, 2000) or latitude (Mizera & Horvath, 2002). Remember that relative velocity means that we are talking about a local kinematic frame of reference—in other words, the speed and direction of fluid flow relative to the object of interest. These drag forces increase with the square of velocity and often dramatically affect performance.

The Drag force on an object has several sources: *surface drag*, *pressure drag*, and *wave drag*. Understanding these drag forces is important for minimizing these resistances in many sports and activities.

Surface drag can be thought of as a fluid friction force, much like solid friction force studied in chapter 6. Surface drag is also commonly called friction drag or skin friction drag. It results from the frictional force between fluid molecules moving past the surface of an object and the frictional force between the various layers of the fluid. Viscosity is the internal resistance of a

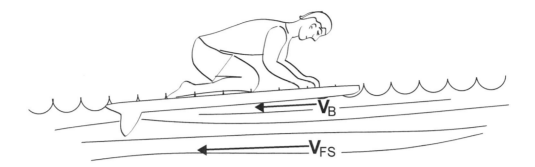

Figure 8.4. The water nearest a surfboard forms a boundary layer that flows more slowly ($\mathbf{V}_B$) past the board than the free stream velocity ($\mathbf{V}_{FS}$) because of friction with the board and fluid friction.

fluid to flow. Air has a lower viscosity than water, which has a lower viscosity than maple syrup.

Suppose a surfer is floating on their board waiting for the right wave (Figure 8.4). The fluid flow below the apparently stationary surfboard creates surface drag from the flow of the ocean under the board. Water molecules immediately adjacent to the board are slowed by shear forces between them and the molecules of the board. So the fluid close to the board moves slower than the ocean water farther from the board. In fact, there is a region of water layers close to the board that moves more slowly because of viscous (fluid friction) forces between the fluid particles. This region of fluid affected by surface drag and viscosity near an object is called the **boundary layer**. Layers of fluid more distant from the object that are not affected by drag forces with the object represent the free stream velocity. Have you ever started driving your car and notice a small insect on the hood or windshield wipers? I am willing to wager that most of you noticed the considerable speed you had to drive to disrupt the boundary layer the insect stood in before it was swept away! We will see that this relative or free stream velocity is one of the most important factors affecting the drag and lift forces between objects and fluids.

Performers cannot change the viscosity of the fluid they move in, but they can modify the roughness of their body or equipment to decrease surface drag. Surfboards and skis are waxed, a swimmer may shave body hair, or very smooth body suits may be worn to decrease surface drag. Some suits actually introduce texture on portions of the fabric to modify both lift and drag forces (Benjanuvatra, Dawson, Blanksby, & Elliott, 2002). While it is important to minimize surface drag, the largest fluid resistance in many sports tends to be from pressure drag.

The second kind of drag force that dominates the fluid resistance in many sports is **pressure drag**. Pressure drag is the resistance force to fluid flow that is created by a pressure differential when the fluid flows around a submerged object. A simplified illustration of this phenomenon is presented in Figure 8.5. The collision of the object and molecules of fluid creates a high pressure on the front of the object, while a lower-pressure region or wake is formed behind the object. The region of higher pressure "upstream" creates a resultant force backward on the object. We will see that the mechanics of this pressure differential is a bit more complicated and related to many factors. Fortunately, many of these factors can be modified to reduce the fluid

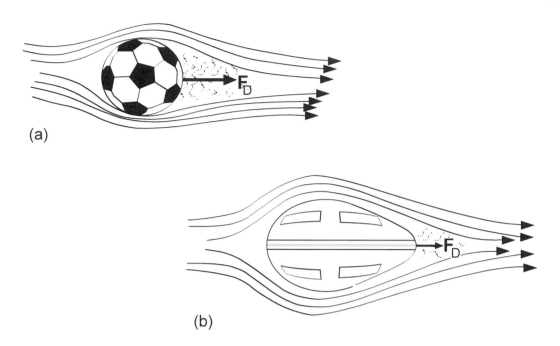

(a)

(b)

Figure 8.5. Form drag forces (**F**₍D₎) result from a vacuum pressure formed in the pocket formed behind a submerged object (a). Decreasing the pressure in this wake is how contouring the rear profile of an object (streamlining) decreases form drag (b).

resistance to many human movements. Some human movements may also use drag as a propulsive force.

To understand the variations in pressure drag, one must differentiate two different kinds of fluid flow in the boundary layer: *laminar* and *turbulent*. The air flow past a tennis ball can be highlighted by smoke introduced into a wind tunnel, depicted in Figure 8.6, which shows both predominantly laminar and turbulent flow. Laminar flow typically occurs in low-velocity conditions with streamlined objects where the fluid particles can flow relatively undisturbed in parallel layers. Turbulent flow occurs when fluid molecules bounce off the object and each other, mixing in chaotic fashion.

The kind of fluid flow over an object also affects pressure drag. At low velocities the boundary layer is laminar and cannot

flow very far around a non-rotating sphere before peeling away from the surface (Figure 8.7a), creating a large form drag. At

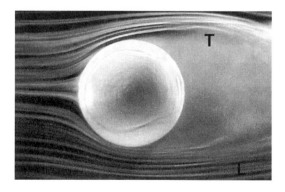

Figure 8.6. The air flow past a tennis ball shows both laminar (L) and turbulent (T) fluid motion. The top-spin on the ball deflects the air flow creating another fluid force called lift. Photo courtesy of NASA Ames Research Center Fluid Mechanics Laboratory and Cislunar Aerospace, Inc.

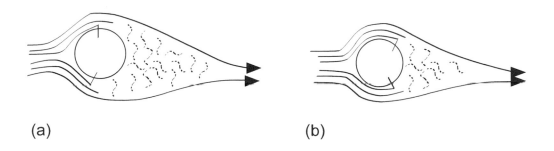

(a)   (b)

Figure 8.7. Spheres like sport balls create different fluid flows and drag force depending on many factors. Primarily laminar flow (a) can result in large pressure drag because of early separation of the boundary layer for a large wake, while turbulent flow (b) will often boundary separation and decrease pressure drag.

higher velocities, the boundary layer flow is turbulent and more resistant to the pressure gradient as it flows around the object. This results in a later point of separation (Figure 8.7b) and lower pressure drag than laminar flow. In most objects there is not a distinct transition from laminar to turbulent flow, but a critical or transition region where flow is unstable can be either laminar or turbulent. This transition region is important in the flight of spherical balls because the coefficient of drag can drop dramatically, creating a "drag crisis." Increasing the roughness of the ball (scuffing a baseball or putting dimples on a golf ball) can decrease the velocity where these lower drag forces occur. Scientists interested in fluid mechanics use a dimensionless ratio (the Reynolds number: Re) to combine the effects of object geometry on fluid flow. This chapter will not go into detail on Reynolds numbers, but interested students can see Mehta (1985) or Mehta and Pallis (2001a,b) for more information on Reynolds numbers related to sports balls.

Much of the variation in the flight characteristics of many sport balls is related to differences in drag and lift forces that are directly related to variations in fluid flow in the transition region of Reynolds numbers. This provides a great opportunity for skill

and coaching to modify the flight characteristics of many shots or throws in sports. Many of these important effects are counterintuitive. For example, slightly increasing the roughness of a sphere (golf or baseball) might decrease drag by promoting a more turbulent boundary layer, while increasing the lift forces generated. Another example is the nature of the felt on tennis balls. The felt has a major influence on the drag coefficient (Mehta & Pallis, 2001a), so professional tennis players when serving select balls in part based on the amount of felt fluff and wear. In the next section we will study how surface roughness of rotating balls can also be used to increase the fluid force of lift.

The two major techniques employed to decrease pressure drag in human movement are (a) decreasing the frontal area and (b) streamlining. The smaller the frontal area, the less the fluid must be accelerated to flow around the object. Extending the downstream lines of an object also decreases pressure drag by delaying separation and decreasing the turbulent wake behind the object. Swimming strokes often strike a balance between maintaining a streamlined body position and a one that maximizes propulsion. The high speeds and large surface areas in cycling make streamlined

Figure 8.8. High-speed sports like cycling (or skiing) use streamlining to decrease speed losses due to drag forces. Image used with permission from Getty Images.

Application: Drafting

Sports with very high relative velocities of fluid flow are strongly affected by drag. One strategy used to minimize drag forces in these sports (cycling, car racing) is drafting. Drafting means following closely behind another competitor, essentially following in their wake. The athlete in front will use more energy against greater pressure drag, while the drafting athlete experiences less fluid resistance and can use less energy while they draft. The strategy of the drafting athlete is often to outsprint the leader near the end of the race. In many team racing sports it is the teammate who expends the extra energy to be in the front early in the race who makes it possible for other team members to finish in a higher final position. Drafting even has advantages in some lower-velocity events. An athlete running about 1 m behind another runner can decrease air resistance, decreasing the metabolic cost of running by about 7% (Pugh, 1971).

equipment and body positions critical (Figure 8.8).

The third kind of drag is **wave drag**. At the surface of a fluid it is possible that disturbances will create waves within the fluid that resist the motion of an object with area projecting at this surface. Wave drag can constitute a major resistance in swimming (Rushall, Sprigings, Holt, & Cappaert, 1994). Triathletes swimming in the open water must overcome wave drag from both the wind and from their fellow competitors. Swimmers in enclosed pools are less affected by wave drag than those swimming in the open water because of lane makers and gutters designed to dampen waves. Small variations in lane placement, however, may affect the wave drag experienced by a swimmer.

Lift

The fluid force acting at right angles to the flow of fluid is called **lift** (Figure 8.9). Just like contact forces are resolved into right-angle components (friction and normal reaction), fluid forces are resolved into the right-angle forces of drag and lift. Since lift acts at right angles to the flow of the fluid, the direction of the lift force in space varies and depends on the shape, velocity, and rotation of the object. It is unwise to assume that the lift always acts upward. For example, the wings on race cars are designed to

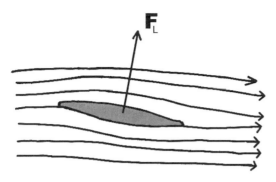

Figure 8.9. The fluid force acting at right angles to the relative flow of fluid is called lift. Lift acts in all directions, not just upward.

create a downward lift force to stabilize the car and keep it in contact with the ground.

The size of the lift force can also be modeled with a coefficient of lift (C_L) and a familiar equation: $F_L = \frac{1}{2}C_L\ \rho\ A_P\ V^2$. Just like drag, lift varies with the square of the relative velocity (V^2) of fluid. Earlier we characterized drag as primarily a fluid resistance. Lift tends to be a fluid force and is often used for propulsion. One of the ear-

ly leaders in swimming research, "Doc" Counsilman at Indiana University, used high-speed films of skilled swimmers to measure the complex patterns of arm and leg motions and was instrumental in demonstrating the importance of lift as a propulsive force in swimming (Counsilman, 1971). Whether lift or drag is the primary propulsive force used in swimming is a controversial issue (Sanders, 1998), and other theories like vortices (Arellano, 1999) and axial fluid flow (Toussaint *et al.*, 2002) are currently being examined. The important thing for swim coaches to realize is that precise arm and leg movements are required to use the hands and feet effectively, and that skilled swimmers learn to use both lift and drag forces for propulsion.

Synchronized swimming and competitive swimming tend to use small "sculling" hand movements to create lift forces for propulsion. A skilled swimmer precisely adjusts the pitch of their hands to maximize the down-the-pool resultant of the lift and drag forces (Figure 8.10). This is much like the high-tech propellers in modern aircraft

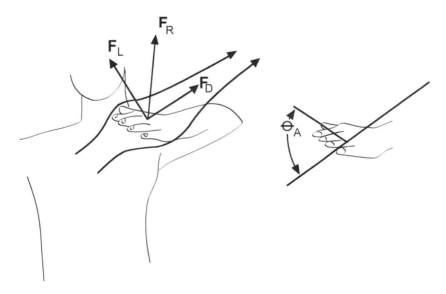

Figure 8.10. The inward sweep skill of a freestyle swimmer's hand may be selected to maximize the down-the-pool resultant of the lift and drag acting on the hand (a). This angle of attack (θ_A) is critical to the lift and drag created (b).

that change the pitch of a blade based on flying conditions. The complexity of fluid flow over the human body has made it difficult to resolve the controversy over which fluid forces are most influential in propulsion. Another example of controversy and potential research is to understand why elite swimmers usually keep their fingers slightly spread. It is unknown if this improves performance from increased surface area for the hand, or that the flow through the fingers acts like a slotted airplane wing in modifying the lift created at lower speeds of fluid flow. Coaches should base their instruction on the kinematics of elite swimmers and allow scholars to sort out whether lift, drag, or a vortex (swirling eddies) modifying the flow of fluid is the primary propulsive mechanism for specific swimming strokes.

Activity

After trying the ball-submersion experiment, try out this little activity using freestyle swimming technique. Compare the number of arm pulls it takes to cross the pool using two extremes in arm pull technique. First, try an arm pull with a primarily paddling motion, straight downward under your shoulder. The next arm pull should be more like the traditional freestyle technique, sculling the hand/arm in a narrow "S" pattern (frontal plane view) down the body. The paddle stroke would use primarily drag for propulsion, while the sculling motion would combine lift and drag for propulsion. Attempt to match the speed/tempo of the pulls and employ a flotation assist (like a pull buoy), and no flutter kick for a true comparison. Which fluid force seems to be most effective in pulling your body through the water with the fewest strokes?

There are two common ways of explaining the cause of lift: Newton's Laws and Bernoulli's Principle. Figure 8.11 shows a side view of the air flow past a discus in flight. The lift force can be understood using Newton's second and third laws. The air molecules striking the undersurface of the discus are accelerated or deflected off its surface. Since the fluid is accelerated in the direction indicated, there must have been a resultant force (F_A) acting in that direction on the fluid. The reaction force (F_R) acting on the discus creates the lift and drag forces on the discus.

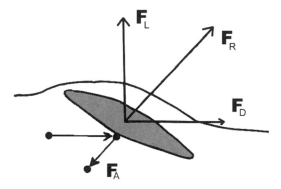

Figure 8.11. The kinetics of the lift and drag forces can be explained by Newton's laws and the interaction of the fluid and the object. The air molecules ( ) deflecting off the bottom of the discus creates the lift (F_L) and drag (F_D) acting on the discus.

The other explanation for lift forces is based on pressure differences in fluids with different velocities discovered by the Swiss mathematician Daniel Bernoulli. **Bernoulli's Principle** states that the pressure in a fluid is inversely proportional to the velocity of the fluid. In other words, the faster the fluid flow, the lower the pressure the fluid will exert. In many textbooks this has been used to explain how lift forces are created on airplane wings. Airplane wings are designed to create lift forces from airflow over the wing (Figure 8.12). Fluid mol-

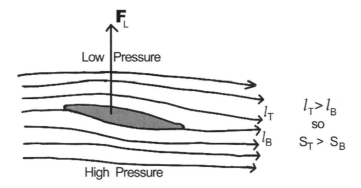

Figure 8.12. The lift force (F_L) acting on a discus or airplane wing can be explained using Bernoulli's Principle. The greater distance (and faster speed of fluid flow) over the top of the wing (l_T) compared to the distance under the bottom (l_B) creates a pressure differential. The high pressure below and lower pressure above the wing lifts the airplane.

ecules passing over the top of the wing cover a greater distance than molecules passing under the wing in the same amount of time and, therefore, have a greater average speed than the airflow under the wing. The lower pressure above the wing relative to below the wing creates a lift force toward the top of the wing. Unfortunately, this simplistic explanation is not technically correct. Rather, it's an oversimplification of a complex phenomenon (visit the NASA Bernoulli vs. Newton webpage, about the competing theories about lift forces in fluids at http://www.grc.nasa.gov/WWW/K-12/airplane/bernnew.html). Bernoulli's equation only accounts for force changes due to fluid pressure (no work, heat, or friction) or a frictionless (inviscid) flow. This is not the case in most fluid dynamics situations (airplane wings or hands in the pool). Unfortunately, Bernoulli's Principle has

also been overgeneralized to lift forces on sport balls.

The Magnus Effect

Lift forces can also be created by the spin imparted to spherical balls. These lift forces arise because of pressure differences and fluid deflection resulting from ball spin. This phenomenon of lift force in spinning balls is called the **Magnus Effect**, after German engineer Gustav Magnus, though it may have been discovered a century earlier (Watts & Bahill, 2000). Sport balls hit or thrown with topspin have trajectories that curve more downward than balls with minimal spin or backspin. This greater downward break comes from the vertical resultant force from gravity and the primarily downward lift force from the Magnus Effect.

Activity: Bernoulli's Principle

An easy way to demonstrate Bernoulli's Principle is to use a small (5 × 10 cm) piece of regular weight paper to simulate an airplane wing. If you hold the sides of the narrow end of the paper and softly blow air over the top of the sagging paper, the decrease in pressure above the paper (higher pressure below) will lift the paper.

Recall that it was noted that Bernoulli's Principle is often overgeneralized to explain the lift force from the Magnus Effect. This oversimplification of a complex phenomenon essentially begins by noting that a rotating sphere affects motion in the boundary layer of air (Figure 8.13) because of the very small irregularities in the surface of the ball and the viscosity of the fluid molecules. Fluid flow past the ball is slowed where the boundary layer rotation opposes the flow, but the free stream fluid flow will be faster when moving in the same direction as the boundary layer. For the tennis ball with topspin illustrated in Figure 8.13, Bernoulli's Principle would say that there is greater pressure above the ball than below it, creating a resultant downward lift force. As direct and appealing as this explanation is, it is incorrect because Bernoulli's Principle does not apply to the

kinds of fluid flow past sport balls since the fluid flow has viscous properties that create a separation of the boundary layer (Figure 8.6). Bernoulli's Principle may only apply to pressure differences away from or outside the boundary layer of a spinning ball.

A better explanation of the lift force is based on how ball spin creates a deflection of the fluid, as evidenced by the shifted separation point of the boundary layer. At the spin rates that occur in sports, the boundary layers cannot stick to the ball all the way around because of an adverse pressure gradient behind the ball. Note how the topspin on the ball in Figure 8.6 creates earlier separation of the boundary layer on the top of the ball, which results in upward deflection of the wake behind the ball (Mehta & Pallis, 2001a). The backward motion of the boundary layer on the bottom of the ball increases the momentum of the boundary

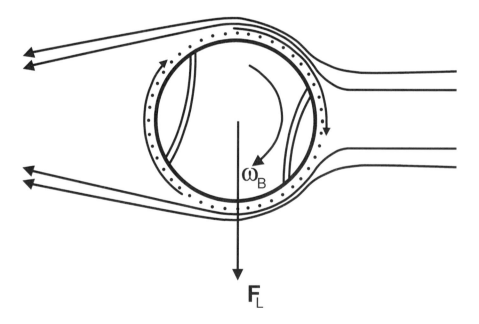

Figure 8.13. The lift forces created on spinning spheres is called the Magnus Effect. An overly simplified application of Bernoulli's Principle is often incorrectly used to explain the cause of this fluid force. Spin on the tennis ball drags the boundary layer of fluid in the direction of the spin. Fluid flow past the ball is slowed where the boundary layer opposes the free stream flow, increasing the fluid pressure. The topspin on this ball (like the ball in figure 8.6) creates a downward lift force that combines with gravity to make a steep downward curve in the trajectory.

layer, allowing it to separate later (downstream), while the boundary layer separates sooner on top of the ball. This asymmetric separation of the boundary layer results in an upward deflection on the wake. An upward force on the fluid means that an equal and opposite downward lift (Newton's third law) is acting on the ball.

The lift force created by the Magnus Effect is apparent in the curved trajectory of many sport balls. Golf balls are hit with backspin to create lift forces in flight that resist gravity and alter the trajectory of shots. Golf balls given sidespin create lift forces that curve a ball's flight more in the horizontal plane. Figure 8.14 renders a schematic of flight for various golf shots created using different sidespins. A tennis player im-

parting sidespin to a ball also creates a lateral lift force that makes the ball curve in flight. The flat trajectory of a fastball pitch in baseball results from an upward component of lift force that decreases the effect of gravity. Lift forces on sport balls are vectorially added to other forces like drag and weight to determine the resultant forces on the ball. The interaction of these forces creates the trajectory or flight path of the ball. The lift forces in a fastball are not larger than the weight of the baseball, so player perceptions of fastballs "rising" is an illusion based on their expectation that the ball will drop more before it crosses the plate. Fastballs don't rise; they just drop less than similar pitches with minimal backspin or topspin.

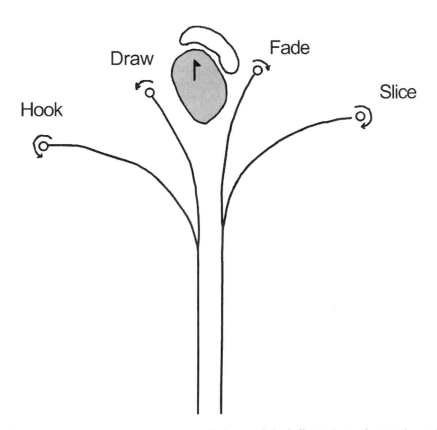

Figure 8.14. Horizontal plane trajectories of various golf shots and the ball spin (curved arrows) creating these curves with Magnus forces.

Much of the skill of baseball pitching relies on a pitcher's ability to vary the speed and spin of pitches. A curveball is pitched with an element of topspin that has a lift component in the same direction as gravity, so there is greater downward break as the ball nears the plate. The steepness of this break has resulted in hitters saying that a good curveball looks like it "drops off the table." Looking at the curveball in baseball (much like topspin shots in other sports, like volleyball and tennis) will provide a nice review of the kinetics of lift forces.

Figure 8.15 shows a three-dimensional reconstruction of a major league curveball from two perspectives. The curveball has a gradual break that looks much steeper from the relatively poor vantage point of the hitter. Why does so much of the ball's break occur late in the trajectory when the hitter is swinging the bat and cannot change their swing? The major factors are the changing direction of the Magnus force in space and the slowing of the ball from drag.

Recall that the Magnus force acts perpendicular to the flow of fluid past the ball. This means that the Magnus force for a curveball primarily acts downward, adding to the drop created by gravity, but the horizontal component of the lift force changes. As the direction of the pitch changes, so does the fluid flow and lift force (Figure 8.16). As the ball breaks downward, the Magnus force has a backward component

that slows the ball even more. This extra slowing and extra downward force contribute to the increasing "break" in the pitch late in its trajectory. Novice golfers can experience the same surprise if they consistently have trouble with "hooking" or "slicing" their drives. A "hooked" drive might initially look quite straight when the moderate sideward force is hard to detect due to the great initial speed of the ball. Unfortunately, as they watch their "nice" drive later in its trajectory, the ball seems to begin curving sideways late in flight. Diagram a transverse plane view of a "hooked" shot in golf. Draw the lift force acting on the ball and note its change in direction as the direction of the ball changes.

The coefficient of lift (C_L) in spinning balls tends to be less sensitive to variations in Reynolds numbers than drag, so the size of the Magnus force depends mostly on spin and ball roughness (Alaways et al., 2001). Athletes can create more break on balls by increasing spin or increasing the surface roughness of the ball. In baseball, pitches can be thrown with four seams perpendicular to the throw, which increases C_L two to three times more than a two-seam rotation (Alaways et al., 2001).

Interesting exceptions to the dominant effect of lift forces on the flight of many sport balls are projections with minimal ball spin. A baseball "knuckleball" and a volleyball "floater" serve are examples of

---

Activity: Lift and Angle of Attack

When driving on an uncrowded road, roll down a window and put your hand out just into the flow of the air rushing past. Drive at a constant speed to standardize air speed and experiment with various hand shapes and angles of attack to the air. Think about the sport balls/objects your hand can simulate and note the drag you (and the simulated object) experience at that relative velocity of air flow. How much does the drag increase as you increase the frontal area of your hand? Can you make the lift force act downward? Find the angle of attack that seems to have the most lift and the least drag (maximum lift/drag ratio). See if your classmates have observed similar results.

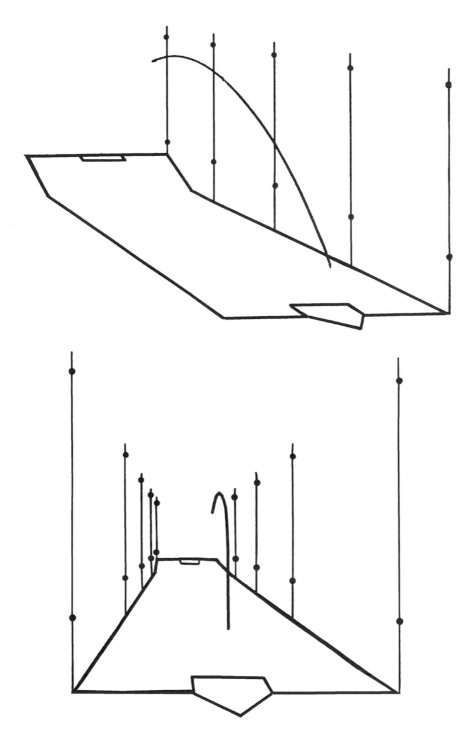

Figure 8.15. Trajectory of the same curveball thrown by a major league player from two perspectives. Note the majority of the downward break occurs late in the trajectory, creating the illusion (to the hitter) of the ball "dropping off the table." Adapted from Allman (1984).

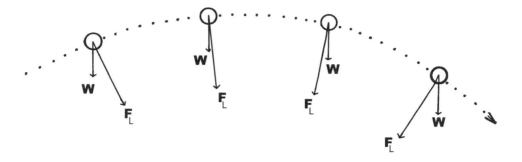

Figure 8.16. The late "break" of a curveball can be explained by the changing direction of the Magnus force (F_L) on the ball and gravity (**W**). The downward deflection of the ball accelerates because the lift forces not only act downward with gravity, but backward (toward the pitcher). Slowing of the ball allows for more downward break.

techniques where the ball is projected with virtually no spin. The erratic trajectory and break of these balls are due to unpredictable variations in air flow past the ball. As the ball gradually rotates, air flow can be diverted by a seam or valve stem, making the ball take several small and unpredictable "breaks" during its trajectory. So spin, and the lack of it, on a sport ball has a major effect on trajectory.

PRINCIPLE OF SPIN

It is clear that fluid forces affect the motion of objects through a fluid. Lift is a key fluid force that can be modified by imparting spin on a projectile. The **Principle of Spin** is related to using the spin on a projectile to obtain an advantageous trajectory or bounce. Kinesiology professionals can use the principle of spin to understand the most successful techniques in many activities. The upward lift force created by backspin in a golf shot increases the distance of a drive (Figure 8.17a), while the backspin on a basketball jump shot is primarily used to keep the ball close to the hoop when impacting the rim or backboard (Figure 8.17b). The bottom of a basketball with backspin is moving faster than the center of the ball be-

cause the ball is rotating. This increases the friction force between the ball and the rim, decreasing the horizontal velocity of the ball, which makes the ball bounce higher. In applying the spin principle, professionals should weigh the trajectory and bounce effects of spin changes.

Applying spin to projectiles by throwing or striking have a key element in common that can be used to teach clients. The body or implement applies force to the ball off-center, creating a torque that produces spin on the projectile. The principles of torque production can be applied to the creation of spin, in that a larger force or a larger moment arm will increase the torque and spin produced. Coaching athletes to project or hit balls with minimal spin to create erratic trajectories requires that the object be in contact with a force in line with the ball's center of gravity. In volleyball, for example, the athlete is taught to strike through the center of the ball with minimal wrist snap. The flat impact through the ball's center of gravity and minimal torque from wrist rotation ensures that the ball will have minimal spin.

Unfortunately, the linear speed of the projectile is inversely proportional to the spin created. In other words, the more spin produced in the throw or hit comes at a cost

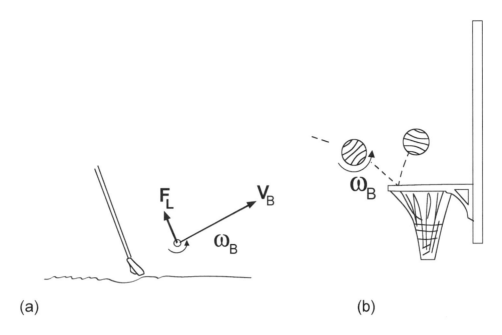

(a)   (b)

Figure 8.17. The principle of spin is used on a golf ball to create lift forces (F_L) that affect ball trajectory, while spin on a basketball is primarily used to modify ball rebound to increase the chance of a made basket.

of lower ball speed. In tennis the lateral break of a ball with slice (sidespin) will not travel as fast as a flat serve (minimal spin) hit with the same effort. Much of the art of teaching and coaching is being able to evaluate a person's performance, diagnosing the factors related to spin and speed production that are appropriate for a specific situation.

There is one more advantage of imparting spin to a projectile that is not related to fluid or contact forces on a surface. This third advantage of projectile spin is related to Newton's laws and conservation of angular momentum. Any object in angular motion without external-acting torques (like a projectile) will conserve angular momentum. This inertia in a rotating object can be used to keep the projectile in a certain orientation. A pass in American football does not create significant lift force, but the spin stabilizes the flight of the ball in a streamlined position. Divers and gymnasts

(human body projectiles) can overcome this inertia and move body parts relative to an axis of rotation with internal muscle forces. In these situations athlete can transfer angular momentum from one axis to another (e.g., add a twist in the middle of somersaults) by asymmetric motions of body parts. Coaches of these sports need to be familiar with this interesting application of the spin principle (see Yeadon, 1991, 1997).

Knowing what advantage of spin in a particular situation is important and how to mechanically create it are important, but this knowledge must be integrated with knowledge from other kinesiology disciplines. A physical educator could ask a junior high student to "use an eccentric force" or "increase the effective moment arm" and may be mechanically correct, but a good teacher selects an appropriate cue that communicates the essential correction without using such technical language. Think about what would be good cues for hitting a sport ball to create topspin, backspin, right or left sidespin. Deciding whether cues about the ball (target) or body action (technique) are most relevant depends on the situation. This is another example of how a biomechanical principle must be integrated in an interdisciplinary fashion with other kinesiology disciplines (e.g., motor development, motor learning, psychology).

SUMMARY

Fluid forces from air and water have a significant effect on human movement. The main fluid forces are buoyancy, lift, and drag. Buoyancy is the supporting or floating force that a fluid exerts on an object as it is submerged in the fluid. The size of the buoyant force can be determined by Archimedes Principle. The fluid force that acts in the same direction as the relative flow of fluid is drag, while the fluid force acting at right angles to the flow is lift. The size of the lift and drag forces depends on many factors, but they vary with the square of the relative velocity of the fluid. Lift forces can be created on spinning spherical balls through the Magnus Effect. Kinesiology professionals can apply the Spin Principle to help performers create spin on projectiles like sport balls. Imparting more spin to a projectile usually comes at the expense of a loss in some linear velocity, but the lift force can be used to gain an advantage from an altered flight or bounce relative to a no-spin projection.

KEY TERMS

Archimedes Principle
Bernoulli's principle
boundary layer
buoyancy
center of buoyancy
drag
lift
Magnus Effect
spin principle

REVIEW QUESTIONS

1. What are the major fluid forces and in what directions do they act?
2. What factors affect the fluid resistance acting on projectiles? What factor is most influential in creating fluid forces?
3. Compare and contrast the motion of the center of gravity and center of buoyancy with various body segment movements.
4. Explain why streamlining decreases fluid resistance.
5. How do fluid forces affect the optimal projection angles proposed earlier in chapter 5?
6. Why do golf balls have dimples?
7. Draw or trace a person and estimate their center of buoyancy. Trace the person two more times with various exercise and swimming flotation devices and re-estimate the likely center of buoyancy.
8. How is the density of water related to whether an object will float in water?
9. Explain why topspin serves in volleyball curve downward?
10. What are the benefits of imparting spin to round balls in sports?
11. How are the spin and the "break" of a ball in flight related?
12. Draw a free-body diagram of a golf ball in flight and explain how the resultant forces on the ball affect its flight.
13. Why do swimmers and cyclists shave their body, but a baseball pitchers illegally roughen the surface of the ball?
14. What forces increase and decrease when exercising in water?

SUGGESTED READING

Adair, R. (1990). *The physics of baseball*. New York: Harper & Row.

Alaways, L. W., Mish, S. P., & Hubbard, M. (2001). Identification of release conditions and aerodynamic forces in pitched-baseball trajectories. *Journal of Applied Biomechanics*, **17**, 63–76.

Arellano, R. (1999). Vortices and propulsion. In R. Sanders & J. Linsten (Eds.), *SWIMMING: Applied proceedings of the xvii international symposium on biomechanics in sports* (Vol. 1, p. 53–66). Perth, WA: Edith Cowan University.

Berger, M. A. M., de Groot, G., & Hollander, A. P. (1995). Hydrodynamic drag and lift force on human hand/arm models. *Journal of Biomechanics*, **28**, 125–133.

Counsilman, J. E. (1971). The application of Bernoulli's Principle to human propulsion in water. In L. Lewillie and J. Clarys (Eds.), *First international symposium on biomechanics of swimming* (pp.59–71). Brussels: Université Libre de Bruxelles.

McLean, S. P., & Hinrichs, R. N. (2000a). Influence of arm position and lung volume on the center of buoyancy of competitive swimmers. *Research Quarterly for Exercise and Sport*, **71**, 182–189.

McLean, S. P., & Hinrichs, R. N. (2000b). Buoyancy, gender, and swimming performance. *Journal of Applied Biomechanics*, **16**, 248–263.

Mehta, R. D., & Pallis, J. M. (2001b). Sports ball aerodynamics: Effects of velocity, spin and surface roughness. In F. H. Froes, & S. J. Haake (Eds.), *Materials and science in sports* (pp. 185–197). Warrendale, PA: The Minerals, Metals and Materials Society [TMS].

Mureika, J. R. (2000). The legality of wind and altitude assisted performances in the sprints. *New Studies in Athletics*, **15**(3/4), 53–58.

Olds, T. (2001). Modelling of human locomotion: Applications to cycling. *Sports Medicine*, **31**, 497–509.

Toussaint, H. M., van den Berg, C., & Beek, W. J. (2002). "Pumped-up propulsion" during front crawl swimming. *Medicine and Science in Sports and Exercise*, **34**, 314–319.

Watts, R. G., & Bahill, A. T. (2000). *Keep your eye on the ball: The science and folklore of baseball* (2nd ed.). New York: W.H. Freeman.

Yeadon, M. R. (1997). The biomechanics of human flight. *American Journal of Sports Medicine*, **25**, 575–580.

WEB LINKS

Aerodynamics in Tennis—NASA/Cislunar Aerospace project to promote science education through sport science.
http://wings.avkids.com/Tennis/Book/index.html

Cycling Aerodynamics—cycling page by Smits and Royce of Princeton University.
http://www.princeton.edu/~asmits/Bicycle_web/bicycle_aero.html

Cycling Aerodynamics and power calculation page.
http://www.exploratorium.edu/cycling/aerodynamics1.html

Floating, density, and air foil simulations—part of Mechanics simulations at explorescience.com.
http://www.explorescience.com/activities/activity_list.cfm?categoryID=10

Bernoulli vs. Newton—NASA webpage on the competing theories for lift forces in fluids.
http://www.grc.nasa.gov/WWW/K-12/airplane/bernnew.html

ISBS Coaching Information Service—select swimming link.
http://www.sportscoach-sci.com/

PART **IV**

APPLICATIONS OF BIOMECHANICS IN QUALITATIVE ANALYSIS

The personal trainer depicted here is using the principles of biomechanics to qualitatively analyze the exercise technique of his client. Biomechanical principles must be integrated with other kinesiology sciences in the qualitative analysis of human movement. The chapters in part IV provide guided examples of applying biomechanics in qualitative analysis for several kinesiology professions: physical education, coaching, strength and conditioning, and sports medicine. A variety of guided examples and questions for discussion are presented. The lab activities related to part IV provide students with opportunities to integrate biomechanical principles with other subdisciplines of kinesiology in the qualitative analysis of human movement. A sample table with the principles of biomechanics for qualitative analysis can be found in Appendix E.

CHAPTER 9

Applying Biomechanics in Physical Education

Physical educators teach a wide variety of human movements, and biomechanics provides a rationale critical for evaluating technique and prescribing intervention to help young people improve. Biomechanics also allows physical educators to identify exercises and physical activities that contribute to the physical development of various muscle groups and fitness components. This chapter illustrates how biomechanical knowledge and the nine principles of biomechanics can be integrated with other sport sciences in qualitative analysis of human movement. Five skills commonly taught in physical education are discussed, and the various tasks of qualitative analysis (Knudson & Morrison, 2002) are emphasized in the examples. Real movement performances and typical teaching cues are used to show how biomechanics is applied to real-world physical education. Qualitative analysis is a critical evaluative and diagnostic skill that can be employed for improvement of movement in physical education.

QUALITATIVE ANALYSIS OF KICKING TECHNIQUE

The primary task of a professional physical educator may be the qualitative analysis of movement technique to facilitate learning of motor skills. Biomechanics is the primary sport science focusing on movement technique, so it is logical that physical educators should use the principles of biome-

chanics in helping students move safely and effectively. Biomechanics provides knowledge relevant to all four tasks of qualitative analysis (Figure 2.9).

Imagine that you are an elementary physical educator planning a lesson on kicking as a lead-up to soccer, so you are involved in the preparatory task of qualitative analysis. In preparing to teach and qualitatively analyze kicking, you list the critical features and teaching points of the movement (Table 9.1). As students practice this skill, you are planning to evaluate these critical features and diagnose student performance using biomechanical principles. Which biomechanical principles seem most relevant to the critical features of high-speed place-kicking?

Five of the critical features presented in Table 9.1 are strongly related to several of the biomechanical principles. The opposition and coordination involved in high-

Table 9.1
CRITICAL FEATURES AND TEACHING CUES FOR FAST PLACE KICKING

Critical feature	Possible teaching /intervention cues
Visual focus	Head down and focus on the ball
Opposition	Turn your hip to the ball
Foot plant	Plant your foot next to the ball
Coordination	Swing your hip and leg
Impact position	Kick the center of the ball
Follow-through	Follow-through toward the target

speed kicking are all strongly influenced by the principles of range of motion, coordination, and segmental interaction. In addition, the force–motion, force–time, and optimal projection principles are important in kicking as well. The teacher might plan to keep the principles of inertia, spin, and balance in the back of their mind, so they will not be a focus of observation. These three principles are not likely to play a significant role in the kicking executed by most primary school children.

A child making a full-effort kick toward a goal is observed to consistently have a technique like that illustrated in Figure 9.1. Remember that good qualitative analysis requires the analyst to observe several performances so that clear trends of strengths and weaknesses can be identified, rather than jumping to conclusions or identifying unimportant "errors" (Knudson & Morrison, 2002). What critical features are strongly and weakly performed? These

judgments are part of the evaluation process within the evaluation/diagnosis task of qualitative analysis.

The child illustrated in Figure 9.1 is clearly at a low developmental level of kicking. The teacher could praise the student's focus on the ball, strong approach, and balance during the kick. The list of biomechanical weaknesses is long at this beginning stage of learning. The biomechanical principles that are weakly incorporated into the kick are force–motion, optimal projection, inertia, range of motion, coordination, and segmental interaction. The student applies a suboptimal force to the ball because they plant the support foot well behind the ball, and impact the ball with their toe rather than the proximal instep (top of the shoelaces). Low-trajectory shots are desirable, but this kick, rolling along the ground, will slow the ball down as it rolls, making it easier for opponents to intercept. Finally, the student needs considerable

Figure 9.1. The technique of a young person making a high-speed soccer kick. The time between images is 0.08 s.

practice to increase the range of motion of the kick and to refine a well-timed sequential coordination that transfers energy through segmental interactions. Highly skilled kickers will approach the ball at an angle to increase the contralateral hip range of motion that can be sequentially combined with the hip and knee motions of the kicking leg. Which of these weaknesses do you think is most important to kicking success? One effective intervention strategy would be to provide a cue to plant their foot next to the ball. This is a simple correction that might be related to other weaknesses and might motivate the student with initial success and improvement.

Toward the end of the lesson you notice another child consistently kicking as in the sequence illustrated in Figure 9.2. What biomechanical principles are strongly or weakly performed in Figure 9.2?

The student depicted in Figure 9.2 is more skilled than the student from the pre-

vious example. Note the more vigorous approach to the ball. The intensity (inertia) of this approach is apparent in the length of the hurdle to the plant leg and the trunk lean used to maintain balance. It is hard to judge from the figure, but the ball is kicked at the desirable low trajectory. Some educators might conclude that all the biomechanical principles were well applied in this kick. The only two principles that might be slightly improved are range of motion and coordination. If the student were to approach the ball from a more oblique angle, the rotation of the pelvis on the left hip could be increased (range of motion) and combined (sequential coordination) with the good coordination of the kicking hip and knee.

Which of these small improvements, range of motion or coordination, do you think could be easily changed by this student in practice? Improvement in what principle would increase performance the

Figure 9.2. The technique of another soccer player kicking for maximum speed. Time between images is 0.08 s.

most? These are the issues that are important for a physical educator to examine in the diagnosis and intervention stages of qualitative analysis. The teacher might review some recent research and review papers on kicking (Barfield, 1998; Davids, Lees, & Burwitz, 2000; Dorge, Bull Andersen, Sorensen, & Simonsen, 2002). The following examples of qualitative analysis will illustrate the use of the biomechanical principles in these more difficult phases of qualitative analysis.

QUALITATIVE ANALYSIS OF BATTING

Imagine you are a physical educator working on batting with young boys and girls. Most primary school children receive some experience intercepting and striking objects from elementary physical education. The difficulty of the skill dramatically increases when these young people move from batting slow-moving or stationary (batting tee) objects, to balls thrown with greater speed and spin. Use the technique points and cues in Table 9.2 to analyze the batting technique of the student illustrated in Figure 9.3. Assume the technique illustrated is representative of most batting attempts off a batting tee by this child. What biomechanical

Table 9.2
CRITICAL FEATURES AND TEACHING CUES FOR BATTING

Critical feature	Possible teaching/ intervention cues
Visual focus	Head down and focus on the ball
Opposition	Sideward stance
Readiness	Bat up and elbow back
Weight shift	Short stride toward the pitch
Coordination	Throw your hands through the ball
Follow-through	Follow-through around your body

principles seem to be well applied by this child, and what principles are poorly applied? More importantly, prioritize the weaknesses in an order that you think would result in the best batting performance if the weaknesses were improved.

Most all of the biomechanical principles are relevant to batting performance. The student in Figure 9.3 strongly incorporates many biomechanical principles into batting. His strengths include balance, inertia, and coordination. He strides into the swing and gets the bat in line with the ball. The force–motion principle could be improved since the bat does not squarely collide with the ball (note the tipping batting tee). The principles of force–time and range of motion may be the major weaknesses that could be improved. The student exaggerates the stride and uses an abbreviated follow-through. The physical educator must diagnose the situation and decide if instruction should be focused on the larger than normal range of motion and time in the stride or on the less than expected time/range of motion in the follow-through. Weighing the importance of these principles so as to lead to potential improvement is very difficult. Remember, we noted that this student would soon be applying this skill in the more dynamic condition of impacting a moving ball.

Since the student has good balance, their long stride (which increases range of motion and time of force application) could generate more speed without adversely affecting accuracy. This is typical for a young person with limited upper body strength trying to clobber a ball off a batting tee. Maintaining a long (time and distance) stride in hitting pitched balls, however, is generally a bad tradeoff. Accuracy in contacting the ball becomes more important in dynamic hitting conditions.

It may even be possible to maintain a similar bat speed with a shorter stride if the student improves his follow-through. An

Figure 9.3. A physical education student batting a ball from a tee. Time between images is 0.1 seconds.

abbreviated follow-through means that the hitter is slowing down the bat before impact. Skilled striking involves generating peak velocity at impact, delaying negative accelerations until that point (Knudson & Bahamonde, 2001). The force–time and range-of-motion principles also imply that a short follow-through may increase the risk of injury since the peak forces slowing the movement must be larger. Since the follow-through is an important strategy for minimizing the risk of injury in many movements, the physical educator should rate this intervention ahead of adjusting the preparatory (stride) range of motion. Once the student gets comfortable swinging through the ball, they may have more bat speed at impact, and might be more willing to reduce their stride and weight shift when hitting pitched balls. More recent research on baseball batting has focused on differ-

ences in various bats (Greenwald, Penna, & Crisco, 2001) and hitting from both sides of the plate (McLean and Reeder, 2000). The next section provides an example of diagnosis using biomechanical principles in basketball shooting.

QUALITATIVE ANALYSIS OF THE BASKETBALL FREE THROW

The previous qualitative analysis examples involved movements that must be matched to unpredictable environmental conditions. Motor learning classifies these movements as *open skills*, while skills with very stable conditions are called *closed skills*. When physical educators teach and analyze closed motor skills, they can be confident that performance is more strongly dependent on stereotypical technique rather than a variety of effective techniques. The standardized conditions of the free throw in basketball mean that the stereotypical techniques of a set shot would be optimal. Table 9.3 lists the key technique points and intervention cues that describe good free throw shooting technique.

Suppose an elementary school student is working on her free throw using modified equipment. Using a smaller ball and lower basket is critical to teaching good shooting technique with young children. At this age, they typically cannot employ good shooting technique using a regular ball and a 10-foot-high basket because of their lack of strength. Suppose that observations of the free throw attempts of a young child shows technique consistent with that illustrated in Figure 9.4. Identify the biomechanical principles that are strengths and weaknesses. Then diagnose the situation to determine what biomechanical principle should be the focus of any intervention.

The principles she can be complimented on are her good balance, simultaneous coordination, and spin on the ball. It is difficult to see in Figure 9.4, but this child used only one hand and one leg to shoot because she stepped into the shot. Weaknesses in her shooting technique are the limited use of range of motion and the force–time principles since she is not easily generating the ball speed needed for the shot. Another weakness is in the principle of optimal trajectory. Biomechanical research on shooting has shown that the optimal angles of projection for most set and jump shots are between 49 and 55° above the horizontal (Knudson, 1993). Young basketball players often select "flat" shooting trajectories, which actually require greater ball speed and often have angles of entry that do not allow the ball to pass cleanly through the hoop. This weighing of potential benefits of working on range of motion or initial shot trajectory is the essential diagnostic deci-

Table 9.3
CRITICAL FEATURES AND TEACHING CUES FOR THE FREE THROW

Critical feature	Possible teaching/ intervention cues
Staggered stance	Shooting side foot forward
Shooting plane	Align your arm with the basket
Height of release	Release high above your head
Coordination	Extend your whole body
Angle of release	Shoot with high arc
Ball rotation	Flip your wrist

sion in this case. There are several biomechanical reasons why it is likely more beneficial to work on shot trajectory than increasing range of motion. First, using the desirable trajectory increases the angle of entry and the probability of a made shot. Second, this slightly higher trajectory requires less ball speed than a very flat one. Third, the young player is likely to increase her strength while the desirable trajectory will remain the same. The interaction of biomechanics and performer characteristics suggests to the teacher that subsequent practice should focus on a slightly higher shot trajectory.

EXERCISE/ACTIVITY PRESCRIPTION

Another important content area of physical education is fitness. Physical educators planning to increase student physical fitness must employ biomechanical knowledge to determine the most effective exercises for various parts of the body and fitness components. Like strength and conditioning professionals, physical educators qualitatively analyze exercise technique to be sure that students are safely training their bodies.

Figure 9.4. An elementary student shooting a free throw at an 8-foot-high hoop. Time between images is 0.07 s.

During the first week of your high school weight-training unit, you notice many students performing their curl-up exercises like the student depicted in Figure 9.5. You want to immediately provide some group feedback to help many students with this exercise technique and reinforce some of the technique points you made earlier. Make a list of the critical features or technique points that are important in the curl-up exercise. What biomechanical principles are most related to the objectives of doing curl-ups for health-related fitness (muscular endurance)? Which of the biomechanical principle(s) seem to be weakly applied in the concentric phase of the curl-up for the student shown in Figure 9.5?

The purpose of curl-up exercises is to focus a conditioning stimulus on the abdominal muscles by limiting the contribution of hip flexors and other muscles. The

biomechanical principles that are important in this objective are Force–Motion, Range of Motion, Inertia, and Force–Time. The inertia of the body provides the resistance for the exercise, and the range of motion for the exercise should focus the stress (force–motion) on the abdominal muscles. The repetitions should be slow and controlled (Force–Time) for safety and to promote training for muscular endurance.

The student in Figure 9.5 has several weaknesses in his curl-up technique. He uses too much range of motion, performing more of a sit-up (hip flexion) than a trunk curl. In a curl-up exercise, the abdominal muscles should raise the shoulders to about a 30 to 40° angle with the hip (Knudson, 1996), just lifting the shoulder blades off the ground. Hip flexion is required if the shoulders are to be raised further. The student also decreases the resistance or inertia by

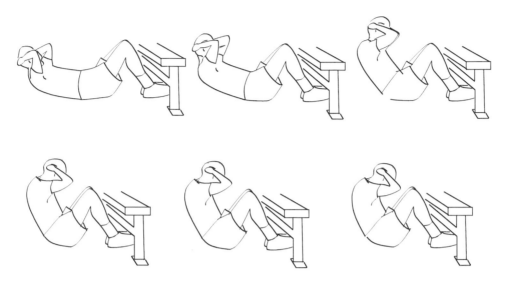

Figure 9.5. Concentric phase technique of a curl-up for a high school student. Time between images is 0.17 s.

keeping the weight of the arms close to the transverse axis of rotation for trunk flexion. The third weakness is in stabilizing his feet with the weight bench. This affects both the Force–Motion Principle and the Principle of Inertia. By stabilizing the feet with the bench, the performer has essentially unlimited inertia for the lower extremities. This allows hip flexor activation to contribute to trunk flexion through the kinematic chain of the lower extremity, so the Force–Motion Principle is not applied well for the training objective of isolating the abdominal muscles. Performing the curl-up without foot stabilization would require greater abdominal activation and stabilization to lift the trunk without hip flexors. The time information in the caption for Figure 9.5 suggests that the student was applying the Force–Time Principle well; in other words, he did not perform the exercise too fast.

The best intervention in this situation is to provide group intervention, reminding all students to perform curl-ups without lower-extremity stabilization. This exercise may feel more difficult, but the teacher can use this opportunity to reinforce the idea that the students are *training* and *teaching* their abdominal muscles an important trunk-stabilizing task. Focusing on using more abdominal muscles for a longer time (Force–Time Principle) better simulates the nearly isometric actions of the muscles in stabilizing the trunk and pelvis. There is a large body of physical therapy literature focused on training specific abdominal muscles so as to stabilize the trunk (McGill, 1998; Vezina & Hubley-Kozey, 2000). The teacher could then provide some individualized intervention for the student. One good strategy would be to compliment (reinforce) the good exercise cadence, but challenge the student to place his hands on top of his head and keep the arms back to increase the resistance for the exercise.

QUALITATIVE ANALYSIS OF CATCHING

Imagine that you are a junior high school physical educator teaching a basketball unit. You have been ingenious in getting the students to realize the rewards of moving without the ball and passing rather than dribbling. There is one small problem that many of the students have poor catching skills. You previously taught students the critical features of catching (Table 9.4) using a variety of cues. In watching a passing drill, you notice a student receiving passes similar to what is illustrated in Figure 9.6. What biomechanical principles are well or poorly incorporated in catching the basketball? Diagnose the situation and prioritize the importance of the biomechanical principles in successful catching for this player and think about what the best intervention would be.

The player has good balance and uses simultaneous coordination in receiving the ball. The Force–Motion Principle was well applied by predicting the location of the ball, intercepting the ball with the hands, and applying the force through the center of gravity of the ball. The two principles

that could be improved are Range of Motion and Force–Time. Since you are a good physical educator, you also note the non-biomechanical factors relevant in this situation: the player appears to visually focus on the ball, is motivated, and is trying her best.

Table 9.4 CRITICAL FEATURES AND TEACHING CUES FOR TWO-HANDED CATCHING	
Critical feature	Possible teaching/ intervention cues
Readiness	Athletic stance
Visual focus	Watch the ball
Intercept	Move and reach towards the ball
Hand position	Thumbs in or thumbs out
Absorption	Give with your hands and arms

Figure 9.6. A junior high school basketball player catching a pass. Time between images is 0.1 seconds.

Diagnosis of this situation is not as difficult as many qualitative analyses because the two weaknesses demonstrated in this example are closely related. Increasing the range of motion in receiving the ball will generally increase the time of force application. You must decide if the player's catching and basketball performance would improve most if her attention were focused on reaching more to intercept the ball or emphasizing how the arms bring the ball in. Both biomechanical principles are important. Can you really say one is more important than the other? The player would clearly improve if she stepped and reached more to intercept the ball earlier and provide more body range of motion to slow the ball down. Increasing range of motion also has a secondary benefit by reducing the risk of a pass being intercepted. How the hand forces opposing ball motion, however, has the most influence on whether a ball is caught or bounces out of a player's grasp. This is a case where some professionals might disagree on the most appropriate intervention. In class, you only have a few seconds and you provide a cue to a student to focus on "giving" with her hands and arms as she receives the ball. You say, "See if you can give with your hands and arms as you catch the ball. Bring that ball in so you barely hear a sound."

SUMMARY

The principles of biomechanics provide a method for physical educators to qualitatively analyze human movement. Several sport and exercise situations commonly faced by physical educators were discussed. The physical educators in the examples employed cue words or phrases to communicate the essence of the biomechanical principles to their students. Physical educators should also integrate the biomechanical principles with their experience, as well as knowledge from other subdisciplines of kinesiology to provide an interdisciplinary approach to qualitative analysis (Knudson & Morrison, 2002).

DISCUSSION QUESTIONS

1. What biomechanical principles are more important in kicking versus trapping a soccer ball?

2. What are the typical teaching points or cues for baseball/softball batting? What biomechanical principles are relevant in these teaching points?

3. How is the application of biomechanical principles different in the free throw versus the jump shot?

4. Which biomechanical principles are relevant to the pushup exercise? How does changing hand position from a wide base of support to a narrow base of support modify the importance of these principles?

5. What biomechanical principles are most relevant to catching a softball? Catching a medicine ball?

6. What are typical teaching points in jumping to rebound a basketball? What points are most important based on the principles of biomechanics?

7. What biomechanical principles are important in throwing a pass in American football?

SUGGESTED READING

Adrian, M. J., & Cooper, J. M. (1995). *Biomechanics of human movement* (2nd ed.). Madison, WI: Brown & Benchmark.

Hay, J. G. (1993). *The biomechanics of sports techniques* (4th. ed.). Englewood Cliffs, NJ: Prentice-Hall.

Knudson, D. (1991). The tennis topspin forehand drive: Technique changes and critical elements. *Strategies*, **5**(1), 19–22.

Knudson, D. (1993). Biomechanics of the basketball jump shot: Six key teaching points. *JOPERD*, **64**(2), 67–73.

Knudson, D., & Morrison, C. (1996). An integrated qualitative analysis of overarm throwing. *JOPERD*, **67**(6), 31–36.

Knudson, D., & Morrison, C. (2002). *Qualitative analysis of human movement* (2nd ed.). Champaign, IL: Human Kinetics.

WEB LINKS

AAHPERD—American Alliance for Health, Physical Education, Recreation, and Dance is the first professional HPERD organization in the United States.
http://www.aahperd.org/

PE Links 4U—website for sharing physical education teaching ideas.
http://www.pelinks4u.org/

PE Central—website for sharing physical education teaching ideas.
http://pe.central.vt.edu/

CHAPTER 10

APPLYING BIOMECHANICS IN COACHING

Coaching athletics also involves teaching motor skills to a wide variety of performers. Traditionally, careers in coaching have focused on working with the physically gifted in interscholastic athletics; however, there are many other levels of coaching: from parents who volunteer to coach their child's team, to the coach of a national team, and to a coach for an individual professional athlete. All these coaching positions benefit from application of biomechanics in coaching decisions. Coaches use biomechanics to analyze technique, determine appropriate conditioning, and treat injuries. Biomechanical knowledge is also important to coaches when coordinating efforts with sports medicine professionals.

QUALITATIVE ANALYSIS OF THROWING TECHNIQUE

Imagine you are a youth softball coach scouting the throwing ability of potential players. You set the players up in the outfield to see how well they can throw the ball to home plate. The technique points for overarm throwing and the cues one would commonly use are listed in Table 10.1. One young person trying out for the team shows a throwing technique like that depicted in Figure 10.1. What are the strengths or weaknesses of their performance in terms of biomechanical principles? Are these weaknesses you are confident can be overcome this season if they become part of your team?

The athlete in Figure 10.1 has a very immature throwing pattern, so he has weaknesses in several biomechanical principles. In fact, the straight arm sling this player uses likely places great stress on the throwing shoulder. The principle most in need of improvement is Range of Motion could improve with a more vigorous approach and a longer stride with the opposite leg. The Inertia of the throwing arm should be reduced in the propulsion phase by flexing the elbow to about 90°. The thrower does rotate their trunk away from and then into the throw, but Sequential Coordination that maximizes Segmental Interaction will require considerable practice. Like many young players, this person throws with a high initial trajectory, violating the Optimal Projection principle. The optimal throwing

Table 10.1
TECHNIQUE POINTS AND CUES
FOR OVERARM THROWING

Technique points	Possible teaching /intervention cues
Approach/stride	Step with the opposite foot toward the target
Opposition & coordination	Turn your side to the target
Arm position	Align your arm with your shoulders
Shoulder internal rotation	Range of motion
Angle of release	Throw the ball low and flat
Relaxation	Be loose and relaxed

Figure 10.1. A softball player throwing with maximum effort to home plate from the outfield.

angles for maximum distance with base-balls and softballs are about 30° (Dowell, 1978).

Some of these weaknesses can be corrected quickly, but some will likely take more than a full season. The athlete should be able to improve his approach, arm action, and angle of projection. Fine-tuning coordination of his throw will likely take longer than a few months. The biomechanics of coordination in overarm throwing is quite complex (Atwater, 1979; Feltner & Dapena, 1986; Fleisig *et al.*, 1999). Consistent practice over a long period of time will gradually build the sequential rotation that optimizes segmental interactions to create a skilled overarm throw. To see if he listens and can easily change aspects of his throwing technique, ask him to step vigorously with his opposite foot and to throw the ball "lower." It is possible that a youth softball coach might select this player for his team

based on other factors. Biomechanical technique in one skill may not be as important as motivational factors or the philosophy employed to help all players develop.

QUALITATIVE ANALYSIS OF DRIBBLING TECHNIQUE

Put yourself in the role of a youth soccer coach. After working on several dribbling drills, you begin a more game-like drill where one player consistently performs as in the illustration in Figure 10.2. Use the technique points and biomechanical principles in Table 10.2 to help guide your observation and qualitative analysis of Figure 10.2. What biomechanical principles are strengths or weaknesses in this performance? Diagnose the performance and decide what would be a good intervention to help this player improve.

Figure 10.2. A soccer player dribbling during a scrimmage.

Table 10.2
TECHNIQUE POINTS AND CUES FOR SOCCER DRIBBLING

Technique points	Possible teaching /intervention cues
Close to body	Keep the ball close to you
Kinesthetic aware-ness/control	Feel the ball on your foot
Awareness of situation	Head up and watch the field
Arch of foot	Push the ball with the arch of the foot
Angle of release	Keep the ball close to the ground

This young player shows good balance in this performance since he does not fall when stumbling over the ball. He has poor control of the ball, which likely contributed to him stepping on the ball. Despite a small stumble, he uses his trail leg to recover the ball. The player needs to adjust their application of the Force–Motion and Range-of-Motion principles to improve their dribbling. Providing a cue that improves one of these principles will likely also improve the angle of release or the Optimal Projection of the ball. Let's diagnose this situation by prioritizing these three weaknesses to provide the best intervention to help this player.

Since this is a young player, you plan to praise his effort and a strong point before focusing attention technique adjustments. Good intervention would be to praise his attention to the ball and recovery from the stumble. It is too early in this player's development to focus intervention on keeping his visual attention on the field. The best intervention may be a cue to "push the ball softly and keep it close to your body." This cue combines the Force–Motion Principle and the Range-of-Motion principles and focuses the player's attention on correct technique. More specific cues on effort or range of motion can follow if future observations of his dribbling yield similar results. Note that a young player is not cognitively ready for complex technique or strategic instruction. The biomechanical complexity of dribbling a soccer ball in the dynamic environment of a game must be appreciated by the coach, but not imposed on a young player too soon.

Table 10.3
TECHNIQUE POINTS AND CUES
FOR BASKETBALL PASSING

Technique points	Possible teaching /intervention cues
Stride	Step toward the target
Speed	Pass quickly
Arm action	Extend arms and thumbs down
Angle of release	Horizontal trajectory

QUALITATIVE ANALYSIS OF CONDITIONING

Junior high and high school coaches often are primarily responsible for developing conditioning programs for their athletes. Coaches must carefully monitor the exercise technique of their athletes to maximize conditioning effects and reduce risk of injury. Suppose you are a junior high basketball coach who has his players perform passing drills with a small medicine ball. The technique points and biomechanical principles you are interested in are listed in Table 10.3. One of your players shows the technique depicted in Figure 10.3. What biomechanical principles are strengths or

weaknesses of their performance, and diagnose the situation to set up intervention.

The weaknesses in this player's exercise technique are related to stride, arm action, and angle of release. The relevant biomechanical principles for these technique points are Inertia, Range of Motion, Coordination, and Optimal Projection. While a variety of passing techniques are used in basketball, the one-handed flip with little weight shift that this player used is not the most desirable technique for high-speed passing. It is hard to judge from the timing information in the figure caption, so we will assume that the athlete used good effort and speed in executing the pass. Motivation clearly affects performance, so the

Figure 10.3. A junior high school basketball player throwing a medicine ball. Time between images is 0.12 s.

weaknesses in some athlete's exercise technique are more related to effort than to neuromuscular errors. The pass will likely have poor speed to the target since only the right arm contributes to the horizontal speed of the pass.

The coach must next diagnose these weaknesses and decide on the best intervention to help this player improve. A good coach would likely focus the player's attention on the correct arm action using both arms (Coordination). The primary reason for this diagnosis is safety, because the use of one arm and trunk twist to propel a heavy object may not be safe loads for poorly trained adolescents. There is also less research on upper body plyometrics than there has been on lower body plyometric exercises (Newton *et al.*, 1997), so what loads and movements are safe is not clear. Cues given for this technique point may also correct the angle of release, increase the speed of the pass, and enhance control of the ball. You decide to work on the stride later for safety reasons. Focusing intervention on the stride does not increase ball speed or decrease the distance (and therefore time) of the pass as much as good coordination with both arms would.

RECRUITMENT

As the golf coach for a university, you have many parents sending you videotapes of their children for potential scholarship consideration. These "daddy" videos can be a nuisance, but you qualitatively analyze the swings of the golfers on them for potential players you might have missed. This information combined with the player's performance in high school and tournaments will help you decide what athletes should be offered scholarships. The technique points and biomechanical principles of the full golf swing you use to analyze swings are presented in Table 10.4. For the player

illustrated in Figure 10.4, evaluate the strengths and weaknesses of their downswing. We will now focus on how the relevant biomechanical principles would help you diagnose the weaknesses of this player and her potential as a golfer on your team.

Table 10.4 TECHNIQUE POINTS AND CUES FOR THE GOLF SWING	
Technique points	Possible teaching /intervention cues
Weight shift	Push with rear, then the front foot
Swing plane	Swing forward and back on same plane
Backswing	Slow and club not past horizontal
Tempo/coordination	Delayed release of the club
Impact/shot trajectory	Divot in front of ball
Follow-through	Long slow finish

This player has an excellent full swing and control of the club. It is difficult to tell from this perspective, but it is likely this player keeps the club in a stable swing plane. The swing has an appropriate range of motion since the backswing terminates with the club virtually horizontal. The player has a good weight shift, hip and trunk twist, and a firm forward leg late in the swing. The follow-through is fine. The two technique points that are difficult to judge from the video (and from the figure) are the Coordination of the swing and the quality of the impact and shot trajectory (Optimal Projection). In short, this particular player has several strengths that suggest she has an excellent golf swing. A good golf coach would be aware of the massive amount of research on the golf swing (Neal & Wilson, 1985; Sprigings & Neal, 2000; Williams & Sih, 2002). There are no obvious

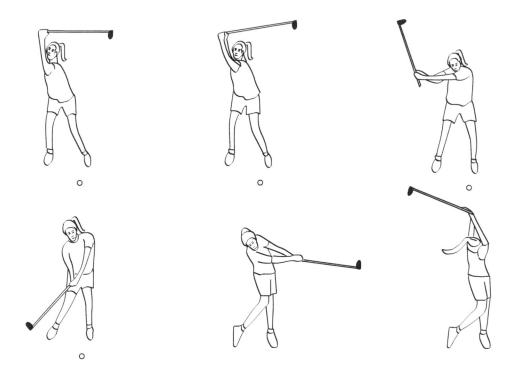

Figure 10.4. The long iron swing of a prospective golf recruit.

warning signs, but a complete diagnosis of this golf swing is difficult to obtain from a single video.

It is possible that the tape was edited to show only the best swings for many shots. To fully diagnose this golf swing, you clearly need to know about impact and shot trajectory relative to the intended target. The sound of the impact might suggest that the ball is well hit; only observation of the ball's flight relative to the intended target will provide clues as to the player's potential and the many subtleties that set high-level golfers apart. A nearly perfect golf swing that strikes the ball with the club face angled away from the target or off-center can produce very poor golf shots. A good golf coach using video for qualitative analysis would get views from several vantage points and gather information on the flight of the ball. This distance and direction information can be written or in recorded form on the audio track of the video. Only an integrated qualitative analysis of all these factors over many strokes would allow the coach to correctly judge this player's potential.

Note how a diagnosis of possible strengths and weaknesses is severely limited when all we have is a single view of a golf swing. Remember that the biomechanical principles related to the golf swing also must be integrated with other kinesiology disciplines. This player might have a flawless swing in practice that turns rough and unpredictable under psychological pressure. If this player's tournament results are good, the coach might invest time talking to their high school coach and plan a trip to see them in action.

QUALITATIVE ANALYSIS OF CATCHING

As a volunteer youth football coach you are working with your receivers on catching passes. Many young players pick up bad habits from playing neighborhood pick-up football games or watching the pros get by with talent rather than optimal technique. The technique points and cues you typically use are listed in Table 10.5. Notice how the critical features are more advanced and specialized than the catching technique points in chapter 9 (e.g., Table 9.4). Which biomechanical principles are strengths and weaknesses in the catching illustrated in Figure 10.5? How would you diagnosis this situation and intervene?

The player in Figure 10.5 made a successful running catch, but the illustration does not show enough of the movement so that we can tell whether the player protect-ed the ball by tucking it into their body. The illustrated view makes it difficult to tell if the player extended his arms (Range of Motion) to intercept the ball and provided time (Force–Time) to absorb the kinetic energy of the ball. Not only is reaching for the ball important in being able to increase the

Table 10.5 TECHNIQUE POINTS AND CUES FOR CATCHING A FOOTBALL PASS	
Technique points	Possible teaching/ intervention cues
Visual focus	Watch the ball, look for the seams
Intercept	Move and reach towards the ball
Hand position	Thumbs in or thumbs out
Absorption	Give with your hands and arms
Protection	Give and tuck the ball away

Figure 10.5. A football player making a catch in practice.

time of force application in order to slow the ball, but visual information on the arms/hands may also help intercept projectiles (van Donkelaar & Lee, 1994). Evaluation of this performance does not clearly identify any weaknesses in application of biomechanical principles.

A good intervention strategy would be to praise the player's effort and visual focus on the ball. Reinforcement of important technique points and motivation are good intervention goals while the coach waits to see if subsequent trials demonstrate no major weaknesses. How might the coach increase the difficulty of the catching drill to see if poor technique develops? Catching in a game situation involves many more environmental distractions. A knowledge of research concerning technique errors (Williams & McCririe, 1988) and environmental constraints (Savelsbergh & Whiting, 1988) in catching is clearly relevant for coaching football. What would be a better perspective for the coach to observe if the player is really reaching away from the body to intercept the ball?

SUMMARY

Coaches employ the principles of biomechanics to qualitatively analyze the movements of their athletes. This chapter explored the use of biomechanical principles in coaching softball, soccer, golf, football, and conditioning for basketball. Like physical educators, coaches often use cue words or phrases to communicate intervention to players. Coaches must integrate biomechanical principles with experience and other kinesiology subdisciplines (Knudson & Morrison, 2002). For example, coaches most often need to take into account conditioning (exercise physiology) and motivational issues (sports psychology) when dealing with athletes.

DISCUSSION QUESTIONS

1. Are certain biomechanical principles more important to the advanced athlete? Which and why?

2. Athletics coaches often have the opportunity of working closely with a smaller number of performers over a greater length of time than other kinesiology professionals. Does this concern for long-term performance increase or decrease the importance of biomechanical principles?

3. Have coaching organizations adequately promoted continuing education in sport sciences like biomechanics?

4. Which biomechanical principles are relevant to athlete quickness? Can biomechanics be used to coach an athlete to be quicker? If so, how does this improvement compare to improvement from conditioning?

5. Are biomechanical principles relevant to talent identification?

6. While the "daddy" videos discussed above might give the coach a general indication of the swings of players, what important aspects of golf competition may not show up on these videos? What important biomechanical issues might be difficult to determine from inadequate camera views?

7. Prioritize the following factors based on their importance in coaching beginning, intermediate, and advanced athletes for a specific sport: biomechanics, maturation, physiology, psychology.

SUGGESTED READING

Brancazio, P. (1984). *Sport science: Physical laws and optimum performance*. New York: Simon & Schuster.

Brody, H. (1987). *Tennis science for tennis players*. Philadelphia: University of Pennsylvania Press.

Dyson, G. (1986). *Mechanics of athletics* (8th ed.). New York: Holmes & Meier.

Ecker, T. (1996). *Basic track and field biomechanics* (2nd ed.). Los Altos, CA: Tafnews Press.

Elliott, B. C., & Mester, J. (Eds.) (1998). *Training in sport: Applying sport science*. New York: John Wiley & Sons.

Farrally, M. R., & Cochran, A. J. (Eds.) (1999). *Science and golf, III*. Champaign, IL: Human Kinetics.

Hay, J. G. (2000). *The biomechanics of sport techniques*, Englewood Cliffs, NJ: Prentice-Hall.

Jorgensen, T. P. (1994). *The physics of golf*. New York: American Institute of Physics.

Knudson, D. (2001, July). Improving stroke technique using biomechanical principles. *Coaching and Sport Science Review*, pp. 11–13.

Knudson, D., & Morrison, C. (2002). *Qualitative analysis of human movement* (2nd ed.). Champaign, IL: Human Kinetics.

Zatsiorsky, V. (Ed.) (2000). *Biomechanics in sport: Performance enhancement and injury prevention*. London: Blackwell Science.

WEB LINKS

ASEP—American Sport Education Program, which provides resources for developing coaching skills.
http://www.asep.com/

CAC—Coaching Association of Canada, which provides coaching development resources.
http://www.coach.ca/

CIS—ISBS Coaching Information Service, which provides articles applying biomechanics for coaches.
http://www.sportscoach-sci.com/

CHAPTER 11

Applying Biomechanics in Strength and Conditioning

Strength and conditioning is a profession in which a great deal of biomechanical research has been conducted recently. The National Strength and Conditioning Association (NSCA) is the leading professional strength and conditioning association in the world, and their journals—*Strength and Conditioning Journal* and *Journal of Strength and Conditioning Research*—have been receptive to articles on the biomechanics of exercise. Traditionally, strength and conditioning careers were limited to coaching the physically gifted in intercollegiate athletics. However, more and more opportunities exist for personal training with a wide variety of clients in the private sector.

Strength coaches and personal trainers are responsible for prescribing exercises that benefit their clients. On the surface this may seem a simple task, but in reality it is quite complicated. Exercises must be selected and exercise technique monitored. Exercises must be relevant, and the intensity must be sufficient for a training response but not too great as to cause overtraining or a high risk of injury. Biomechanics helps strength and conditioning professionals to assess these risk:benefit ratios, determine the most appropriate (sport-specific) exercises, and evaluate technique during training. As in teaching and coaching, biomechanical knowledge is important for the strength and conditioning professional so they can coordinate their efforts with sports medicine professionals.

QUALITATIVE ANALYSIS OF SQUAT TECHNIQUE

One of the most common and important exercises in athletic conditioning is the parallel squat. The squat is a functional exercise used for a wide variety of sports and other fitness objectives. The squat is usually performed as a free-weight exercise, making movement technique critical to overloading the target muscle groups and minimizing the risk of injury. Exacting technique in free-weight exercises is necessary because small variations allow other muscles to contribute to the lift, diminishing overload of the muscles or movements of interest. What are the main technique points of the squat often emphasized by strength and conditioning experts? Which biomechanical principles are most strongly related to those technique points?

Table 11.1 presents some of the typical technique points and cues for the parallel or front squat. Evaluate the strengths and weaknesses in the biomechanical principles related to the eccentric phase of the squat illustrated in Figure 11.1. Again, assume the lifter has performed a couple of repetitions this way and you are confident you can identify stable strengths and weaknesses in application of the principles.

The lifter depicted in Figure 11.1 has very good squat technique, so there are virtually no weaknesses in application of biomechanical principles. His stance width

Figure 11.1. The eccentric phase of a person doing a squat. Time between images is 0.2 seconds.

Table 11.1	
TECHNIQUE POINTS AND CUES FOR THE SQUAT	
Technique points	Possible intervention cues
Stance	Athletic position
Extended/neural spine	Slight arch
Slow, smooth movement	Slow and smooth
Keep thighs above horizontal	Thighs parallel to the ground

is appropriate, and there is no indication of difficulties in terms of control of the body or the bar (Balance). The images suggest that the motion was smooth, with simultaneous Coordination. The timing information in the caption indicates the squat was slow, maximizing the time the muscles were stressed (Force–Time Principle). This lifter also keeps his spine straight with normal lordosis, so the spinal loads are primarily compression and are evenly applied across the disks. This more axial loading be- tween the spinal segments is safest for the spine. Recent research has shown that spinal flexion reduces the extensor muscle component of force resisting anterior shear in the spine (McGill, Hughson, & Parks, 2000), making it more difficult for the muscles to stabilize the spine. Strength and conditioning coaches would also need to be familiar with research on the effect of weight belts in squats and other heavy lifting exercises.

Our lifter completed this exercise with the appropriate full Range of Motion, while not hyperflexing the knee. There is good trunk lean, which distributes the load on both the hip and knee extensors. The amount of trunk lean (hip flexion) in a squat is the primary factor in determining the distribution of joint moments that contribute to the exercise (Escamilla, 2001; Hay, Andrews, Vaughan, & Ueya, 1983; Mc-Laughlin, Lardner, & Dillman, 1978). The more upright posture in the front squat decreases the hip and lumbar extensor torques, while increasing the knee extensor torques required in the exercise.

A large part of the strength and conditioning professional's job is motivating and monitoring athletes. The coach needs to look for clues to the athlete's effort or a change in their ability to continue training. Some of these judgments involve application of biomechanical principles. How an athlete's Balance changes over a practice or several sets of an exercise could give a strength coach clues about fatigue. Since the figure and introduction give no clues to this aspect of performance, the best intervention in this situation is to praise the good technique of the athlete and possibly provide encouragement to motivate them.

Strength and conditioning professionals also must integrate sport-specific training with other practice and competition. The next example will focus on the sport-specificity of a plyometric training exercise.

QUALITATIVE ANALYSIS OF DROP JUMPS

Plyometrics are common exercises for improving speed and muscular power movements in athletes. Plyometric exercises use weights, medicine balls, and falls to exaggerate stretch-shortening-cycle muscle actions. Considerable research has focused on drop jumps as a lower-body plyometric exercise for improving jumping ability (Bobbert, 1990). Recent research has shown that drop jump exercise programs can increase bone density in children (Fuchs, Bauer, & Snow, 2001). Qualitative analysis of drop jumps is important in reducing the risk of injury in these exercises and monitoring technique that has been observed to vary between subjects (Bobbert *et al.*, 1986). Qualitative analysis is also important because drop jumping and resistance training can affect the technique used in various jumping movements (Hunter & Marshall, 2002). Table 11.2 presents important technique points and cues for drop jumps.

Table 11.2
TECHNIQUE POINTS AND CUES FOR DROP JUMPS

Technique points	Possible intervention cues
Landing position	Toe-heel landing
Rapid rebound	Quick bounce
Minimize countermovement	Range of motion
Arm integration	Arms down and up

What are the strengths and weaknesses in the drop jump performance illustrated in Figure 11.2?

The athlete doing the drop jump illustrated in Figure 11.2 has several good technique points, and possibly one weakness. The strong points of her technique are good lower-extremity positioning before touchdown, moderate countermovement, and a nearly vertical takeoff. This indicates good Balance during the exercise. It is difficult to evaluate the speed or quickness of the performance from the drawings with no temporal information in the caption. This athlete did have a short eccentric phase with a quick reversal into the concentric phase. Occasionally subjects will have a longer eccentric phase that minimizes the stretch-shortening-cycle effect of drop jumps (Bobbert *et al.*, 1986). The Force–Time Principle applied to plyometric exercises explains why large forces and high rates of force development are created over the short time of force application in plyometrics.

The obvious weakness is not using her arms in the exercise. Most athletes should strive to utilize an arm swing with coordination similar to jumping or the specific event for which they are training. If the arms are accelerated downward as the athlete lands, this will decrease eccentric loading of the lower extremities. For jump-specific training, cue the athletes to swing their arms downward in the drop so the arms are

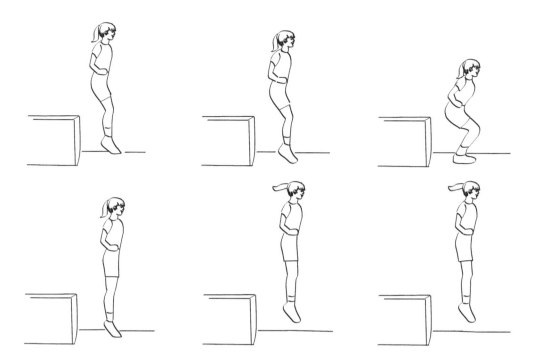

Figure 11.2. An athlete doing a drop jump exercise.

swinging behind them during the loading phase, increasing the intensity of eccentric loading of the lower extremities. The vigorous forward and upward swing of the arms from this position increases the vertical ground-reaction force through segmental interaction (Feltner *et al.*, 1999). The cue "arms down and up" could be used to remind an athlete of the technique points she should be focusing on in the following repetitions. A key conditioning principle is that the exercises selected for training should closely match the training objectives or movement that is to be improved. This matching of the exercise conditions to performance conditions is the conditioning principle of specificity. Exercise specificity will also be examined in the next example.

EXERCISE SPECIFICITY

In the past, exercise specificity was often based on a functional anatomical analysis (chapter 3) of the movement of interest. Exercises were selected that supposedly trained the muscles hypothesized to contribute to the movement. We saw in chapters 3 and 4 that biomechanics research has demonstrated that this approach to identifying muscle actions often results in incorrect assumptions. This makes biomechanical research on exercise critical to the strength and conditioning field. The strength and conditioning professional can also subjectively compare the principles of biomechanics in the exercise and the movement of interest to examine the potential specificity of training.

Suppose you are a strength and conditioning coach working with the track and field coach at your university to develop a training program for javelin throwers. You search SportDiscus for biomechanical research on the javelin throw and the conditioning literature related to overarm throwing patterns. What biomechanical principles are most relevant to helping you qualitatively analyze the javelin throw? The technique of a javelin throwing drill is illustrated in Figure 11.3. These principles would then be useful for examining potential exercises that would provide specificity for javelin throwers. Let's see how the principles of biomechanics can help you decide which exercise to emphasize more in the conditioning program: the bench press or pullovers. We will be limiting our discussion to technique specificity.

The principles most relevant to the javelin throw are Optimal Projection, Inertia, Range of Motion, Force–Motion, Force–Time, Segmental Interaction, and Coordination Continuum. Athletes throw the javelin by generating linear momentum (using Inertia) with an approach that is transferred up the body in a sequential

overarm throwing pattern. These principles can be used in the qualitative analysis of the throwing performances of the athletes by coaches, while the strength and conditioning professional is interested in training to improve performance and prevent injury. The fast approach (Range of Motion) and foul line rules make the event very hard on the support limb, which must stop and transfer the forward momentum to the trunk (Morriss, Bartlett, & Navarro, 2001). This Segmental Interaction using energy from the whole body focuses large forces (Force–Motion) in the upper extremity. The size and weight of the javelin also contribute to the high stresses on the shoulder and elbow joints. While some elastic cord exercises could be designed to train the athlete to push in the direction of the throw (Optimal Projection), this section will focus

Figure 11.3. Typical technique for the javelin throw drill.

on the specificity of two exercises: the bench press and pullovers. Space does not permit a discussion of other specificity issues, like eccentric training for the plant foot or training for trunk stability.

For specificity of training, the exercises prescribed should match these principles and focus on muscles that contribute (Force–Motion) to the joint motions (Range of Motion), and those which might help stabilize the body to prevent injury. While much of the energy to throw a javelin is transferred up the trunk and upper arm, a major contributor to shoulder horizontal adduction in overarm patterns is likely to be the pectoralis major of the throwing arm. The question then becomes: which exercises most closely match Range of Motion and Coordination in the javelin throw? Matching the speed of movement and determining appropriate resistances are also specificity issues that biomechanics would help inform.

Biomechanical research on the javelin can then help select the exercise and customize it to match pectoralis major function during the event. EMG and kinetic studies can be used to document the temporal location and size of muscular demands. Kinematic research help identify the shoulder range and speed of shoulder motion in the javelin throw. A good strength and conditioning coach would review this research on the javelin throw with the track coach (Bartlett & Best, 1988; Bartlett et al., 1996).

If the bench press and pullover exercise techniques remain in their traditional (supine) body position and joint ranges of motion, the bench press may provide the most activity-specific training for the javelin throw. The bench press typically has the shoulder in 90° of abduction, matching its position in the javelin throw. The bench press could be performed (assuming adequate spotting and safety equipment) with a fast speed to mimic the SSC of the javelin throw. This would also mimic the muscle

actions and rate of force development (Force–Time). Even greater sport specificity may be achieved by using plyometric bench presses with medicine balls. The plyometric power system (Wilson et al., 1993) is a specialized piece of equipment that would also allow for dynamic bench press throws.

Pullovers often have greater shoulder abduction that is unlike the range of motion in the event. Pullovers also have a range of motion that requires greater scapular upward rotation and shoulder extension, which tends to compress the supraspinatus below the acromion process of the scapula. Athletes in repetitive overarm sports often suffer from this impingement syndrome, so pullovers may be a less safe training exercise than the bench press.

The other training goal that is also related to movement specificity is prevention of injury. What muscles appear to play more isometric roles in stabilizing the lower extremity, the shoulder, and elbow? What research aside from javelin studies could be used to prescribe exercises that stabilize vulnerable joints? What muscles are likely to have eccentric actions to "put on the brakes" after release? What exercises or movements are best for training to reduce the risk of injury? Why might training the latissimus dorsi potentially contribute to the performance and injury prevention goals of training for the javelin throw?

INJURY RISK

Imagine you are a strength coach at a junior college. You closely watch many of the young men in your preseason conditioning program because they have had little serious weight training in their high schools, and others may be pushing themselves too hard to meet team strength standards to qualify for competition. Suppose you see a player performing the bench press using

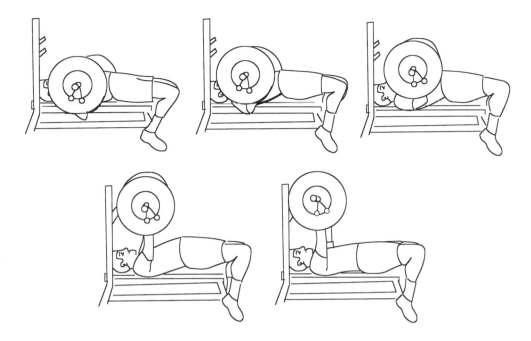

Figure 11.4. The concentric phase of a bench press from an athlete struggling to make a weight goal.

the technique illustrated in Figure 11.4. What are the strengths and weaknesses of performance? How would you diagnosis this performance and what intervention would you use?

The biomechanical principles relevant to the bench press are Balance, Coordination Continuum, Force–Time, and Range of Motion. When training for strength, resistance is high, the athlete must have good control of the weight (Balance), and coordination during the lift will be simultaneous. The force–time profile of strength training attempts to maintain large forces applied to the bar through as much of the range of motion as possible. The SSC nature of the movement should be minimized. This keeps the movement slow and force output near the weight of the bar. High initial forces applied to the ball results in lower forces applied to the bar later in the range of motion (Elliott *et al.*, 1989). The principle of Range of Motion in strength training tends

toward one of two extremes. First, minimize the range of motion of joints that do not contribute to the movement and of those that allow other muscles to contribute to the movement. Second, the range of motion for joint movements or muscles that are targeted by the exercise should be maximized.

The two principles most strongly related to exercise safety in the bench press are Balance and Range of Motion. Athletes must control the weight of the bar at all times, and a lack of control will affect the range of motion used in the exercise. The athlete in Figure 11.3 shows weaknesses in both balance and range of motion. Since the athlete is struggling to "make weight," the difference in strength between the sides of the body manifests as uneven motion of the bar and poor balance. The athlete also hyperextended his lumbar spine in straining to lift the weight.

Several aspects of this performance may have a strength coach thinking about a risk of immediate and future injury: lateral strength imbalance, poor control of bar motion, and hyperextension of the lumbar spine. Since the athlete is "maxing-out," some of these weaknesses can be expected, but safety is the greatest concern. Spotters can assist lifters with poor bar control, or who can complete the lift with only one side of their body. Hyperextension of the spine, however, is an immediate risk to the athlete's low-back health. Hyperextension of the lumbar spine under loading is dangerous because of uneven pressures on the intervertebral disks and greater load bearing on the facet joints. The best intervention here is to terminate the lift with assistance from a spotter and discontinue testing only when the athlete maintains a neutral and supported spinal posture on the bench. Here the immediate risk of injury is more important than balance, skill in the exercise, or passing a screening test.

EQUIPMENT

Equipment can have quite a marked influence on the training effect of an exercise. Exercise machines, "preacher" benches, and "Smith" machines are all examples how equipment modifies the training stimulus of weight-training exercises. Strength and conditioning catalogues are full of specialized equipment and training aids; unfortunately, most of these devices have not been biomechanically studied to determine their safety and effectiveness. Garhammer (1989) provides a good summary of the major kinds of resistance exercise machines in his review of the biomechanics of weight training.

Let's revisit the squat exercise using one of these training devices. This device is a platform that stabilizes the feet and lower legs. A person performing the eccentric phase of a front squat with this device is depicted in Figure 11.5. Compare the squat technique of this subject with the technique in the traditional squat (Figure 11.1). What biomechanical principles are affected most by the use of this device?

Inspection of Figure 11.5 shows that there are several Range-of-Motion differences between the two squat exercises. Squatting with the device results in less knee flexion and ankle dorsiflexion. Note how the lower leg remains nearly vertical, and how the center of mass of the athlete/bar is shifted farther backward in this squat. There does not appear to be any obvious differences in trunk lean between the two devices with these performers. What do you think are the training implications for these small differences? Which body position at the end of the eccentric phase seems to be more specific to football, skiing, or volleyball: this or the front squat?

Using the device makes balancing easier, although it puts the line of gravity of the body/bar well behind the feet. The larger base of support and Inertia (body and stand) stabilizes the exerciser in the squat. It is not possible to compare the kinetics of the two exercises from qualitative analysis of the movements, but it is likely there are differences in the loading of the legs and back (Segmental Interaction). What joints do you think are most affected (think about the moment arm for various body segment and shearing forces in the knee)? What kinds of biomechanical studies would you like to see if you were advising the company on improving the device?

SUMMARY

Strength and conditioning professionals use the principles of biomechanics to qualitatively analyze the technique of exercises, evaluate the appropriateness of exercises, and reduce the risk of injury from dangerous exercise technique. Qualitative analysis of several free weight exercises was pre-

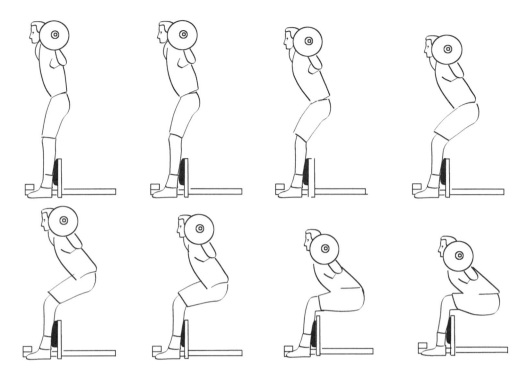

Figure 11.5. The eccentric phase of a person doing a squat using a foot and leg stabilizing stand.

sented, and we examined the biomechanical principles in the qualitative analysis of exercises machines. Strength and conditioning professionals also must integrate physiological and psychological knowledge with biomechanical principles to maximize client improvement. Since strength training utilizes loads closer to the ultimate mechanical strength of tissues, professionals need to keep safety and exacting exercise technique in mind.

DISCUSSION QUESTIONS

1. The squat and various leg-press exercise stations are often used interchangeably. What biomechanical principles are more important in the squat than in the leg press, and how would you educate lifters who think that the exercises do the same thing?

2. An athlete back in the weight room after initial rehabilitation from an injury is apprehensive about resuming their conditioning program. What biomechanical principles can be modified in adapting exercises for this athlete? Suggest specific exercises and modifications.

3. What aspect of exercise specificity (muscles activated or joint motions) do you think is most important in training for sports? Why? Does analysis of the biomechanical principles of exercises and sport movement help you with this judgment?

4. If an athlete uses unsafe technique in the weight room, should the coach's response be swift and negative for safety's sake, or should they take a positive (teachable moment) approach in teaching safer technique? Are there athlete (age, ability, etc.) or exercise factors that affect the best approach?

5. Athletes train vigorously, pushing their limits, treading a fine line between training safely and overtraining. Are there biomechanical indicators that could help the strength and conditioning professional recognize when training intensity has moved beyond overload to dangerous? Why?

6. For a specific sport movement, determine if conditioning exercises should emphasize Force-Time or Force-Motion to be more activity-specific.

7. What biomechanical principles are relevant to training overarm-throwing athletes with upper-body plyometric exercises? Be sure to integrate the muscle mechanics knowledge summarized in chapter 4 in your answer.

8. Strength training resistances are often expressed as percentages of maximum strength (1RM). If loads on the musculoskeletal system were also expressed as percentages of mechanical strength, what training loads do you think would be safe (acceptable risk) or unsafe (unacceptable risk)?

9. Which is most important in selecting weight training resistances: training studies or biomechanical tissue tolerances? Why?

SUGGESTED READING

Atha, J. (1981). Strengthening muscle. *Exercise and Sport Sciences Reviews*, **9**, 1–73.

Baechle, T. R., & Earle, R. W. (Eds.) (2000). *Essentials of strength training and conditioning* (2nd ed.). Champaign, IL: Human Kinetics.

Bartlett, R. M., & Best, R. J. (1988). The biomechanics of javelin throwing: A review. *Journal of Sports Sciences*, **6**, 1–38.

Garhammer, J. (1989). *Weight lifting and training*. In C. Vaughan (Ed.), *Biomechanics of sport* (pp. 169–211). Boca Raton, FL: CRC Press.

Knudson, D., & Morrison, C. (2002). *Qualitative analysis of human movement* (2nd ed.). Champaign, IL: Human Kinetics.

Knuttgen, H. G., & Kraemer, W. J. (1987). Terminology and measurement in exercise performance. *Journal of Applied Sport Science Research*, **1**, 1–10.

Komi, P. V. (Ed.) (1992). *Strength and power in sport*. London: Blackwell Science.

Stone, M., Plisk, S., & Collins, D. (2002). Training principles: Evaluation of modes and methods of resistance training—A coaching perspective. *Sports Biomechanics*, **1**, 79–103.

Wilson, G. J. (1994). Strength and power in sport. In J. Bloomfield, T. R. Ackland, & B. C. Elliott (Eds.) *Applied anatomy and biomechanics in sport* (pp. 110–208). Melbourne: Blackwell Scientific Publications.

Zatsiorsky, V. (1995). *Science and practice of strength training*. Champaign, IL: Human Kinetics.

WEB LINKS

NSCA—National Strength and Conditioning Association.
  http://www.nsca-lift.org/menu.htm

PCPFS Research Digest—research reviews published by the President's Council on Physical Fitness and Sports.
  http://www.fitness.gov/Reading_Room/Digests/digests.html

CHAPTER 12

Applying Biomechanics in Sports Medicine and Rehabilitation

Biomechanics also helps professionals in clinical settings to determine the extent of injury and to monitor progress during rehabilitation. Many sports medicine programs have specific evaluation and diagnostic systems for identification of musculoskeletal problems. The physical therapist and athletic trainer analyzing walking gait or an orthopaedic surgeon evaluating function after surgery all use biomechanics to help inform decisions about human movement.

These clinical applications of biomechanics in qualitative analysis tend to focus more on localized anatomical issues than the examples in the previous three chapters. This chapter cannot replace formal training in gait analysis (Perry, 1992), injury identification (Shultz, Houglum, and Perrin, 2000), or medical diagnosis (Higgs & Jones, 2000). It will, however, provide an introduction to the application of biomechanical principles in several sports medicine professions. Biomechanical principles must be integrated with the clinical training and experience of sports medicine professionals

INJURY MECHANISMS

Most sports medicine professionals must deduce the cause of injuries from the history presented by patients or clients. Occasionally athletic trainers may be at a practice or competition where they witness an injury. Knowledge of the biomechanical causes of certain injuries can assist an athletic trainer in these situations, in that diagnosis of the particular tissues injured is facilitated. Imagine you are an athletic trainer walking behind the basket during a basketball game. You look onto the court and see one of your athletes getting injured as she makes a rebound (see Figure 12.1). What kind of injury do you think occurred? What about the movement gave you the clues that certain tissues would be at risk of overload?

The athlete depicted in Figure 12.1 likely sprained several knee ligaments. Landing from a jump is a high-load event for the lower extremity, where muscle activity must be built up prior to landing. It is likely the awkward landing position, insufficient pre-impact muscle activity, and twisting (internal tibial rotation) contributed to the injury. It is also likely that the anterior (ACL) and posterior (PCL) cruciate ligaments were sprained. The valgus deformation of the lower leg would also suggest potential insult to the tibial (medial) collateral ligament. Female athletes are more likely to experience a non-contact ACL injury than males (Malone, Hardaker, Garrett, Feagin, & Bassett, 1993), and the majority of ACL injuries are non-contact injuries (Griffin *et al.*, 2000). There are good recent reviews of knee ligament injury mechanisms (Bojsen-Moller & Magnusson, 2000; Whiting & Zernicke, 1998).

You rush to the athlete with these injuries in mind. Unfortunately, any of these sprains are quite painful. Care must be taken to comfort the athlete, treat pain

Figure 12.1. A basketball player injuring her knee during a rebound.

and inflammation, and prevent motion that would stress the injured ligaments. Joint tests and diagnostic imaging will eventually be used to diagnosis the exact injury. What biomechanical issue or principle do you think was most influential in this injury?

EXERCISE SPECIFICITY

The principle of specificity also applies to therapeutic exercise in rehabilitation settings. The exercises prescribed must match the biomechanical needs of the healing patient. Exercises must effectively train the muscles that have been weakened by injury and inactivity. Biomechanical research on therapeutic exercise is even more critical since therapists need to know when inter-

nal loadings may exceed the mechanical strengths of normal and healing tissues.

Imagine that you are a physical therapist treating a runner with patellofemoral pain syndrome. Patellofemoral pain syndrome (PFPS) is the current terminology for what was commonly called chondromalacia patella (Thomee, Agustsson, & Karlsson, 1999). PFPS is likely inflammation of the patellar cartilage since other knee pathologies have been ruled out. It is believed that PFPS may result from misalignment of the knee, weakness in the medial components of the quadriceps, and overuse. If the vastus medialis and especially the vastus medialis obliquus (VMO) fibers are weak, it is hypothesized that the patella may track more laterally on the femur and irritate either the patellar or femoral cartilage. The exercises commonly prescribed to focus activation on the VMO are knee extensions within 30° of near complete extension, similar short-arc leg presses/squats, and isometric quadriceps setting at complete extension, and these exercises with combined hip adduction effort. While increased VMO activation for these exercises is not conclusive (see Earl, Schmitz, and Arnold, 2001), assume you are using this therapeutic strategy when evaluating the exercise technique in Figure 12.2. What biomechanical principles are strengths and weaknesses in this exercise.

Most biomechanical principles are well performed. Balance is not much of an issue in a leg press machine because mechanical restraints and the stronger limb can compensate for weakness in the affected limb. There is simultaneous Coordination, and there appears to be slow, smooth movement (Force–Time).

The principle that is the weakest for this subject is the large knee flexion Range of Motion. This subject has a knee angle of about 65° at the end of the eccentric phase

of the exercise. This very flexed position puts the quadriceps at a severe mechanical disadvantage, which results in very large muscle forces and the consequent large stresses on the patellofemoral and tibiofemoral joints. This exercise technique can irritate the PFPS and does not fit the therapeutic strategy, so the therapist should quickly instruct this person to decrease the range of motion. Providing a cue to only slightly lower the weight or keeping the knees extended to at least 120° would be appropriate for a patient with PFPS.

A better question would be: should this person even be on this leg press machine? Would it be better if they executed a different exercise? A leg press machine requires less motor control to balance the resistance than a free-weight squat exercise, so a leg press may be more appropriate than a squat. Maybe a more appropriate exercise would be a leg press machine or a cycle that allows the subject to keep the hip extended (reducing hip extensor contributions and increasing quadriceps demand) and limit

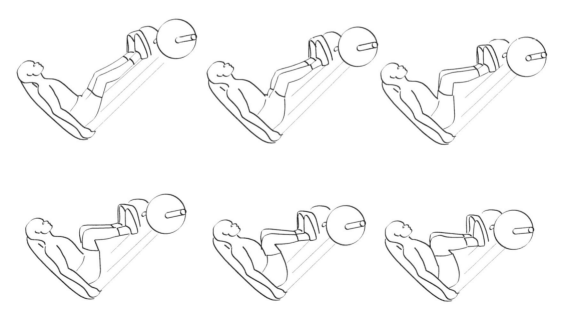

Figure 12.2. The leg press technique of a person trying to remediate patellofemoral pain.

the amount of knee flexion allowed. The differences in muscle involvement are likely similar to upright versus recumbent cycling (Gregor, Perell, Rushatakankovit, Miyamoto, Muffoletto, & Gregor, 2002). These subtle changes in body position and direction of force application (Force–Motion) are very important in determining the loading of muscles and joints of the body. Good therapists are knowledgeable about the biomechanical differences in various exercises, and prescribe specific rehabilitation exercises in a progressive sequence to improve function.

EQUIPMENT

Sports medicine professionals often prescribe prosthetics or orthotics to treat a variety of musculoskeletal problems. Prosthetics are artificial limbs or body parts. Orthotics are devices or braces that support, cushion, or guide the motion of a body. Shoe inserts and ankle, knee, or wrist braces are examples of orthotics. Orthotics can be bought "off the shelf" or custom-build for a particular patient.

Shoe inserts are a common orthotic treatment for excessive pronation of the subtalar joint. One origin of excessive pronation is believed to be a low arch or flat foot. A person with a subtalar joint axis below 45° in the sagittal plane will tend to have more pronation from greater eversion and adduction of the rear foot. It has been hypothesized that the medial support of an orthotic will decrease this excessive pronation.

Figure 12.3 illustrates a rear frontal plane view of the maximum pronation position in running for an athlete diagnosed with excessive rear-foot pronation. The two images show the point of maximum pronation when wearing a running shoe (a) and when wearing the same shoe with a custom semirigid orthotic (b). Imag-

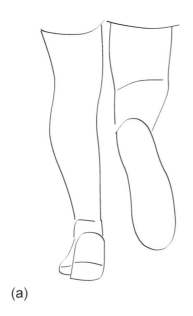
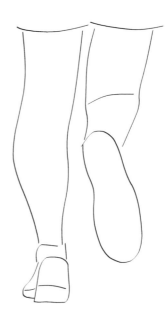

(a)   (b)

Figure 12.3. Rear frontal plane view of the positions of maximum pronation in running in shoes (a) and shoes with a semi-rigid orthotic (b) on a treadmill at 5.5 m/s.

ine that you are the athletic trainer working with this runner. The runner reports that it is more comfortable to run with the orthotic, an observation that is consistent with decreased pain symptoms when using orthotics (Kilmartin & Wallace, 1994). You combine this opinion with your visual and videotaped observations of the actions of her feet in running.

Inspection of Figure 12.3 suggests that there is similar or slightly less pronation when the runner is wearing an orthotic. Biomechanical research on orthotics and rear-foot motion have not as of yet determined what amount of pronation or speed of pronation increases the risk of lower-extremity injuries. The research on this intervention is also mixed, with little evidence of the immediate biomechanical effects of orthotics on rear-foot motion and the hypothesized coupling with tibial internal rotation (Heiderscheit, Hamill, & Tiberio, 2001). In addition, it is unclear if the small decrease in pronation (if there was one) in this case is therapeutic. The comfort and satisfaction perceived by this runner would also provide some support for continued use of this orthotic.

READINESS

Orthopaedic surgeons and athletic trainers must monitor rehabilitation progress before clearing athletes to return to their practice routine or competition. Recovery can be documented by various strength, range-of-motion, and functional tests. Subjective measures of recovery include symptoms reported by the athlete and qualitative analyses of movement by sports medicine professionals. Athletes will often be asked to perform various movements of increasing demands, while the professional qualitatively evaluates the athlete's control of the injured limb. A couple of common functional tests for athletes with knee injuries

are multiple hops for distance or time (Fitzgerald et al., 2001).

Imagine you are an athletic trainer working with an athlete rehabilitating an ACL injury in her right knee. You ask the athlete to perform a triple hop for maximum distance. The technique of the first hop is illustrated in Figure 12.4. As you measure the distance hopped, you go over the strengths and weaknesses in terms of the biomechanical principles of the hop in your mind. Later you will combine this assessment with the quantitative data. The distance hopped on the injured limb should not be below 80% of the unaffected limb (Fitzgerald et al., 2001). What biomechanical principles are strength and weaknesses, and what does a diagnosis of this hopping performance tell you about her readiness to return to practice? Biomechanical technique is just one aspect of many areas that must be evaluated in making decisions on returning athletes to play (Herring et al., 2002).

Most all biomechanical principles are well performed by this athlete. This athlete is showing good hopping technique with nearly Optimal Projection for a long series of hops. She shows good Coordination of arm swing, integrated with good simultaneous flexion and extension of the lower extremity. She appears to have good Balance, and her application of the Range-of-Motion and Force–Time principles in the right leg shows good control of eccentric and concentric muscle actions. There are no apparent signs of apprehension or lack of control of the right knee. If these qualitative observations are consistent with the distance measured for the three hops, it is likely the athletic trainer would clear this athlete to return to practice. The therapist might ask the coach to closely monitor the athlete's initial practices for signs of apprehension, weakness, or poor technique as she begins more intense and sport-specific movements.

Figure 12.4. An athlete doing a triple hop test.

SUMMARY

Sports medicine professionals use biomechanical principles to understand injury mechanisms, select appropriate injury prevention and rehabilitation protocols, and monitor recovery. In the specificity example, we saw that qualitative analysis of exercise technique can help sports medicine professionals ensure that the client's technique achieves the desired training effect. Qualitative analysis in sports medicine often focuses on an anatomical structure level more often than other kinesiology professions. Qualitative analysis of therapeutic exercise also requires an interdisciplinary approach (Knudson & Morrison, 2002), especially integrating clinical training and experience with biomechanics. Other issues sports medicine professionals must take into account beyond biomechan-

ical principles are pain, fear, motivation, and competitive psychology.

DISCUSSION QUESTIONS

1. What biomechanical principle do you think is more important in rehabilitating from an ankle sprain, Balance or Range of Motion?

2. Patients recovering from knee injuries are often given braces to prevent unwanted movement and to gradually increase allowable motion. What movement characteristics would indicate that a patient is ready to exercise or function without a brace?

3. Sports medicine professionals looking for the causes of overuse injuries often evaluate joints distant from the affected area (Kibler & Livingston, 2001) because of

Segmental Interaction through the kinematic chain. What biomechanical principles can provide cues to potential overuse injuries in other parts of the body?

4. A major injury in athletic and sedentary populations is low-back pain. What abdominal and back muscles are most specific to injury prevention for an office worker and a tennis player?

5. Athletes using repetitive overarm throwing often suffer from impingement syndrome. What biomechanical principles can be applied to the function of the shoulder girdle and shoulder in analyzing the exercise and throwing performance of an injured athlete?

6. You are an trainer working with an athlete recovering from a third-degree ankle sprain. You and the athlete are deciding whether to use athletic tape or an ankle brace. What does a qualitative biomechanical analysis suggest in the better of these two options? What biomechanical studies would you suggest to investigate the clinical efficacy of these options?

7. What biomechanical principles should be focused on when a therapist or trainer is working with elderly clients to prevent falls?

8. An adapted physical educator has referred a young person who might have Developmental Coordination Disorder (DCD) to a physician. Before various imaging and neurological tests are performed, what biomechanical principles should be the focus of observation, and what simple movement tests would be appropriate in the initial physical/orthopaedic exam?

SUGGESTED READING

Dvir, Z. (Ed.) (2000). *Clinical biomechanics*. New York: Churchill Livingstone.

Eriksson, E. (1996). The scientific basis of rehabilitation. *American Journal of Sports Medicine, 24*, S25–S26.

Fitzgerald, G. K., Lephart, S. M., Hwang, J. H., & Wainner, R. S. (2001). Hop tests as predictors of dynamic knee stability. *Journal of Orthopaedic and Sports Physical Therapy, 31*, 588–597.

Hawkins, D., & Metheny, J. (2001). Overuse injuries in youth sports: Biomechanical considerations. *Medicine and Science in Sports and Exercise, 33*, 1701–1707.

Kibler, W. B., and Livingston (2001). Closed chain rehabilitation of the upper and lower extremity. *Journal of the American Academy of Orthopaedic Surgeons, 9*, 412–421.

Knudson, D., & Morrison, C. (2002). *Qualitative analysis of human movement* (2nd ed.). Champaign, IL: Human Kinetics.

Nordin, M., & Frankel, V. (2001). *Basic biomechanics of the musculoskeletal system* (3rd ed.). Baltimore: Williams & Wilkins.

Smith, L. K., Weiss, E. L., & Lehmkuhl, L. D. (1996). *Brunnstrom's clinical kinesiology* (5th ed.). Philadelphia: F. A. Davis.

Whiting, W. C., & Zernicke, R. F. (1998). *Biomechanics of musculoskeletal injury*. Champaign, IL: Human Kinetics.

WEB LINKS

ACSM—The American College of Sports Medicine is a leader in the clinical and scientific aspects of sports medicine and exercise. ACSM provides the leading professional certifications in sports medicine.
http://acsm.org/

APTA—American Physical Therapy Association
http://www.apta.org/

CGA—International Clinical Gait Analysis website, which posts interesting case studies, discussions, and learning activities.
http://aisr1.lib.tju.edu/cga/

FIMS—International Federation of Sports Medicine
http://www.fims.org/

Gillette Children's Hospital Videos and CDROMs
http://www.gillettechildrens.com/programs-services/motion-analysis/motionanalysisvideos.html

GCMAS—North American organization called the Gait and Clinical Movement Analysis Society
http://www.gcmas.org/

ISB Technical Group on footwear biomechanics
http://www.staffs.ac.uk/isb-fw/

NATA—National Athletic Trainers' Association
http://www.nata.org/

References

Aagaard, P., Simonsen, E. B., Magnusson, S. P., Larsson, B., & Dyhre-Poulsen, P. (1998). A new concept for isokinetics hamstring: quadriceps muscle strength ratio. *American Journal of Sports Medicine*, **26**, 231–237.

Aagaard, P., Simonsen, E. B., Andersen, J. L., Magnusson, S. P., Bojsen-Moller, F., & Dyhre-Poulsen, P. (2000). Antagonist muscle coactivation during isokinetic knee extension. *Scandinavian Journal of Medicine and Science in Sports*, **10**, 58–67.

Abe, T., Kumagai, K., & Brechue, W. F. (2000). Fasicle length of leg muscles is greater in sprinters than distance runners. *Medicine and Science in Sports and Exercise*, **32**, 1125–1129.

Abernethy, P., Wilson, G., & Logan, P. (1995). Strength and power assessment: Issues, controversies and challenges. *Sports Medicine*, **19**, 401–417.

Abraham, L. D. (1991). *Lab manual for KIN 326: Biomechanical analysis of movement.* Austin, TX.

Adair, R. (1990). *The physics of baseball.* New York: Harper & Row.

Adamson, G. T., & Whitney, R. J. (1971). Critical appraisal of jumping as a measure of human power. In J. Vredenbregt & J. Wartenweiler (Eds.), *Biomechanics II* (pp. 208–211). Baltimore: University Park Press.

Adrian, M. J., & Cooper, J. M. (1995). *Biomechanics of human movement* (2nd ed.). Madison, WI: Brown & Benchmark.

Ait-Haddou, R., Binding, P., & Herzog, W. (2000). Theoretical considerations on co-contraction of sets of agonist and antagonist muscles. *Journal of Biomechanics*, **33**, 1105–1111.

Akeson, W. H., Amiel, D., Abel, M. F., Garfin, S. R., & Woo, S. L. (1987). Effects of immobilization on joints. *Clinical Orthopaedics*, **219**, 28–37.

Alaways, L. W., Mish, S. P., & Hubbard, M. (2001). Identification of release conditions and aerodynamic forces in pitched-baseball trajectories. *Journal of Applied Biomechanics*, **17**, 63–76.

Aleshinsky, S. Y. (1986a). An energy "sources" and "fractions" approach to the mechanical energy expenditure problem, I: Basic concepts, description of the model, analysis of one–link system movement. *Journal of Biomechanics*, **19**, 287-293.

Aleshinsky, S. Y. (1986b). An energy "sources" and "fractions" approach to the mechanical energy expenditure problem, IV: Critcism of the concepts of "energy" transfers within and between segments. *Journal of Biomechanics*, **19**, 307–309.

Alexander, R. M. (1992). *The human machine.* New York: Columbia University Press.

Allard, P., Stokes, I. A., & Blanchi, J. (Eds.) (1995). *Three-dimensional analysis of human motion.* Champaign, IL: Human Kinetics.

Allman, W. F. (1984). Pitching rainbows: The untold physics of the curve ball. In E. W. Schrier & W. F. Allman (Eds.), *Newton at the bat: The science in sports* (pp. 3–14). New York: Charles Scribner & Sons.

Anderson, F. C., & Pandy, M. G. (1993). Storage and utilization of elastic strain energy during jumping. *Journal of Biomechanics*, **26**, 1413–1427.

Aragon-Vargas, L. F., & Gross, M. M. (1997a). Kinesiological factors in vertical jumping performance: Differences among individuals. *Journal of Applied Biomechanics*, **13**, 24–44.

Aragon-Vargas, L. F., & Gross, M. M. (1997b). Kinesiological factors in vertical jumping performance: Differences within individuals. *Journal of Applied Biomechanics*, **13**, 45–65.

Arellano, R. (1999). Vortices and propulsion. In R. Sanders & J. Linsten (Eds.), *SWIMMING: Applied proceedings of the xvii international symposium on biomechanics in sports* (Vol. 1, p. 53–66). Perth, WA: Edith Cowan University.

Ariel, G. B. (1983). Resistive training. *Clinics in Sports Medicine*, **2**, 55–69.

Arndt, A. N., Komi, P. V., Bruggemann, G. P., & Lukkariniemi, J. (1998). Individual muscle contributions to the *in vivo* achilles tendon force. *Clinical Biomechanics*, **13**, 532–541.

Arokoski, J. P. A., Jurvelin, J. S., Vaatainen, U., & Helminen, H. J. (2000). Normal and pathological adaptations of articular cartilage to joint loading. *Scandinavian Journal of Medicine and Science in Sports*, **10**, 186–198.

Aruin, A. S. (2001). Simple lower extremity two-joint synergy. *Perceptual and Motor Skills*, **92**, 563–568.

Ashton-Miller, J. A., & Schultz, A. B. (1988). Biomechanics of the human spine and trunk. *Exercise and Sport Sciences Reviews*, **16**, 169–204.

Atha, J. (1981). Strengthening muscle. *Exercise and Sport Sciences Reviews*, **9**, 1–73.

Atwater, A. E. (1979). Biomechanics of overarm throwing movements and of throwing injuries. *Exercise and Sport Sciences Reviews*, **7**, 43–85.

Atwater, A. E. (1980). Kinesiology/biomechanics: Perspectives and trends. *Research Quarterly for Exercise and Sport*, **51**, 193–218.

Atwater, A. E. (1982). Basic biomechanics applications for the coach. In J. Terauds (Ed.), *Biomechanics in sports: Proceedings of the international symposium of biomechanics in sports* (pp. 21–30). Del Mar, CA: Academic Publishers.

Babault, N., Pousson, M., Ballay, Y., & Van Hoecke, J. (2001). Activation of human quadriceps femoris during isometric, concentric, and eccentric contractions. *Journal of Applied Physiology*, **91**, 2628–2634.

Baechle, T. R., & Earle, R. W. (Eds.) (2000). *Essentials of strength training and conditioning* (2nd ed.). Champaign, IL: Human Kinetics.

Bahill, A. T., & Freitas, M. M. (1995). Two methods of recommending bat weights. *Annals of Biomedical Engineering*, **23**, 436–444.

Barclay, C. J. (1997). Initial mechanical efficiency in cyclic contractions of mouse skeletal muscle. *Journal of Applied Biomechanics*, **13**, 418–422.

Barfield, W. R. (1998). The biomechanics of kicking in soccer. *Clinics in Sports Medicine*, **17**, 711–728.

Barlow, D. A. (1971). Relation between power and selected variables in the vertical jump. In J. M. Cooper (Ed.), *Selected topics on biomechanics* (pp. 233–241). Chicago: Athletic Institute.

Barr, A. E., & Barbe, M. F. (2002). Pathophysiological tissue changes associated with repetitive movement: A review of the evidence. *Physical Therapy*, **82**, 173–187.

Bartlett, R. (1999). *Sports biomechanics: Reducing injury and improving performance*. London: E&FN Spon.

Bartlett, R., Muller, E., Lindinger, S., Brunner, F., & Morriss, C. (1996). Three-dimensional evaluation of the kinematics release parameters for javelin throwers of different skill levels. *Journal of Applied Biomechanics*, **12**, 58–71.

Bartlett, R. M., & Best, R. J. (1988). The biomechanics of javelin throwing: A review. *Journal of Sports Sciences*, **6**, 1–38.

Basmajian, J. V., & De Luca, C. J. (1985). *Muscles alive: Their functions revealed by electromyography* (5th. ed.). Baltimore: Williams & Wilkins.

Basmajian, J. V., & Wolf, S. L. (Eds.) (1990). *Therapeutic exercise* (5th ed.). Baltimore: Williams & Wilkins.

Bates, B. T., Osternig, L. R., Sawhill, J. A., & Janes, S. L. (1983). An assessment of subject variability, subject-shoe interaction, and the evaluation of running shoes using ground reaction forces. *Journal of Biomechanics*, **16**, 181–191.

Bauer, J. J., Fuchs, R. K., Smith, G. A., & Snow, C. M. (2001). Quantifying force magnitude and loading rate from drop landings that induce osteogenesis. *Journal of Applied Biomechanics*, **17**, 142–152.

Bawa, P. (2002). Neural control of motor output: Can training change it? *Exercise and Sport Sciences Reviews*, **30**, 59–63.

Beattie, P. F., & Meyers, S. P. (1998). Magnetic resonance imaging in low back pain: General pinciples and clinical issues. *Physical Therapy*, **78**, 738–753.

Beer, F. P., & Johnston, E. R. (1984). *Vector mechanics for engineers: Statics and dynamics* (4th ed.). New York: McGraw-Hill.

Benjanuvatra, N. Dawson, B., Blanksby, B. A., & Elliott, B. C. (2002). Comparison of buoyancy, passive and net active drag forces between Fastskin™ and standard swimsuits. *Journal of Science and Medicine in Sport*, **5**, 115–123.

Berger, M. A. M., de Groot, G., & Hollander, A. P. (1995). Hydrodynamic drag and lift force on human hand/arm models. *Journal of Biomechanics*, **28**, 125–133.

Bernstein, N. A. (1967). *The co-ordination and regulation of movements*. Oxford: Pergamon Press.

Berthoin, S., Dupont, G., Mary, P., & Gerbeaux, M. (2001). Predicting sprint kinematic parameters from anaerobic field tests in physical education students. *Journal of Strength and Conditioning Research*, **15**, 75–80.

Biewener, A. A. (1998). Muscle function *in vivo*: A comparison of muscles used for elastic energy savings versus muscles used to generate mechanical power. *American Zoologist*, **38**, 703–717.

Biewener, A. A., & Blickhan, R. (1988). Kangaroo rat locomotion: Design for elastic energy storage or acceleration. *Journal of Experimental Biology*, **140**, 243–255.

Biewener, A. A., & Roberts, T. J. (2000). Muscle and tendon contributions to force, work, and elastic energy savings: A comparative perspective. *Exercise and Sport Sciences Reviews*, **28**, 99–107.

Bird, M., & Hudson, J. (1998). Measurement of elastic-like behavior in the power squat. *Journal of Science and Medicine in Sport*, **1**, 89–99.

Bjerklie, D. (1993). High-tech olympians. *Technology Review*, **96**(1), 22–30.

Blackard, D. O., Jensen, R. L., & Ebben, W. P. (1999). Use of EMG analysis in challenging kinetic chain terminology. *Medicine and Science in Sports and Exercise*, **31**, 443–448.

Bloomfield, J., Ackland, T. R., & Elliott, B. C. (1994). *Applied anatomy and biomechanics in sport*. Melbourne: Blackwell Scientific Publications.

Bobbert, M. F. (1990). Drop jumping as a training method for jumping ability. *Sports Medicine*, **9**, 7–22.

Bobbert, M. F., & van Ingen Schenau, G. J. (1988). Coordination in vertical jumping. *Journal of Biomechanics*, **21**, 249–262.

Bobbert, M. F., & van Soest, A. J. (1994). Effects of muscle strengthening on vertical jump height: A simulation study. *Medicine and Science in Sports and Exercise*, **26**, 1012–1020.

Bobbert, M. F., & van Zandwijk, J. P. (1999). Dynamics of force and muscle stimulation in human vertical jumping. *Medicine and Science in Sports and Exercise*, **31**, 303–310.

Bobbert, M. F., Makay, M., Schinkelshoek, D., Huijing, P. A., & van Ingen Schenau, G. J. (1986). Biomechanical analysis of drop and countermovement jumps. *European Journal of Applied Physiology*, **54**, 566–573.

Bobbert, M. F., Gerritsen, K. G. M., Litjens, M. C. A., & van Soest, A. J. (1996). Why is countermovement jump height greater than squat jump height? *Medicine and Science in Sports and Exercise*, **28**, 1402–1412.

Boden, B. P., Griffin, L. Y., & Garrett, W. E. (2000). Etiology and prevention of noncontact ACL injury. *Physician and Sportsmedicine*, **28**(4), 53–60.

Bojsen-Moller, F., & Magnusson, S. P. (Eds.) (2000). Basic science of knee joint injury mechanisms [special issue]. *Scandinavian Journal of Medicine and Science in Sports*, **10**(2).

Bouisset, S. (1973). EMG and muscle force in normal motor activities. In J. E. Desmedt (Ed.), *New developments in electromyography and clinical neurophysiology* (pp. 502–510). Basel: Karger.

Brancazio, P. (1984). *Sport science: Physical laws and optimum performance*. New York: Simon & Schuster.

Bressler, B., & Frankel, J. P. (1950). The forces and moments in the leg during level walking. *Transactions of the American Society of Mechanical Engineers*, **72**, 27–36.

Brody, H. (1987). *Tennis science for tennis players*. Philadelphia: University of Pennsylvania Press.

Broer, M. R., & Zernicke, R. F. (1979). *Efficiency of human movement* (4th ed.). Philadelphia: W.B. Saunders.

Brondino, L., Suter, E., Lee, H., & Herzog, W. (2002). Elbow flexor inhibition as a function of muscle length. *Journal of Applied Biomechanics*, **18**, 46–56.

Brown, I. E., & Loeb, G. L. (1998). Post-activation potentiation—A clue for simplifying models of muscle dynamics. *American Zoologist*, **38**, 743–754.

Brown, L. E. (Ed.) (2000). *Isokinetics in human performance*. Champaign, IL: Human Kinetics.

Buckalew, D. P., Barlow, D. A., Fischer, J. W., & Richards, J. G. (1985). Biomechanical profile of elite women marathoners. *International Journal of Sport Biomechanics*, **1**, 330–347.

Bunn, J. W. (1972). *Scientific principles of coaching* (2nd ed.). Englewood Cliffs, NJ: Prentice-Hall.

Burke, R. E. (1986). The control of muscle force: Motor unit recruitment and firing patterns. In N. L. Jones, N. McCartney, & A. J. McComas (Eds.), *Human muscle power* (pp. 97–106). Champaign, IL: Human Kinetics.

Burkholder, T. J., Fingado, B., Baron, S., & Lieber, R. L. (1994). Relationship between muscle fiber types and sizes and muscle architectural properties in the mouse hindlimb. *Jounal of Morphology*, **221**, 177–190.

Callaghan, M. J., & Oldham, J. A. (1996). The role of quadricepts exercise in the treatment of patellofemoral pain syndrome. *Sports Medicine*, **21**, 384–391.

Calvano, N. J., & Berger, R. E. (1979). Effects of selected test variables on the evaluation of football helmet performance. *Medicine and Science in Sports*, **11**, 293–301.

Cappozzo, A., Marchetti, M., & Tosi, V. (1992). *Biolocomotion: A century of research using moving pictures*. Rome: Promograph.

Carlstedt, C. A., & Nordin, M. (1989). Biomechanics of tendons and ligaments. In M. Nordin & V. H. Frankel (Eds.), *Basic biomechanics of the musculoskeletal system* (2nd ed., pp. 59–74). Philadelphia: Lea & Febiger.

Cavagna, G. A., Saibene, P. F., & Margaria, R. (1965). Effect of negative work on the amount of positive work performed by an isolated muscle. *Journal of Applied Physiology*, **20**, 157–158.

Cavanagh, P. R. (1988). On "muscle action" vs "muscle contraction." *Journal of Biomechanics*, **21**, 69.

Cavanagh, P. R. (1990). Biomechanics: A bridge builder among the sport sciences. *Medicine and Science in Sports and Exercise*, **22**, 546–557.

Cavanagh, P. R., & Kram, R. (1985). The efficiency of human movement—A statement of the problem. *Medicine and Science in Sports and Exercise*, **17**, 304–308.

Cavanagh, P. R., & LaFortune, M. A. (1980). Ground reaction forces in distance running. *Journal of Biomechanics*, **15**, 397–406.

Cavanagh, P. R., Andrew, G. C., Kram, R., Rogers, M. M., Sanderson, D. J., & Hennig, E. M. (1985). An approach to biomechanical profiling of elite distance runners. *International Journal of Sport Biomechanics*, **1**, 36–62.

Chaffin, B. D., Andersson, G. B. J., & Martin, B. J. (1999). *Occupational biomechanics* (3rd ed.). New York: Wiley.

Challis, J. H. (1998). An investigation of the influence of bi-lateral deficit on human jumping. *Human Movement Science*, **17**, 307–325.

Chalmers, G. (2002). Do golgi tendon organs really inhibit muscle activity at high force levels to save muscles from injury, and adapt with strength training? *Sports Biomechanics*, **1**, 239–249.

Chaloupka, E. C., Kang, J., Mastrangelo, M. A., Scibilia, G., Leder, G. M., & Angelucci, J. (2000). Metabolic and cardiorespiratory responses to continuous box lifting and lowering in non-impaired subjects. *Journal of Orthopaedic and Sports Physical Therapy*, **30**, 249–262.

Chapman, A. E. (1985). The mechanical properties of human muscle. *Exercise and Sport Sciences Reviews*, **13**, 443–501.

Chapman, A. E., & Caldwell, G. E. (1983). Kinetic limitations of maximal sprinting speed. *Journal of Biomechanics*, **16**, 79–83.

Chapman, A. E., & Sanderson, D. J. (1990). Muscular coordination in sporting skills. In J. M. Winters & S. L. Y. Woo (Eds.), *Multiple muscle systems: Biomechanics and movement organization* (pp. 608–620). New York: Springer-Verlag.

Charlton, W. P. H., St. John, T. A., Ciccotti, M. G., Harrison, N., & Schweitzer, M. (2002). Differences in femoral notch anatomy between men and women: A magnetic resonance imaging study. *American Journal of Sports Medicine*, **30**, 329–333.

Cholewicki, J., & McGill, S. M. (1992). Lumbar posteriour ligament involvement during extremely heavy lifts estimated from fluoroscopic measurements. *Journal of Biomechanics*, **25**, 17–28.

Chow, J. W., Darling, W. G., & Ehrhardt, J. C. (1999). Determining the force–length–velocity relations of the quadriceps muscles, I: Anatomical and geometric parameters. *Journal of Applied Biomechanics*, **15**, 182–190.

Chowdhary, A. G., & Challis, J. H. (2001). The biomechanics of an overarm throwing task: A simulation model examination of optimal timing of muscle activations. *Journal of Theoretical Biology*, **211**, 39–53.

Ciccone, C. D. (2002). Evidence in practice. *Physical Therapy*, **82**, 84–88.

Ciullo, J. V., & Zarins, B. (1983). Biomechanics of the musculotendinous unit: Relation to athletic performance and injury. *Clinics in Sports Medicine*, **2**, 71–86.

Corbin, C. B., & Eckert, H. M. (Eds.) (1990). *Resolution: The Evolving undergraduate major*. AAPE Academy Papers No. 23 (p. 104). Champaign, IL: Human Kinetics.

Corcos, D. M., Jaric, S., Agarwal, G. C., & Gottlieb, G. L. (1993). Principles for learning single-joint movements, I: Enhanced performance by practice. *Experimental Brain Research*, **94**, 499–513.

Cornbleet, S. & Woolsey, N. (1996). Assessment of hamstring muscle length in school-aged children using the sit-and-reach test and the inclinometer measure of hip joint angle. *Physical Therapy*, **76**, 850–855.

Counsilman, J. E. (1971). The application of Bernoulli's Principle to human propulsion in water. In L. Lewillie and J. Clarys (Eds.), *First international symposium on biomechanics of swimming* (pp.59–71). Brussels: Université Libre de Bruxelles.

Cox, V. M., Williams, P. E., Wright, H., James, R. S., Gillott, K. L., Young, I. S., & Goldspink, D. F. (2000). Growth induced by incremental static stretch in adult rabbit latissimus dorsi muscle. *Experimental Physiology*, **85**(3), 193–202.

Cronin, J., McNair, P. J., & Marshall, R. N. (2001a). Developing explosive power: A comparison of technique and training. *Journal of Science and Medicine in Sport*, **4**, 59–70.

Cronin, J. B., McNair, P. J., & Marshall, R. N. (2001b). Magnitude and decay of stretch-induced enhancement of power output. *European Journal of Applied Physiology*, **84**, 575–581.

Cross, R. (2000). The coefficient of restitution for collisions of happy balls, unhappy balls, and tennis balls. *American Journal of Physics*, **68**, 1025–1031.

Cross, R. (2002). Measurements of the horizontal coefficient of restitution for a super ball and a tennis ball. *American Journal of Physics*, **70**, 482–489.

Daish, C. B. (1972). *The physics of ball games*. London: The English Universities Press.

Darling, W. G., & Cooke, J. D. (1987). Linked muscle activation model for movement generation and control. *Journal of Motor Behavior*, **19**, 333–354.

Davids, K., Lees, A., & Burwitz, L. (2000). Understanding and measuring coordination and control in kicking skills in soccer: Implications for talent identification and skill acquisition. *Journal of Sports Sciences*, **18**, 703–714.

Davis, K. G., & Marras, W. S. (2000). Assessment of the relationship between box weight and trunk kinematics: Does a reduction in box weight necessarily correspond to a decrease in spinal loading? *Human Factors*, **42**, 195–208.

De Deyne, P. G. (2001). Application of passive stretch and its implications for muscle fibers. *Physical Therapy*, **81**, 819–827.

De Koning, F. L., Binkhorst, R. A., Vos, J. A., & van Hof, M. A. (1985). The force–velocity relationship of arm flexion in untrained males and females and arm-trained athletes. *European Journal of Applied Physiology*, **54**, 89–94.

De Koning, J. J., Houdijk, H., de Groot, G., & Bobbert, M. F. (2000). From biomechanical theory to application in top sports: The Klapskate story. *Journal of Biomechanics*, **33**, 1225–1229.

Delp, S. L., Loan, J. P., Hoy, M. G., Zajac, F. E., & Rosen, J. M. (1990). An interactive graphic-based model of the lower extremity to study orthopaedic surgical procedures. *Journal of Biomedical Engineering*, **37**, 757–767.

De Luca, C. J. (1997). The use of surface electromyography in biomechanics. *Journal of Applied Biomechanics*, **13**, 135–163.

Denny-Brown, D., & Pennybacker, J. B. (1938). Fibrillation and fasciculation in voluntary muscle. *Brain*, **61**, 311–344.

Desmedt, J. E., & Godaux, D. (1977). Ballistic contractions in man: Characteristic recruitment pattern of single motor units of the tibialis anterior muscle. *Journal of Physiology* (London), **264**, 673–694.

DeVita, P., & Skelly, W. A. (1992). Effect of landing stiffness on joint kinetics and energenetics in the lower extremity. *Medicine and Science in Sports and Exercise*, **24**, 108–115.

Dickerman, R. D., Pertusi, R., & Smith, G. H. (2000). The upper range of lumbar spine bone mineral density? An examination of the current world record holder in the squat lift. *International Journal of Sports Medicine*, **21**, 469–470.

Di Fabio, R. P. (1999). Making jargon from kinetic and kinematic chains. *Journal of Orthopaedic and Sports Physical Therapy*, **29**, 142–143.

Dillman, C. J., Murray, T. A., & Hintermeister, R. A. (1994). Biomechanical differences of open and closed chain exercises with respect to the shoulder. *Journal of Sport Rehabilitation*, **3**, 228–238.

Dixon, S. J., Batt, M. E., & Collop, A. C. (1999). Artificial playing surfaces research: A review of medical, engineering and biomechanical aspects. *International Journal of Sports Medicine*, **20**, 209–218.

Doorenbosch, C. A. M., Veeger, D., van Zandwij, J., & van Ingen Schenau, G. (1997). On the effectiveness of force application in guided leg movements. *Journal of Motor Behavior*, **29**, 27–34.

Dorge, H. C., Bull Andersen, T., Sorensen, H., & Simonsen, E. B. (2002). Biomechanical differences in soccer kicking with the preferred and the non–preferred leg. *Journal of Sports Sciences*, **20**, 293-299.

Dotson, C. O. (1980). Logic of questionable density. *Research Quarterly for Exercise and Sport*, **51**, 23–36.

Dowell, L. J. (1978). Throwing for distance: Air resistance, angle of projection, and ball size and weight. *Motor Skills: Theory into Practice*, **3**, 11–14.

Dowling, J. J., & Vamos, L. (1993). Identification of kinetic and temporal factors related to vertical jump performance. *Journal of Applied Biomechanics*, **9**, 95–110.

Duck, T. A. (1986). Analysis of symmetric vertical jumping using a five segment computer simulation. In M. Adrian & H. Deutsch (Eds.), *Biomechanics: The 1984 olympic scientific congress proceedings* (pp. 173–178). Eugene, OR: Microform Publications.

Dufek, J. S., Bates, B. T., Stergiou, N., & James, C. R. (1995). Interactive effects between group and single–subject response patterns. *Human Movement Science*, **14**, 301-323.

Dvir, Z. (Ed.) (2000). *Clinical biomechanics*. New York: Churchill Livingstone.

Dyson, G. (1986). *Mechanics of athletics* (8th ed.). New York: Holmes & Meier.

Earl, J. E., Schmitz, R. J., & Arnold, B. L. (2001). Activation of the VMO and VL during dynamic mini-squat exercises with and without isometric hip adduction. *Journal of Electromyography and Kinesiology*, **11**, 381–386.

Ebenbichler, G. R., Oddsson, L., Kollmitzer, J., & Erim, Z. (2001). Sensory-motor control of the lower back: Implications for rehabilitation. *Medicine and Science in Sports and Exercise*, **33**, 1889–1898.

Ecker, T. (1996). *Basic track and field biomechanics* (2nd ed.). Los Altos, CA: Tafnews Press.

Edgerton, V. R., Roy, R. R., Gregor, R. J., & Rugg, S. (1986). Morphological basis of skeletal muscle power output. In N. L. Jones, N. McCartney, & A. J. McComas (Eds.), *Human muscle power* (pp. 43–64). Champaign, IL: Human Kinetics.

Edman, K. A., Mansson, A., & Caputo, C. (1997) The biphasic force velocity relationship in force muscle fibers and its evaluation in terms of cross-bridge function. *Journal of Physiology*, **503**, 141–156.

Eksstrand, J., Wiktorsson, M., Oberg, B., & Gillquist, J. (1982). Lower extremity goniometric measurements: A study to determine their reliability. *Archives of Physical Medicine and Rehabilitation*, **63**, 171–175.

Elby, A. (2001). Helping physics students learn how to learn. *American Journal of Physics*, **69**, S54–S64.

Elftman, H. (1939). Forces and energy changes in the leg during walking. *American Journal of Physiology*, **125**, 339–356.

Elliott, B. (1981). Tennis racquet selection: A factor in early skill development. *Australian Journal of Sport Sciences*, **1**, 23–25.

Elliott, B. (1983). Spin and the power serve in tennis. *Journal of Human Movement Studies*, **9**, 97–104.

Elliott, B. (1999). Biomechanics: An integral part of sport science and sport medicine. *Journal of Science and Medicine and Sport*, **2**, 299–310.

Elliott, B. C., & Mester, J. (Eds.) (1998). *Training in sport: Applying sport science*. New York: John Wiley & Sons.

Elliott, B. C., Wilson, G. J., & Kerr, G. K. (1989). A biomechanical analysis of the sticking region in the bench press. *Medicine and Science in Sports and Exercise*, **21**, 450–462.

Elliott, B. C., Baxter, K. G., & Besier, T. F. (1999). Internal rotation of the upper-arm segment during a stretch-shortening cycle movement. *Journal of Applied Biomechanics*, **15**, 381–395.

Eng, J. J., Winter, D. A., & Patla, A. E. (1997). Intralimb dynamics simplify reactive control strategies during locomotion. *Journal of Biomechanics*, **30**, 581–588.

Englehorn, R. A. (1983). Agonist and antagonist muscle EMG activity changes with skill acquisition. *Research Quarterly for Exercise and Sport*, **54**, 315–323.

Enoka, R. (2002). *The neuromechanical basis of human movement* (3rd ed.). Champaign, IL: Human Kinetics.

Enoka, R. M. (1996). Eccentric contractions require unique activation strategies by the nervous system. *Journal of Applied Physiology*, **81**, 2339–2346.

Eriksson, E. (1996). The scientific basis of rehabilitation. *American Journal of Sports Medicine*, **24**, S25–S26.

Escamilla, R. F. (2001). Knee biomechanics of the dynamic squat exercise. *Medicine and Science in Sports and Exercise*, **33**, 127–141.

Escamilla, R. F., Speer, K. P., Fleisig, G. S., Barrentine, S. W., & Andrews, J. R. (2000). Effects of throwing overweight and underweight baseballs on throwing velocity and accuracy. *Sports Medicine*, **29**, 259–272.

Ettema, G. J. C. (2001). Muscle efficiency: The controversial role of elasticity and mechanical energy conversion in stretch–shortening cycles. *European Journal of Applied Physiology*, **85**, 457–465.

Farrally, M. R., & Cochran, A. J. (Eds.) (1999). *Science and golf, III*. Champaign, IL: Human Kinetics.

Federative Committee on Anatomical Terminology (1998). *Terminologia anatomica*. Stuttgart: Thieme.

Feldman, A. G., Levin, M. F., Mitnitski, A. M., & Archambault, P. (1998). Multi-muscle control in human movements. *Journal of Electromyography and Kinesiology*, **8**, 383–390.

Feltner, M. (1989). Three-dimensional interactions in a two-segment kinetic chain, II: Application to the throwing arm in baseball pitching. *International Journal of Sport Biomechanics*, **5**, 420–450.

Feltner, M., & Dapena, J. (1986). Dynamics of the shoulder and elbow joints of the throwing arm during a baseball pitch. *International Journal of Sport Biomechanics*, **2**, 235–259.

Feltner, M. E., Fraschetti, D. J., & Crisp, R. J. (1999). Upper extremity augmentation of lower extremity kinetics during countermovement vertical jumps. *Journal of Sports Sciences*, **17**, 449–466.

Finni, T., Ikegawa, S., & Komi, P. V. (2001). Concentric force enhancement during human movement. *Acta Physiological Scandinavica,* **173,** 369–377.

Finni, T., Komi, P. V., & Lepola, V. (2000). In vivo human triceps surae and quadriceps femoris muscle function in a squat jump and counter movement jump. *European Journal of Applied Physiology,* **83,** 416–426.

Fitts, R. H., & Widrick, J. L. (1996). Muscle mechanics: Adaptations with exercise-training. *Exercise and Sport Sciences Reviews,* **24,** 427–473.

Fitzgerald, G. K., Lephart, S. M., Hwang, J. H., & Wainner, R. S. (2001). Hop tests as predictors of dynamic knee stability. *Journal of Orthopaedic and Sports Physical Therapy,* **31,** 588–597.

Fleisig, G. S., Andrews, J. R., Dillman, C. J., & Escamilla, R. F. (1995). Kinetics of baseball pitching with implications about injury mechanisms. *American Journal of Sports Medicine,* **23,** 233–239.

Fleisig, G. S., Barrentine, S. W., Zheng, N., Escamilla, R. F., & Andrews, J. R. (1999). Kinematic and kinetic comparison of baseball pitching among various levels of development. *Journal of Biomechanics,* **32,** 1371–1375.

Fowles, J. R., Sale, D. G., & MacDougall, J. D. (2000a). Reduced strength after passive stretch of the human plantar flexors. *Journal of Applied Physiology,* **89,** 1179–1188.

Fowles, J. R., MacDougall, J. D., Tarnopolsky, M. A., Sale, D. G., Roy, B. D., & Yarasheski, K. E. (2000b). The effects of acute passive stretch on muscle protein syntheis in humans. *Canadian Journal of Applied Physiology,* **25,** 165–180.

Frank, K., Baratta, R. V., Solomonow, M., Shilstone, M., & Riche, K. (2000). The effect of strength shoes on muscle activity during quiet standing. *Journal of Applied Biomechanics,* **16,** 204–209.

Frederick, E. C. (1986). Biomechanical consequences of sport shoe design. *Exercise and Sport Sciences Reviews,* **14,** 375–400.

Fuchs, R. K., Bauer, J. J., & Snow, C. M. (2001). Jumping improves hip and lumbar spine bone mass in prepubescent children: A randomized control trial. *Journal of Bone and Mineral Research,* **16,** 148–156.

Fujii, N., & Hubbard, M. (2002). Validation of a three-dimensional baseball pitching model. *Journal of Applied Biomechanics,* **18,** 135–154.

Fukashiro, S., & Komi, P. V. (1987). Joint moment and mechanical power flow of the lower limb during vertical jump. *International Journal of Sports Medicine,* **8**(Supp.), 15–21.

Fukunaga, T., Ichinose, Y., Ito, M., Kawakami, Y., & Fukashiro, S. (1997). Determinination of fasicle length and pennation in a contracting human muscle in vivo. *Journal of Applied Physiology,* **82,** 354–358.

Fukunaga, T., Kawakami, Y., Kubo, K., & Keneshisa, H. (2002). Muscle and tendon interaction during human movements. *Exercise and Sport Sciences Reviews,* **30,** 106–110.

Funato, K., Matsuo, A., & Fukunaga, T. (1996). Specific movement power related to athletic performance in weight lifting. *Journal of Applied Biomechanics,* **12,** 44–57.

Fung, Y. C. (1981). *Biomechanics: Mechanical properties of living tissues.* New York: Springer-Verlag.

Gabriel, D. A., & Boucher, J. P. (2000). Practicing a maximal performance task: A cooperative strategy for muscle activity. *Research Quarterly for Exercise and Sport,* **71,** 217–228.

Gajdosik, R. & Lusin, G. (1983). Hamstring muscle tightness: Reliability of an active–knee-extension test. *Physical Therapy,* **63,** 1085-1088.

Gambetta, V. (1995, April). Following a functional path. *Training & Conditioning,* pp. 25–30.

Gambetta, V. (1997, Winter). Leg strength for sport performance. *Sports Coach,* pp. 8–10

Gandevia, S. C. (1999). Mind, muscles and motoneurons. *Journal of Science and Medicine in Sport,* **2,** 167–180.

Gareis, H., Solomonow, M., Baratta, R., Best, R., & D'Ambrosia, R. (1992). The isometric length-force models of nine different skeletal muscles. *Journal of Biomechanics,* **25,** 903–916.

Garhammer, J. (1989). *Weight lifting and training.* In C. Vaughan (Ed.), *Biomechanics of sport* (pp. 169–211). Boca Raton, FL: CRC Press.

Garrett, W. E. (1996). Muscle strain injuries. *American Journal of Sports Medicine*, **24**(6), S2–S8.

Gatton, M. L., & Pearcy, M. J. (1999). Kinematics and movement sequencing during flexion of the lumbar spine. *Clinical Biomechanics*, **14**, 376–383.

Gielen, S. (1999). What does EMG tell us about muscle function? *Motor Control*, **3**, 9–11.

Gleim, G. W., & McHugh, M. P. (1997). Flexibility and its effects on sports injury and performance. *Sports Medicine*, **24**, 289–299.

Gottlieb, G. L. (1996). Muscle compliance: Implications for the control of movement. *Exercise and Sport Sciences Reviews*, **24**, 1–34.

Gottlieb, G. L., Corcos, D. M., & Agarwal, G. C. (1989). Strategies for the control of single mechanical degree of freedom voluntary movements. *Behavior and Brain Sciences*, **12**, 189–210.

Grace, T. G. (1985). Muscle imbalance and extremity injury: A perplexing relationship. *Sports Medicine*, **2**, 77–82.

Greenwald, R. M., Penna, L. H., & Crisco, J. J. (2001). Differences in batted ball speed with wood and aluminum baseball bats: A batting cage study. *Journal of Applied Biomechanics*, **17**, 241–252.

Gregor, R. J. (Ed.) (1997). Mechanics and energetics of the stretch-shorening cycle [special issue]. *Journal of Applied Biomechanics*, **13**(4).

Gregor, R. J., & Abelew, T. A. (1994). Tendon force measurments and movement control: A review. *Medicine and Science in Sports and Exercise*, **26**, 1359–1372.

Gregor, S. M., Perell, K. L., Rushatakankovit, S., Miyamoto, E., Muffoletto, R., & Gregor, R. J. (2002). Lower extremity general muscle moment patterns in healthy individuals during recumbent cycling. *Clinical Biomechanics*, **17**, 123–129.

Grieco, A., Molteni, G., DeVito, G., & Sias, N. (1998). Epidemiology of musculoskeletal disorders due to biomechanical overload. *Ergonomics*, **41**, 1253–1260.

Griffin, L. Y., Agel, J., Albohm, M. J., *et al.* (2000). Noncontact anterior cruciate ligament injuries: Risk factors and prevention strategies. *Journal of the American Academy of Orthopaedic Surgeons*, **8**, 141–150.

Griffiths, R. I. (1989). The mechanics of the medial gastrocnemius muscle in the freely hopping wallaby. *Journal of Experimental Biology*, **147**, 439–456.

Griffiths, R. I. (1991). Shortening of muscle fibers during stretch of the active cat medial gastrocnemius muscle: The role of tendon compliance. *Journal of Physiology* (London), **436**, 219–236.

Gruen, A. (1997). Fundamentals of videography—A review. *Human Movement Science*, **16**, 155–187.

Gulch, R. W. (1994). Force–velocity relations in human skeletal muscle. *International Journal of Sports Medicine*, **15**, S2–S10.

Hagen, K. B., Hallen, J., & Harms-Ringdahl, K. (1993). Physiological and subjective responses to maximal repetitive lifting employing stoop and squat technique. *European Journal of Applied Physiology*, **67**, 291–297.

Hakkinen, K., Komi, P., & Alen, M. (1985). Effect of explosive type strength training on isometric force– and relaxation–time, electromyographic and muscle fibre characteristics of leg extensor muscles. *Acta Physiologiica Scandinavica*, **125**, 587–600.

Hamill, J., & Knutzen, K. M. (1995). *Biomechanical basis of human movement*. Baltimore: Williams & Wikins.

Harman, E. A., Rosenstein, M. T., Frykman, P. N., & Rosenstein, R. M. (1990). The effects of arm and countermovement on vertical jumping. *Medicine and Science in Sports and Exercise*, **22**, 825–833.

Harris, J. C. (1993). Using kinesiology: A comparison of applied veins in the subdisciplines. *Quest*, **45**, 389–412.

Hatze, H. (1974). The meaning of the term: "Biomechanics." *Journal of Biomechanics*, **7**, 189–190.

Hatze, H. (1993). The relationship between the coefficient of restitution and energy losses in tennis rackets. *Journal of Applied Biomechanics*, **9**, 124–142.

Hatze, H. (2000). The inverse dynamics problem of neuromuscular control. *Biological Cybernetics*, **82**, 133–141.

Hawkins, D., & Metheny, J. (2001). Overuse injuries in youth sports: Biomechanical considerations. *Medicine and Science in Sports and Exercise*, **33**, 1701–1707.

Hay, J. G. (1987). *A bibliography of biomechanics literature* (5th ed.). Iowa City: University of Iowa Press.

Hay, J. G. (1993). *The biomechanics of sports techniques* (4th. ed.). Englewood Cliffs, NJ: Prentice-Hall.

Hay, J. G. (2000). *The biomechanics of sport techniques*, Englewood Cliffs, NJ: Prentice-Hall.

Hay, J. G., & Reid, J. G. (1982). *Anatomy, mechanics, and human motion* (2nd ed.). Englewood Cliffs, NJ: Prentice-Hall.

Hay, J. G., Vaughan, C. L., & Woodworth, G. G. (1981). Technique and performance: Identifying the limiting factors. In A. Morecki (Ed.), *Biomechanics VII-B* (pp. 511–520). Baltimore: University Park Press.

Hay, J. G., Andrews, J. G., Vaughan, C. L., & Ueya, K. (1983). Load, speed and equipment effects in strength-training exercises. In H. Matsui & K. Kobayashi (Eds.), *Biomechanics III-B* (pp. 939–950). Champaign, IL: Human Kinetics.

Hay, J. G., Miller, J. A., & Canterna, R. W. (1986). The techniques of elite male long jumpers. *Journal of Biomechanics*, **19**, 855–866.

Hayes, W. C. (1986). Bone mechanics: From tissue mechanical properties to an assessment of structural behavior. In G. W. Schmid-Schonbein, S. L.-Y. Woo, & B. W. Zweifach (Eds.), *Frontiers in biomechanics* (pp. 196–209). New York: Springer-Verlag.

Heckman, C. J., & Sandercock, T. G. (1996). From motor unit to whole muscle properties during locomotor movements. *Exercise and Sport Sciences Reviews*, **24**, 109–133.

Heerkens, Y. F., Woittiez, R. D., Huijing, P. A., Huson, A. van Ingen Schenau, G. J., & Rozendal, R. H. (1986). Passive resistance of the human knee: The effect of immobilization. *Journal of Biomedical Engineering*, **8**, 95–104.

Heiderscheit, B., Hamill, J., & Tiberio, D. (2001). A biomechanical perspective: Do foot orthoses work? *British Journal of Sports Medicine*, **35**, 4–5.

Hellebrandt, F. A. (1963). Living anatomy. *Quest*, **1**, 43–58.

Henneman, E., Somjen, G., & Carpenter, D. O. (1965). Excitability and inhibitability of motoneurons of different sizes. *Journal of Neurophysiology*, **28**, 599–620.

Herbert, R., Moore, S., Moseley, A., Schurr, K., & Wales, A. (1993). Making inferences about muscles forces from clinical observations. *Australian Journal of Physiotherapy*, **39**, 195–202.

Herring, S. A., Bergfeld, J. A., Boyd, J., *et al.* (2002). The team physician and return-to-play issues: A consensus statement. *Medicine and Science in Sports and Exercise*, **34**, 1212–1214.

Herring, S. W., Grimm, A. F., & Grimm, B. R. (1984). Regulation of sarcomere number in skeletal muscle: A comparison of hypotheses. *Muscle and Nerve*, **7**, 161–173.

Herzog, W. (1996a). Muscle function in movement and sports. *American Journal of Sports Medicine*, **24**, S14–S19.

Herzog, W. (1996b). Force-sharing among synergistic muscles: Theoretical considerations and experimental approaches. *Exercise and Sport Sciences Reviews*, **24**, 173–202.

Herzog, W. (2000). Muscle properties and coordination during voluntary movement. *Journal of Sports Sciences*, **18**, 141–152.

Herzog, W., & Leonard, T. R. (2000). The history dependence of force production in mammalian skeletal muscle following stretch-shortening and shortening-stretch cycles. *Journal of Biomechanics*, **33**, 531–542.

Herzog, W., Koh, T., Hasler, E., & Leonard, T. (2000). Specificity and plasticity of mammalian skeletal muscles. *Journal of Applied Biomechanics*, **16**, 98–109.

Higgs, J., & Jones, M. (2000). *Clinical reasoning in the health professions*. London: Butterworth-Heinemann.

Hill, A. V. (1926, April). The scientific study of athletics. *Scientific American*, pp. 224–225.

Hill, A. V. (1927, August). Are athletes machines? *Scientific American*, pp. 124–126.

Hill, A. V. (1970). *The first and last experiments in muscle mechanics*. Cambridge: Cambridge University Press.

Hinson, M. N., Smith, W. C., & Funk, S. (1979). Isokinetics: A clairfication. *Research Quarterly*, **50**, 30–35.

Hintermeister, R. A., Bey, M. J., Lange, G. W., Steadman, J. R., & Dillman, C. J. (1998). Quantification of elastic resistance knee rehabilitation exercises. *Journal of Orthopaedic and Sports Physical Therapy*, **28**, 40–50.

Hirashima, M., Kadota, H., Sakurai, S., Kudo, K., & Ohtsuki, T. (2002). Sequential muscle activity and its functional role in the upper extremity and trunk during overarm throwing. *Journal of Sports Sciences*, **20**, 301–310.

Hochmuth, G., & Marhold, G. (1978). The further development of biomechanical principles. In E. Asmussen & K. Jorgensen (Eds.), *Biomechanics VI-B* (pp. 93–106). Baltimore: University Park Press.

Hof, A. L. (2001). The force resulting from the action of mono- and biarticular muscle in a limb. *Journal of Biomechanics*, **34**, 1085–1089.

Holder-Powell, H. M., & Rutherford, O. M. (1999). Reduction in range of movement can increase maximum voluntary eccentric forces for the human knee extensor muscles. *European Journal of Applied Physiology*, **80**, 502–504.

Holt, J., Holt, L. E., & Pelham, T. W. (1996). Flexibility redefined. In T. Bauer (Ed.), *Biomechanics in Sports XIII* (pp. 170–174). Thunder Bay, Ontario: Lakehead University.

Hong, D., & Roberts, E. M. (1993). Angular movement characteristics of the upper trunk and hips in skilled baseball pitching. In J. Hamill, T. R. Derrick, & E. H. Elliott (Eds.), *Biomechanics in Sports XI* (pp. 339–343). Amherst, MA: International Society of Biomechanics in Sports.

Hong, D., Cheung, T. K., & Roberts, E. M. (2000). A three-dimensional, six-segment chain analysis of forceful overarm throwing. *Journal of Electromyography and Kinesiology*, **11**, 95–112.

Hore, J., Watts, S., & Martin, J. (1996). Finger flexion does not contribute to ball speed in overarm throws. *Journal of Sports Sciences*, **14**, 335–342.

Horrigan, J. M., Shellock, F. G., Mink, J. H., & Deutsch, A. L. (1999). Magnetic resonance imaging evaluation of muscle usage associated with three exercises for rotator cuff rehabilitation. *Medicine and Science in Sports and Exercise*, **31**, 1361–1366.

Houston, C. S., & Swischuk, L. E. (1980). Varus and valgus—No wonder they are confused. *New England Journal of Medicine*, **302**(8), 471–472.

Hubbard, M. (1989). The throwing events in track and field. In C. Vaughan (Ed.), *Biomechanics of sport* (pp. 213–238). Boca Raton, FL: CRC Press.

Hubbard, M. (1993). Computer simulation in sport and industry. *Journal of Biomechanics*, **26**(S1), 53–61.

Hubbard, M., & Alaways, L. W. (1987). Optimum release conditions for the new rules javelin. *International Journal of Sport Biomechanics*, **3**, 207–221.

Hubbard, M., de Mestre, N. J., & Scott, J. (2001). Dependence of release variables in the shot put. *Journal of Biomechanics*, **34**, 449–456.

Hubley, C. L., & Wells, R. P. (1983). A work–energy approach to determine individual joint contributions to vertical jump performance. *European Journal of Applied Physiology*, **50**, 247–254.

Hudson, J. L. (1986). Coordination of segments in the vertical jump. *Medicine and Science in Sports and Exercise*, **18**, 242–251.

Hudson, J. L. (1989). *Guidelines and standards: Application of concepts.* Paper presented to the AAHPERD National Convention, Boston.

Hudson, J. L. (1990). The value visual variables in biomechanical analysis. In E. Kreighbaum & A. McNeill (Eds.), *Proceedings of the VIth international symposium of the international society of biomechanics in sports* (pp. 499–509). Bozeman: Montana State University.

Hudson, J. L. (1995). Core concepts in kinesiology. *JOPERD*, **66**(5), 54–55, 59–60.

Hudson, J. L. (1997). The biomechanics body of knowledge. In J. Wilkerson, K. Ludwig, & M. Bucher (Eds.), *Proceedings of the IVth national symposium on teaching biomechanics* (pp. 21–42). Denton: Texas Woman's University Press.

Hudson, J. L. (2000). Sports biomechanics in the year 2000: Some observations on trajectory. In Y. Hong & D. P. Johns (Eds.), *Proceedings of the XVIth international symposium of the international society of biomechanics in sports* (Vol. 2, pp. 525–528). Hong Kong: The Chinese University of Hong Kong.

Huijing, P. A., & Baan, G. C. (2001). Extramuscular myofacial force transmission within the rat anterior tibial compartment: Proximo-distal differences in muscle force. *Acta Physiological Scandinavica*, **173**, 297–311.

Hukins, D. W. L., Kirby, M. C., Sikoryu, T. A., Aspden, R. M., & Cox, A. J. (1990). Comparison of structure, mechanical properties, and functions of lumbar spinal ligaments. *Spine*, **15**, 787–795.

Hunter, J. P., & Marshall, R. N. (2002). Effects of power and flexibility training on vertical jump technique. *Medicine and Science in Sports and Exercise*, **34**, 478–486.

Hutton, R. S. (1993). Neuromuscular basis of stretching exercise. In P. Komi (Ed.), *Strength and power in sports* (pp. 29–38). London: Blackwell Scientific Publications.

Huxham, F. E., Goldie, P. A., & Patla, A. E. (2001). Theoretical considerations in balance assessment. *Australian Journal of Physiotherapy*, **47**, 89–100.

Ichinose, Y., Kawakami, Y., Ito, M., Kanehisa, H., & Fukunaga, T. (2000). In vivo extimation of contraction velocity of human vastus lateralis muscle during "isokinetic" action. *Journal of Applied Physiology*, **88**, 851–856.

International Society of Biomechanics Subcommittee on Standards and Terminology (1987, Winter). *ISB Newsletter*, p. 9.

Iossifidou, A. N., & Baltzopoulos, V. (2000). Inertial effects on moment development during isokinetic concentric knee extension testing. *Journal of Orthopaedic and Sports Physical Therapy*, **30**, 317–327.

Ito, M., Kawakami, Y., Ichinose, Y., Fukashiro, S., & Fukunaga, T. (1998). Nonisometric behavior of fascicles during isometric contractions of a human muscle. *Journal of Applied Physiology*, **85**, 1230–1235.

Izquierdo, M., Ibanez, J., Gorostiaga, E., Gaurrues, M., Zuniga, A., Anton, A., Larrion J. L., & Hakkinen, K. (1999). Maximal strength and power characteristics in isometric and dynamic actions of the upper and lower extremities in middle-aged and older men. *Acta Physiological Scandinavica*, **167**, 57–68.

Jackson, A. S., & Frankiewicz, R. J. (1975). Factorial expressions of muscular strength. *Research Quarterly*, **46**, 206–217.

Jakobi, J. M., & Cafarelli, E. (1998). Neuromuscular drive and force production are not altered during bilateral contractions. *Journal of Applied Physiology*, **84**, 200–206.

James, C. R., Dufek, J. S., & Bates, B. T. (2000). Effects of injury proneness and task difficulty on joint kinetic variability. *Medicine and Science in Sports and Exercise*, **32**, 1833–1844.

Jensen, J. L., Phillips, S. J., & Clark, J. E. (1994). For young jumpers, differences are in movement's control, not its coordination. *Research Quarterly for Exercise and Sport*, **65**, 258–268.

Johnson, R. (1990). *Effects of performance principle training upon skill analysis competency*. Ph.D. dissertation. Ohio State University.

Johnson, D. L., & Bahamonde, R. (1996). Power output estimate in university athletes. *Journal of Strength and Conditioning Research*, **10**, 161–166.

Jorgensen, T. P. (1994). *The physics of golf*. New York: American Institute of Physics.

Kamibayashi, L. K., & Richmond, F. J. R. (1998). Morphometry of human neck muscles. *Spine*, **23**, 1314–1323.

Kaneko, M., Fuchimoto, T., Toji, H., & Suei, K. (1983). Training effect of different loads on the force–velocity and mechanical power output in human muscle. *Scandinavian Journal of Sports Sciences*, **5**, 50–55.

Kasprisin, J. E., & Grabiner, M. D. (2000). Joint angle-dependence of elbow flexor activation levels during isometric and isokinetics maximum voluntary contractions. *Clinical Biomechanics*, **15**, 743–749.

Kawakami, Y., Ichinose, Y., & Fukunaga, T. (1998). Architectural and functional features of human triceps surae muscles during contraction. *Journal of Applied Physiology*, **85**, 398–404.

Kawakami, Y., Ichinose, Y., Kubo, K., Ito, M., Imai, M., & Fukunaga, T. (2000). Architecture of contracting human muscles and its functional significance. *Journal of Applied Biomechanics*, **16**, 88–98.

Keating, J. L., & Matyas, T. A. (1998). Unpredictable error in dynamometry measurements: A quantitative analysis of the literature. *Isokinetics and Exercise Science*, **7**, 107–121.

Kellis, E., & Baltzopoulos, V. (1997). The effects of antagonistic moment on the resultant knee joint moment during isokinetics testing of the knee extensors. *European Journal of Applied Physiology*, **76**, 253–259.

Kellis, E., & Baltzopoulos, V. (1998). Muscle activation differences between eccentric and concentric isokinetic exercise. *Medicine and Science in Sports and Exercise*, **30**, 1616–1623.

Kelly, B. T., Kadrmas, W. R., & Speer, K. P. (1996). The manual muscle examination for rotator cuff strength: An electromyographic investigation. *American Journal of Sports Medicine*, **24**, 581–588.

Kendall, F. P., McCreary, E. K., & Provance, P. G. (1993). *Muscles: Testing and function* (4th ed.). Baltimore: Williams & Wilkins.

Khan, K. M., Cook, J. L., Taunton, J. E., & Bonar, F. (2000). Overuse tendinosis, not tendinitis, I: A new paradigm for a difficult clinical problem. *Physician and Sportsmedicine*, **28**(5), 38–48.

Kibler, W. B., & Livingston, B. (2001). Closed-chain rehabilitation of the upper and lower extremity. *Journal of the American Academy of Orthopaedic Surgeons*, **9**, 412–421.

Kilmartin, T. E., & Wallace, A. (1994). The scientific basis for the use of biomechanical foot orthoses in the treatment of lower limb sports injuries—A review of the literature. *British Journal of Sports Medicine*, **28**, 180–184.

Kim, E., & Pack, S. (2002). Students do not overcome conceptual difficulties after solving 1000 traditional problems. *American Journal of Physics*, **70**, 759–765.

Kinesiology Academy (1992, Spring). Guidelines and standards for undergraduate kinesiology/biomechanics. *Kinesiology Academy Newsletter*, pp. 3–6.

Kingma, I., Toussaint, H. M., Commissaris, D. A. C. M., Hoozemans, M. J. M., & Ober, M. J. (1995). Optimizing the determination of the body center of mass. *Journal of Biomechanics*, **28**, 1137–1142.

Kirkendall, B. T., & Garrett, W. E. (1997). Function and biomechanics of tendons. *Scandinavian Journal of Medicine and Sciences in Sports*, **7**, 62–66.

Klee, A., Jollenbeck, T., & Weimann, K. (2000). Correlation between muscular function and posture—Lowering the degree of pelvic inclination with exercise. In Y. Hong & D. P. Johns (Eds.), *Proceedings of XVIIIth international symposium on biomechanics in sports* (Vol. 1, pp. 162–165). Hong Kong: The Chinese University of Hong Kong.

Kleissen, R. F. M., Burke, J. H., Harlaar, J., & Zilvold, G. (1998). Electromyography in the biomechanical analysis of human movement and its clinical application. *Gait and Posture*, **8**, 143–158.

Klute, G. K., Kallfelz, C. F., & Czerniecki, J. M. (2001). Mechanical properties of prosthetic limbs: Adapting to the patient. *Journal of Rehabilitation Research and Development*, **38**, 299–307.

Knudson, D. (1991). The tennis topspin forehand drive: Technique changes and critical elements. *Strategies*, **5**(1), 19–22.

Knudson, D. (1993). Biomechanics of the basketball jump shot: Six key teaching points. *JOPERD*, **64**(2), 67–73.

Knudson, D. (1996). A review of exercises and fitness tests for the abdominal muscles. *Sports Medicine Update*, **11**(1), 4–5, 25–30.

Knudson, D. (1997). The Magnus Effect in baseball pitching. In J. Wilkerson, K. Ludwig, & M. Butcher (Eds.), *Proceedings of the 4th national symposium on teaching biomechanics* (pp. 121–125). Denton: Texas Woman's University Press.

Knudson, D. (1998). Stretching: Science to practice. *JOPERD*, **69**(3), 38–42.

Knudson, D. (1999a). Issues in abdominal fitness: Testing and technique. *JOPERD*, **70**(3), 49–55, 64.

Knudson, D. (1999b). Stretching during warm-up: Do we have enough evidence? *JOPERD*, **70**(7), 24–27, 51.

Knudson, D. (1999c). Validity and reliability of visual ratings of the vertical jump. *Perceptual and Motor Skills*, **89**, 642–648.

Knudson, D. (2000). What can professionals qualitatively analyze? *JOPERD*, **71**(2), 19–23.

Knudson, D. (2001a, July). Improving stroke technique using biomechanical principles. *Coaching and Sport Science Review*, pp. 11–13.

Knudson, D. (2001b). A lab to introduce biomechanics using muscle actions. In J. R. Blackwell and D. V. Knudson (Eds.), *Proceedings: Fifth national symposium on teaching biomechanics in sports* (pp. 36–41). San Francisco: University of San Francisco.

Knudson, D. (2001c). Accuracy of predicted peak forces during the power drop exercise. In J. R. Blackwell (Ed.) *Proceedings of oral sessions: XIX international symposium on biomechanics in sports* (pp. 135–138). San Francisco: University of San Francisco.

Knudson, D., & Bahamonde, R. (2001). Effect of endpoint conditions on position and velocity near impact in tennis. *Journal of Sports Sciences*, **19**, 839–844.

Knudson, D., & Blackwell, J. (1997). Upper extremity angular kinematics of the one-handed backhand drive in tennis players with and without tennis elbow. *International Journal of Sports Medicine*, **18**, 79–82.

Knudson, D., & Kluka, D. (1997). The impact of vision and vision training on sport performance. *JOPERD*, **68**(4),17–24.

Knudson, D., & Morrison, C. (1996). An integrated qualitative analysis of overarm throwing. *JOPERD*, **67**(6), 31–36.

Knudson, D., & Morrison, C. (2002). *Qualitative analysis of human movement* (2nd ed.). Champaign, IL: Human Kinetics.

Knudson, D., Magnusson, P., & McHugh, M. (2000, June). Current issues in flexibility fitness. *The President's Council on Physical Fitness and Sports Research Digest*, pp. 1-8.

Knuttgen, H. G., & Kraemer, W. J. (1987). Terminology and measurement in exercise performance. *Journal of Applied Sport Science Research*, **1**, 1–10.

Koh, T. J. (1995). Do adaptations in serial sarcomere number occur with strength training? *Human Movement Science*, **14**, 61–77.

Komi, P. V. (1984). Physiological and biomechanical correlates of muscle function: Effects of muscle structure and stretch-shortening cycle on force and speed. *Exercise and Sport Sciences Reviews*, **12**, 81–121.

Komi, P. V. (1986). The stretch-shortening cycle and human power output. In N. L. Jones, N. McCartney, & A. J. McComas (Eds.), *Human muscle power* (pp. 27–39). Champaign, IL: Human Kinetics.

Komi, P. V. (Ed.) (1992). *Strength and power in sport*. London: Blackwell Science.

Komi, P. V., & Gollhofer, A. (1997). Stretch reflexes can have an important role in force enhancement during the SSC exercise. *Journal of Applied Biomechanics*, **13**, 451–460.

Komi, P. V., Belli, A., Huttunen, V., Bonnefoy, R., Geyssant, A., & Lacour, J. R. (1996). Optic fibre as a transducer of tendomuscular forces. *European Journal of Applied Physiology*, **72**, 278–280.

Kornecki, S., Kebel, A., & Siemienski, A. (2001). Muscular co-operation during joint stabilization, as reflected by EMG. *European Journal of Applied Physiology*, **84**, 453–461.

Kouzaki, M., Shinohara, M., Masani, K., Kanehisa, H., & Fukunaga, T. (2002). Alternate muscle activity observed between knee extensor synergists during low-level sustained contractions. *Journal of Applied Physiology*, **93**, 675–684.

Kovacs, I., Tihanyi, J., DeVita, P., Racz, L., Barrier, J., & Hortobagyi, T. (1999). Foot placement modifies kinematics and kinetics during drop jumping. *Medicine and Science in Sports and Exercise*, **31**, 708–716.

Kraan, G. A., van Veen, J., Snijders, C. J., & Storm, J. (2001). Starting from standing: Why step backwards? *Journal of Biomechanics*, **34**, 211–215.

Kraemer, W. J., Fleck, S. J., & Evans, W. J. (1996). Strength and power training: Physiological mechanisms of adaptation. *Exercise and Sport Sciences Reviews*, **24**, 363–397.

Kreighbaum, E., & Barthels, K. M. (1996). *Biomechanics: A qualitative approach to studying human movement*. Boston: Allyn & Bacon.

Kreighbaum, E. F., & Smith, M. A. (Eds.) (1996). *Sports and fitness equipment design*. Champaign, IL: Human Kinetics.

Krugh, J., & LeVeau, B. (1999). Michael Jordan's vertical jump [abstract]. *Journal of Orthopaedic and Sports Physical Therapy*, **20**, A10.

Kubo, K. Kawakami, Y., & Fukunaga, T. (1999). Influence of elastic properties of tendon structures on jump performance in humans. *Journal of Applied Physiology*, **87**, 2090–2096.

Kubo, K., Kanehisa, H., Kawakami, Y., & Fukunaga, T. (2000a). Elasticity of tendon structures of the lower limbs in sprinters. *Acta Physiological Scandinavica*, **168**, 327–335.

Kubo, K., Kanehisa, H., Takeshita, D., Kawakami, Y., Fukashiro, S., & Fukunaga, T. (2000b). *In vivo* dynamics of human medial gastrocnemius muscle–tendon complex during stretch-shortening cycle exercise. *Acta Physiological Scandinavica*, **170**, 127–135.

Kubo, K., Kanehisa, H., Kawakami, Y., & Fukunaga, T. (2001a). Influence of static stretching on viscoelastic properties of human tendon *in vivo*. *Journal of Applied Physiology*, **90**, 520–527.

Kubo, K., Kanehisa, H., Ito, M., & Fukunaga, T. (2001b). Effects of isometric training on elasticity of human tendon structures *in vivo*. *Journal of Applied Physiology*, **91**, 26–32.

Kubo, K., Kanehisa, H., & Fukunaga, T. (2002). Effect of stretching training on the viscoelastic properties of human tendon structures *in vivo*. *Journal of Applied Physiology*, **92**, 595–601.

Kulig, K., Andrews, J. G., & Hay, J. G. (1984). Human strength curves. *Exercise and Sport Sciences Reviews*, **12**, 417–466.

Kumagai, K., Abe, T., Brechue, W. F., Ryushi, T., Takano, S., & Mizuno, M. (2000). Sprint performance is related to muscle fascicle length in male 100-m sprinters. *Journal of Applied Physiology*, **88**, 811–816.

Kumar, S. (1999). *Biomechanics in ergonomics*. London: Taylor & Francis.

Kurokawa, S., Fukunaga, T., & Fukashiro, S. (2001). Behavior of fasicles and tendinous structures of human gastrocnemius during vertical jumping. *Journal of Applied Physiology*, **90**, 1349–1358.

Lafortune, M. A., & Hennig, E. M. (1991). Contribution of angular motion and gravity to tibial acceleration. *Medicine and Science in Sports and Exercise*, **23**, 360–363.

Lake, M. J., & Cavanagh, P. R. (1996). Six weeks of training does not change running mechanics or improve running economy. *Medicine and Science in Sports and Exercise*, **28**, 860–869.

Lamontagne, A., Malouin, F., & Richards, C. L. (2000). Contribution of passive stiffness to ankle plantar flexor moment during gait after stroke. *Archives of Physical Medicine and Rehabilitation*, **81**, 351–358.

Larkins, C., & Snabb, T. E. (1999). Positive versus negative foot inclination for maximum height two–leg vertical jumps. *Clinical Biomechanics*, **14**, 321-328.

Latash, M. L., & Zatsiorsky, V. M. (Eds.) (2001). *Classics in movement science*. Champaign, IL: Human Kinetics.

Lawson, R. A., & McDermott, L. C. (1987). Student understanding of work–energy and impulse–momentum theorems. *American Journal of Physics*, **55**, 811–817.

Lees, A. (1999). Biomechanical assessment of individual sports for improved performance. *Sports Medicine*, **28**, 299–305.

Lees, A., & Barton, G. (1996). The interpretation of relative momentum data to assess the contribution of the free limbs to the generation of vertical velocity in sports activities. *Journal of Sports Sciences*, **14**, 503–511.

Lees, A., & Fahmi, E. (1994). Optimal drop heights for plyometric training. *Ergonomics*, **37**, 141–148.

Legwold, G. (1984). Can biomechanics produce olympic medals? *Physician and Sportsmedicine*, **12**(1), 187–189.

Lehman, G. J., & McGill, S. M. (2001). Quantification of the differences in electromyographic activity magnitude between the upper and lower portions of the rectus abdominis muscle during selected trunk exercises. *Physical Therapy*, **81**, 1096–1101.

LeVeau, B. (1992). *Williams and Lissner's: Biomechanics of human motion* (3rd ed.). Philadelphia: W. B. Sanders.

Lieber, R. L., & Bodine-Fowler, S. C. (1993). Skeletal muscle mechanics: Implications for rehabilitation. *Physical Therapy*, **73**, 844–856.

Lieber, R. L., & Friden, J. (1993). Muscle damage is not a function of muscle force but active muscle strain. *Journal of Applied Physiology*, **74**, 520–526.

Lieber, R. L., & Friden, J. (1999). Mechanisms of muscle injury after eccentric contration. *Journal of Science and Medicine in Sport*, **2**, 253–265.

Lieber, R. L., & Friden, J. (2000). Functional and clinical significance of skeletal muscle architecture. *Muscle and Nerve*, **23**, 1647–1666.

Lindbeck. L., & Kjellberg, K. (2000). Gender differences in lifting technique. *Ergonomics*, **44**, 202–214.

Luthanen, P., & Komi, P. V. (1978a). Segmental contributions to forces in vertical jump. *European Journal of Applied Physiology*, **38**, 181–188.

Luthanen, P., & Komi, P. V. (1978b). Mechanical factors influencing running speed. In E. Asmussen & K. Jorgensen (Eds.), *Biomechanics VI–B* (pp. 23-29). Baltimore: University Park Press.

Luttgens, K., & Wells, K. F. (1982). *Kinesiology: Scientific basis of human motion* (7th ed.). Philadelphia: Saunders.

Lutz, G. J., & Lieber, R. L. (1999). Skeletal muscle myosin II—Structure and function. *Exercise and Sport Sciences Reviews*, **27**, 63–77.

Lutz, G. E., Palmitier, R. A., An, K. N., & Chao, E. Y. S. (1993). Comparison of tibiofemoral joint forces during open-kinetic-chain and closed-kinetic-chain exercises. *Journal of Bone and Joint Surgery*, **75A**, 732–739.

Maganaris, C. N. (2001). Force–length characteristics of *in vivo* human skeletal muscle. *Acta Physiologica Scandinavica*, **172**, 279–285.

Maganaris, C. N., & Paul, J. P. (2000). Load-elongation characteristics of the *in vivo* human tendon and aponeurosis. *Journal of Experimental Biology*, **203**, 751–756.

Magnusson, S. P., Aagaard, P., Simonsen, E. B., & Bojsen-Moller, F. (2000). Passive tensile stress and energy of the human hamstring muscles *in vivo*. *Scandinavian Journal of Medicine and Science in Sports*, **10**, 351–359.

Maisel, I. (1998, August). MadDash. *Sports Illustrated*, pp. 39–43.

Malinzak, R. A., Colby, S. M., Kirkendall, D. T., Yu, B., & Garrett, W. E. (2001). A comparison of knee joint motion patterns between men and women in selected athletic tasks. *Clinical Biomechanics*, **16**, 438–445.

Malone, T. R., Hardaker, W. T., Garrett, W. E., Feagin, J. A., & Bassett, F. H. (1993). Relationship of gender to anterior cruciate ligament injuries in intercollegiate basketball players. *Journal of the Southern Orthopaedic Association*, **2**, 36–39.

Mann, R. V. (1981). A kinetic analysis of sprinting. *Medicine and Science in Sports and Exercise*, **13**, 325–328.

Margaria, R., Aghemo, P., & Rovelli, E. (1966). Measurement of muscular power (anaerobic) in man. *Journal of Applied Physiology*, **21**, 1662–1664.

Marshall, R. N., & Elliott, B. C. (2000). Long-axis rotation: The missing link in proximal-to-distal segmental sequencing. *Journal of Sports Sciences*, **18**, 247–254.

Matanin, M. J. (1993). *Effects of performance principle training on correct analysis and diagnosis of motor skills*. Ph.D. dissertation. Ohio State University. Dissertation Abstracts International, 54, 1724A.

McBride, J. M., Triplett-McBride, T., Davie, A., & Newton, R. U. (1999). A comparison of strength and power characteristics between power lifters, olympic lifters, and sprinters. *Journal of Strength and Conditioning Research*, **13**, 58–66.

McGill, S. M. (1998). Low back exercises: Evidence for improving exercise regimens. *Physical Therapy*, **78**, 754–765.

McGill, S. M. (2001). Low back stability: From formal description to issues for performance and rehabilitation. *Exercise and Sport Sciences Reviews*, **29**, 26–31.

McGill, S. M., & Norman, R. W. (1985). Dynamically and statically determined low back moments during lifting. *Journal of Biomechanics*, **18**, 877–886.

McGill, S. M., Hughson, R. L., & Parks, K. (2000). Changes in lumbar lordosis modify the role of the extensor muscles. *Clinical Biomechanics*, **15**, 777–780.

McGinnis, P., & Abendroth-Smith, J. (1991). Impulse, momentum, and water balloons. In J. Wilkerson, E. Kreighbaum, & C. Tant, (Eds.), *Teaching kinesiology and biomechanics in sports* (pp. 135–138). Ames: Iowa State University.

McGregor, R. R., & Devereux, S. E. (1982). *EEVeTeC*. Boston: Houghton Mifflin Co.

McLaughlin, T. M., Lardner, T. J., & Dillman, C. J. (1978). Kinetics of the parallel squat. *Research Quarterly*, **49**, 175–189.

McLean, S. P., & Hinrichs, R. N. (2000a). Influence of arm position and lung volume on the center of buoyancy of competitive swimmers. *Research Quarterly for Exercise and Sport*, **71**, 182–189.

McLean, S. P., & Hinrichs, R. N. (2000b). Buoyancy, gender, and swimming performance. *Journal of Applied Biomechanics*, **16**, 248–263.

McLean, S. P., & Reeder, M. S. (2000). Upper extremity kinematics of dominant and non–dominant side batting. *Journal of Human Movement Studies*, **38**, 201-212.

McNair, P. J., Prapavessis, H., & Callender, K. (2000). Decreasing landing forces: Effect of instruction. *British Journal of Sports Medicine*, **34**, 293–296.

McPoil, T. G., Cornwall, M. W., & Yamada, W. (1995). A comparison of two in-shoe plantar pressure measurement systems. *The Lower Extremity*, **2**, 95–103.

Mehta, R. D. (1985). Aerodynamics of sports balls. *Annual Review of Fluid Mechanics*, **17**, 151–189.

Mehta, R. D., & Pallis, J. M. (2001a). The aerodynamics of a tennis ball. *Sports Engineering*, **4**, 177–189.

Mehta, R. D., & Pallis, J. M. (2001b). Sports ball aerodynamics: Effects of velocity, spin and surface roughness. In F. H. Froes, & S. J. Haake (Eds.), *Materials and science in sports* (pp. 185–197). Warrendale, PA: The Minerals, Metals and Materials Society [TMS].

Meinel, K., & Schnabel, G. (1998). *Bewegungslehre— Sportmotorik*. Berlin: Sportverlag Berlin.

Mena, D., Mansour, J. M., & Simon, S. R. (1981). Analysis and synthesis of human swing leg motion during gait and its clinical applications. *Journal of Biomechanics*, **14**, 823–832.

Mero, A., Luthanen, P., Viitasalo, J. T., & Komi, P. V. (1981). Relationships between the maximal running velocity, muscle fiber characteristics, force production, and force relaxation of sprinters. *Scandinavian Journal Sports Sciences*, **3**, 16–22.

Mero, A., Komi, P. V., & Gregor, R. J. (1992). Biomechanics of sprint running: A review. *Sports Medicine*, **13**, 376–392.

Messier, S. P., & Cirillo, K. J. (1989). Effects of verbal and visual feedback system on running technique, perceived exertion and running economy in female novice runners. *Journal of Sports Sciences*, **7**, 113–126.

Miller, D. I. (1980). Body segment contributions to sport skill performance: Two contrasting approaches. *Research Quarterly for Exercise and Sport*, **51**, 219–233.

Miller, D. I. (2000). *Biomechanics of competitive diving*. Fort Lauderdale, FL: U.S. Diving.

Minozzi, S., Pistotti, V., & Forni, M. (2000). Searching for rehabilitation articles on Medline and Embase: An example with cross-over design. *Archives of Physical Medicine and Rehabilitation*, **81**, 720–722.

Mirka, G., Kelaher, D., Baker, A., Harrison, H., & Davis, J. (1997). Selective activation of the external oblique musculature during axial torque production. *Clinical Biomechanics*, **12**, 172–180.

Mizera, F., & Horvath, G. (2002). Influence of environmental factors on shot put and hammer throw range. *Journal of Biomechanics*, **35**, 785–796.

Montgomery, J., & Knudson, D. (2002). A method to determine stride length for baseball pitching. *Applied Research in Coaching and Athletics Annual*, **17**, 75–84.

Moore, S. P., & Marteniuk, R. G. (1986). Kinematic and electromyographic changes that occur as a function of learning a time-constrained aiming task. *Journal of Motor Behavior*, **18**, 397–426.

Morgan, D. L., Whitehead, N. P., Wise, A. K., Gregory, J. E., & Proske, U. (2000). Tension changes in the cat soleus muscle following slow stretch or shortening of the contracting muscle. *Journal of Physiology* (London), **522**, 503–513.

Moritani, T. (1993). Neuromuscular adaptations during the acquisition of muscle strength, power, and motor tasks. *Journal of Biomechanics*, **26**(S1), 95–107.

Moritani, T., & Yoshitake, Y. (1998). The use of electromyography in applied physiology. *Journal of Electromyography and Kinesiology*, **8**, 363–381.

Morrish, G. (1999). Surface electromyography: Methods of analysis, reliability, and main applications. *Critical Reviews in Physical and Rehabilitation Medicine*, **11**, 171–205.

Morriss, C., Bartlett, R., & Navarro, E. (2001). The function of blocking in elite javelin throws: A re–evaluation. *Journal of Human Movement Studies*, **41**, 175-190.

Mow, V. C., & Hayes, W. C. (Eds.) (1991). *Basic orthopaedic biomechanics*. New York: Raven Press.

Mureika, J. R. (2000). The legality of wind and altitude assisted performances in the sprints. *New Studies in Athletics*, **15**(3/4), 53–58.

Murray, M. P., Seireg, A., & Scholz, R. C. (1967). Center of gravity, center of pressure, and supportive forces during human activities. *Journal of Applied Physiology*, **23**, 831–838.

Murray, W. M., Buchanan, T. S., & Delp, S. L. (2000). The isometric functional capacity of muscles that cross the elbow. *Journal of Biomechanics*, **33**, 943–952.

Nagano, A., & Gerritsen, K. G. M. (2001). Effects of neuromuscular strength training on vertical jumping performance—A computer simulation study. *Journal of Applied Biomechanics*, **17**, 113–128.

Nagano, A., Ishige, Y., & Fukashiro, S. (1998). Comparison of new approaches to estimate mechanical output of individual joints in vertical jumps. *Journal of Biomechanics*, **31**, 951–955.

Nakazawa, K., Kawakami, Y., Fukunaga, T., Yano, H., & Miyashita, M. (1993). Differences in activation patterns in elbow flexor muscles during isometric, concentric, and eccentric contractions. *European Journal of Applied Physiology*, **66**, 214–220.

Neal, R. J., & Wilson, B. D. (1985). 3D kinematics and kinetics of the golf swing. *International Journal of Sport Biomechanics*, **1**, 221–232.

Neptune, R. R., & Kautz, S. A. (2001). Muscle activation and deactivation dynamics: The governing properties in fast cyclical human movement performance? *Exercise and Sport Sciences Reviews*, **29**, 76–81.

Neptune, R. R., & van den Bogert, A. J. (1998). Standard mechanical energy analyses do not correlate with muscle work in cycling. *Journal of Biomechanics*, **31**, 239–245.

Newell, K. M., Kugler, P. N., van Emmerick, R. E. A., & McDonald, P. V. (1989). Search strategies and the acquisition of coordination. In S. A. Wallace (Ed.), *Perspectives on the coordination of movement* (pp. 85–122). New York: Elsevier.

Newton, R. U., Kraemer, W. J., Hakkinen, K., Humphries, B. J., & Murphy, A. J. (1996). Kinematics, kinetics, and muscle activation during explosive upper body movements. *Journal of Applied Biomechanics*, **12**, 31–43.

Newton, R. U., Murphy, A. J., Humphries, B. J., Wilson, G. J., Kraemer, W. J., & Hakkinen, K. (1997). Influence of load and stretch shortening cycle on the kinematics, kinetics and muscle activation that occurs during explosive upper-body movements. *European Journal of Applied Physiology*, **75**, 333–342.

Nichols, T. R. (1994). A biomechanical perspective on spinal mechanisms of coordinated muscular action: An architecture principle. *Acta Anatomica*, **151**, 1–13.

Nielsen, A. B., & Beauchamp, L. (1992). The effect of training in conceptual kinesiology on feedback provision patterns. *Journal of Teaching in Physical Education*, **11**, 126–138.

Nigg, B. M. (Ed.) (1986). *Biomechanics of running shoes*. Champaign, IL: Human Kinetics.

Nigg, B. M., Luthi, S. M., & Bahlsen, H. A. (1989). The tennis shoe—Biomechanical design criteria. In B. Segesser & W. Pforringer (Eds.), *The shoe in sport* (pp. 39–46). Chicago: Year Book Medical Publishers.

Nordin, M., & Frankel, V. (2001). *Basic biomechanics of the musculoskeletal system* (3rd ed.). Baltimore: Williams & Wilkins.

Norkin, C. C., & White, D. J. (1995). *Measurement of joint motion—A guide to goniometry* (2nd ed.). Philadelphia: F. A. Davis.

Norman, R. (1975). Biomechanics for the community coach. *JOPERD*, **46**(3), 49–52.

Norman, R. W. (1983). Biomechanical evaluation of sports protective equipment. *Exercise and Sport Sciences Reviews*, **11**, 232–274.

Novacheck, T. F. (1998). The biomechanics of running. *Gait and Posture*, **7**, 77–95.

Nunome, H., Asai, T., Ikegami, Y., & Sakurai, S. (2002). Three-dimensional kinetic analysis of side-foot and instep soccer kicks. *Medicine and Science in Sports and Exercise*, **34**, 2028–2036.

Nussbaum, M. A., & Chaffin, D. B. (1997). Pattern classification reveals intersubject group differences in lumbar muscle recruitment during static loading. *Clinical Biomechanics*, **12**, 97–106.

Olds, T. (2001). Modelling of human locomotion: Applications to cycling. *Sports Medicine*, **31**, 497–509.

Owens, M. S., & Lee, H. Y. (1969). A determination of velocities and angles of projection for the tennis serve. *Research Quarterly*, **40**, 750–754.

Pandy, M. G. (1999). Moment arm of a muscle force. *Exercise and Sport Sciences Reviews*, **27**, 79–118.

Pandy, M. G., Zajac, F. E., Sim, E., & Levine, W. S. (1990). An optimal control model for maximum–height human jumping. *Journal of Biomechanics*, **23**, 1185-1198.

Panjabi, M. M., & White, A. A. (2001). *Biomechanics in the musculoskeletal system*. New York: Churchill Livingstone.

Pappas, G. P., Asakawa, D. S., Delp, S. L., Zajac, F. E., & Drace, J. E. (2002). Nonuniform shortening in the biceps brachii during elbow flexion. *Journal of Applied Physiology*, **92**, 2381–2389.

Patel, T. J., & Lieber, R. L. (1997). Force transmission in skeletal muscle: From actomyosin to external tendons. *Exercise and Sport Sciences Reviews*, **25**, 321–364.

Patla, A. E. (1987). Some neuromuscular strategies characterising adaptation process duirng prolonged activity in humans. *Canadian Journal of Sport Sciences*, **12**(Supp.), 33S–44S.

Paton, M. E., & Brown, J. M. N. (1994). An electomyographic analysis of functional differentiation in human pectoralis major muscle. *Journal of Electromyography and Kinesiology*, **4**, 161–169.

Paton, M. E., & Brown, J. M. N. (1995). Functional differentiation within latissimus dorsi. *Electromyography and Clinical Neurophysiology*, **35**, 301–309.

Peach, J. P., Sutarno, C. G., & McGill, S. M. (1998). Three-dimensional kinematics and trunk mucle myoelectric activity in the young lumbar spine: A database. *Archives of Physical Medicine and Rehabilitation*, **79**, 663–669.

Pedotti, A., & Crenna, P. (1999). Individual strategies of muscle recruitment in complex natural movements. In J. M. Winters & S. L. Woo (Eds.), *Multiple muscle systems* (pp. 542–549). Berlin: Springer-Verlag.

Perrin, D. H. (1993). *Isokinetic excise and assessment*. Champaign, IL: Human Kinetics.

Perry, J. (1992). *Gait analysis: Normal and pathological function*. Thorofare, NJ: Slack Inc.

Phillips, B. A., Lo, S. K., & Mastaglia, F. L. (2000). Muscle force measured using "break" testing with a hand-held myometer in normal subjects aged 20 to 69 years. *Archives of Physical Medicine and Rehabilitation*, **81**, 653–661.

Phillips, S. J., Roberts, E. M., & Huang, T. C. (1983). Quantification of intersegmental reactions during rapid swing motion. *Journal of Biomechanics*, **16**, 411–417.

Piazza, S. J., & Cavanagh, P. R. (2000). Measurement of the screw-home mechanism of the knee is sensitive to errors in axis alignment. *Journal of Biomechanics*, **33**, 1029–1034.

Pinniger, G. J., Steele, J. R., Thorstensson, A., & Cresswell, A. G. (2000). Tension regulation during lengthening and shortening actions of the human soleus muscle. *European Journal of Applied Physiology*, **81**, 375–383.

Plagenhoef, S. (1971). *Patterns of human motion: A cinematographic analysis*. Englewood Cliffs, NJ: Prentice-Hall.

Plagenhoef, S., Evans, F. G., & Abdelnour, T. (1983). Anatomical data for analyzing human motion. *Research Quarterly for Exercise and Sport*, **54**, 169–178.

Poyhonen, T., Kryolainen, H., Keskien, K. L., Hautala, A., Savolainen, J., & Malkia, E. (2001). Electromyographic and kinematic analysis of therapeutic knee exercises under water. *Clinical Biomechanics*, **16**, 496–504.

Prilutsky, B. I. (2000). Coordination of two– and one-joint muscles: Functional consequences and implications for motor control. *Motor Control*, **4**, 1-44.

Prilutsky, B. I., & Zatsiorsky, V. M. (1994). Tendon action of two-joint muscles: Transfer of mechanical energy between joints during jumping, landing, and running. *Journal of Biomechanics*, **27**, 25–34.

Pugh, L. (1971). The influence of wind resistance in running and walking and the mechanical efficiency of work against horizontal or vertical forces. *Journal of Physiology*, **213**, 255–276.

Putnam, C. (1983). Interaction between segments during a kicking motion. In H. Matsui, & K. Kobayashi (Eds.), *Biomechanics VIII-B* (pp. 688–694). Champaign, IL: Human Kinetics.

Putnam, C. (1991). A segment interaction analysis of proximal-to-distal sequential segment motion patterns. *Medicine and Science in Sports and Exercise*, **23**, 130–144.

Putnam, C. (1993). Sequential motions of the body segments in striking and throwing skills: Descriptions and explanations. *Journal of Biomechanics*, **26**(S1), 125–135.

Rassier, D. E., MacIntosh, B. R., & Herzog, W. (1999). Length dependence of active force production in skeletal muscle. *Journal of Applied Physiology*, **86**, 1445–1457.

Ravn, S., Voigt, M., Simonsen, E. B., Alkjaer, T., Bojsen-Moller, F., & Klausen, K. (1999). Choice of jumping strategy in two standard jumps, squat and countermovement jump—Effect of training background or inherited preference? *Scandinavian Journal of Medicine and Science in Sports*, **9**, 201–208.

Richmond, F. J. R. (1998). Elements of style in neuromuscular architecture. *American Zoologist*, **38**, 729–742.

Ridderikhoff, A., Batelaan, J. H., & Bobbert, M. F. (1999). Jumping for distance: Control of the external force in squat jumps. *Medicine and Science in Sports and Exercise*, **31**, 1196–1204.

Rigby, B. J., Hirai, N., Spikes, J. D., & Eyring, H. (1959). The mechanical properties of rat tail tendon. *Journal of General Physiology*, **43**, 265–283.

Roberton, M. A., & Halverson, L. E. (1984). *Developing children—Their changing movement*. Philadelphia: Lea & Febiger.

Roberts, E. M. (1991). Tracking velocity in motion. In C. L. Tant, P. E. Patterson, & S. L. York (Eds.), *Biomechanics in Sports IX* (pp. 3–25). Ames: Iowa State University.

Roberts, T. J., Marsh, R. L., Weyand, P. G., & Taylor, D. R. (1997). Muscular force in running turkeys: The economy of minimizing work. *Science*, **275**, 1113–1115.

Robinovitch, S. N., Hsiao, E. T., Sandler, R., Cortez, J., Liu, Q., & Paiement, G. D. (2000). Prevention of falls and fall–related fractures through biomechanics. *Exercise and Sport Sciences Reviews*, **28**, 74-79.

Rodano, R., & Squadrone, R. (2002). Stability of selected lower limb joint kinetic parameters during vertical jump. *Journal of Applied Biomechanics*, **18**, 83–89.

Rogers, M. M., & Cavanagh, P. R. (1984). Glossary of biomechanical terms, concepts, and units. *Physical Therapy*, **64**, 82–98.

Roncesvalles, M. N. C., Woollacott, M. H., & Jensen, J. L. (2001). Development of lower extremity kinetics for balance control in infants and young children. *Journal of Motor Behavior*, **33**, 180–192.

Ross, A. L., & Hudson, J. L. (1997). Efficacy of a mini-trampoline program for improving the vertical jump. In J. D. Wilkerson, K. Ludwig, & W. Zimmerman (Eds.), *Biomechanics in Sports XV* (pp. 63–69). Denton: Texas Woman's University.

Rowlands, L. K., Wertsch, J. J., Primack, S. J., Spreitzer, A. M., Roberts, D. O., Spreitzer, A. M., & Roberts, M. M. (1995). Kinesiology of the empty can test. *American Journal of Physical Medicine and Rehabilitation*, **74**, 302–304.

Rushall, B. S., Sprigings, E. J., Holt, L. E., & Cappaert, J. M. (1994). A re-evauation of forces in swimming. *Journal of Swimming Research*, **10**, 6–30.

Sale, D. (1987). Influence of exercise and training on motor unit activation. *Exercise and Sport Sciences Reviews*, **16**, 95–151.

Sale, D. (1992). Neural adaptation to strength training. In P. Komi (Ed.), *Strength and Power in Sport* (pp. 249–265). London: Blackwell Scientific Publications.

Sale, D. (2002). Posactivation potentiation: Role in human performance. *Exercise and Sport Sciences Reviews*, **30**, 138–143.

Salsich, G. B., Brown, M., & Mueller, M. J. (2000). Relationships between plantar flexor muscle stiffness, strength, and range of motion in subjects with diabetes-peripheral neuropathy compared to age–matched controls. *Journal of Orthopaedic and Sports Physical Therapy*, **30**, 473-483.

Sandercock, T. G. (2000). Nonlinear summation of force in cat soleus muscle results primarily from stretch of the common–elastic elements. *Journal of Applied Physiology*, **89**, 2206-2214.

Sanders, R. H. (1998). Lift or drag? Let's get skeptical about freestyle swimming. *Sportscience*, http://www.sportsci.org/jour/9901/med.html

Sanders, R. H., & Allen, J. B. (1993). Changes in net joint torques during accommodation to change in surface compliance in a drop jumping task. *Human Movement Science*, **12**, 299–326.

Sargent, D. A. (1921). The physical test of a man. *American Physical Education Review*, **26**, 188–194.

Savelsbergh, G. J. P., & Whiting, H. T. A. (1988). The effect of skill level, external frame of reference and environmental changes on one–handed catching. *Ergonomics*, **31**, 1655-1663.

Sayers, S. P., Harackiewicz, D. V., Harman, E. A., Frykman, P. N., & Rosenstein, M. T. (1999). Cross-validation of three jump power equations. *Medicine and Science in Sports and Exercise*, **31**, 572–577.

Schieb, D. A. (1987, January). The biomechanics piezoelectric force plate. *Soma*, 35–40.

Schmidt, R. A., & Wrisberg, C. A. (2000). *Motor learning and performance* (2nd ed.). Champaign, IL: Human Kinetics.

Scott, W., Stevens, J., & Binder-Macleod, S. A. (2001). Human skeletal muscle fiber type classifications. *Physical Therapy*, **81**, 1810–1816.

Segesser, B., & Pforringer, W. (Eds.) (1989). *The shoe in sport*. Chicago: Year Book Medical Publishers.

Semmler, J. G. (2002). Motor unit synchronization and neuromuscular performance. *Exercise and Sport Sciences Reviews*, **30**, 8–14.

Shadwick, R. E., Steffensen, J. F., Katz, S. L., & Knower, T. (1998). Muscle dynamics in fish during steady swimming. *American Zoologist*, **38**, 755–770.

Sheard, P. W. (2000). Tension delivery from short fibers in long muscles. *Exercise and Sport Sciences Reviews*, **28**, 51–56.

Shirier, I. (1999). Stretching before exercise does not reduce the risk of local muscle injury: A critical review of the clinical and basic science literature. *Clinical Journal of Sport Medicine*, **9**, 221–227.

Shultz, S. J., Houglum, P. A., & Perrin, D. H. (2000). *Assessment of athletic injuries*. Champaign, IL: Human Kinetics.

Simonsen, E. B., Dyhre-Poulsen, P., Voigt, M., Aagaard, P., & Fallentin, N. (1997). Mechanisms contributing to different joint moments observed during human walking. *Scandinavian Journal of Medicine and Science in Sports*, **7**, 1–13.

Siegler, S., & Moskowitz, G. D. (1984). Passive and active components of the internal moment developed about the ankle joint during human ambulation. *Journal of Biomechanics*, **17**, 647–652.

Slifkin, A. B., & Newell, K. M. (2000). Variability and noise in continuous force production. *Journal of Motor Behavior*, **32**, 141–150.

Smith, F. (1982). Dynamic variable resistance and the Universal system. *National Strength and Conditioning Association Journal*, **4**(4), 14–19.

Smith, L. K., Weiss, E. L., & Lehmkuhl, L. D. (1996). *Brunnstrom's clinical kinesiology* (5th ed.). Philadelphia: F. A. Davis.

Soderberg, G. L., & Knutson, L. M. (2000). A guide for use and interpretation of kinesiologic eletromyographic data. *Physical Therapy*, **80**, 485–498.

Sorensen, H., Zacho, M., Simonsen, E. B., Dyhre-Poulsen, P., & Klausen, K. (1996). Dynamics of the martial arts high front kick. *Journal of Sports Sciences*, **14**, 483–495.

Sorensen, H., Zacho, M., Simonsen, E. B., Dyhre-Poulsen, P., & Klausen, K. (2000). Biomechanical reflections on teaching fast motor skills. In Y. Hong & D. P. Johns (Eds.), *Proceedings of XVIIIth international symposium on biomechanics in sports* (Vol. 1, pp. 84–88). Hong Kong: The Chinese University of Hong Kong.

Sprigings, E., Marshall, R., Elliott, B., & Jennings, L. (1994). A three–dimensional kinematic method for determining the effectiveness of arm segment rotations in producing racquet-head speed. *Journal of Biomechanics, 27*, 245-254.

Sprigings, E. J., & Neal, R. J. (2000). An insight into the importance of wrist torque in driving the golfball: A simulation study. *Journal of Applied Biomechanics, 16*, 156–166.

Steindler, A. (1955). *Kinesiology of the human body under normal and pathological conditions.* Springfield, IL: Charles C. Thomas.

Stone, M., Plisk, S., & Collins, D. (2002). Training principles: Evaluation of modes and methods of resistance training—A coaching perspective. *Sports Biomechanics, 1*, 79–103.

Subic, A. J., & Haake, S. J. (2000). *The engineering of sport: Research, development and innovation.* London: Blackwell Scientific.

Talbot, J. A., & Morgan, D. L. (1996). Quantitative analysis of sarcomere non-uniformities in active muscle following a stretch. *Journal of Muscle Research and Cell Motility, 17*, 261–268.

Taylor, A., & Prochazka, A. (Eds.) (1981). *Muscle receptors and movement.* New York: Oxford University Press.

Thomas, J. S., Corcos, D. M., & Hasan, Z. (1998). The influence of gender on spine, hip, knee, and ankle motions during a reaching task. *Journal of Motor Behavior, 30*, 98–103.

Thomee, R., Augustsson, J., & Karlsson, J. (1999). Patellofemoral pain syndrome: A review of current issues. *Sports Medicine, 28*, 245–262.

Tol, J. L., Slim, E., van Soest, A. J., & van Dijk, C. N. (2002). The relationship of the kicking action in soccer and anterior ankle impingement syndrome: A biomechanical analysis. *American Journal of Sports Medicine, 30*, 45–50.

Tomioka, M., Owings, T. M., & Grabiner, M. D. (2001). Lower extremity strength and coordination are independent contributors to maximum vertical jump height. *Journal of Applied Biomechanics, 17*, 181–187.

Torg, J. S. (1992). Epidemiology, pathomechanics, and prevention of football-induced cervical spinal cord trauma. *Exercise and Sport Sciences Reviews, 20*, 321–338.

Toussaint, H. M., van den Berg, C., & Beek, W. J. (2002). "Pumped-up propulsion" during front crawl swimming. *Medicine and Science in Sports and Exercise, 34*, 314–319.

Tsaousidis, N., & Zatsiorsky, V. (1996). Two types of ball–effector interaction and their relative contribution to soccer kicking. *Human Movement Science, 15*, 861–876.

U.S. Department of Health and Human Services. *Physical activity and health: A report of the surgeon general.* Atlanta: U.S. Department of Health and Human Services, Centers for Disease Control and Prevention, National Center for Chronic Disease Prevention and Health Promotion, 1996.

van Dieen, J. H., Hoozemans, M. J. M., & Toussaint, H. M. (1999). Stoop or squat: A review of biomechanical studies on lifting technique. *Clinical Biomechanics, 14*, 685–696.

van Donkelaar, P., & Lee, R. G. (1994). The role of vision and eye motion during reaching to intercept moving targets. *Human Movement Science, 13*, 765–783.

van Ingen Schenau, G. J., & Cavanagh, P. R. (1999). Power equations in endurance sports. *Journal of Biomechanics, 23*, 865–881.

van Ingen Schenau, G. J., De Boer, R. W., & De Groot, B. (1989). Biomechanics of speed skating. In C. Vaughan (Ed.), *Biomechanics of sport* (pp. 121–167). Boca Raton, FL: CRC Press.

van Ingen Schenau, G. J., van Soest, A. J., Gabreels, F. J. M., & Horstink, M. (1995). The control of multi–joint movements relies on detailed internal representations. *Human Movement Science, 14*, 511-538.

van Ingen Schenau, G. J., Bobbert, M. F., & de Haan, A. (1997). Does elastic energy enhance work and efficiency in the stretch–shortening cycle? *Journal of Applied Biomechanics, 13*, 389-415.

van Zandwijk, J. P., Bobbert, M. F., Munneke, M., & Pas, P. (2000). Control of maximal and submaximal vertical jumps. *Medicine and Science in Sports and Exercise, 32*, 477–485.

Veloso, A., & Abrantes, J. (2000). Estimation of power output from leg extensor muscles in the acceleration phase of the sprint. In Y. Hong & D. P. Johns (Eds.), *Proceedings of XVIIIth international symposium on biomechanics in sports* (Vol. 1, pp. 97–101). Hong Kong: The Chinese University of Hong Kong.

Vezina, M. J., & Hubley-Kozey, C. L. (2000). Muscle activation in therapeutic exercises to improve trunk stability. *Archives of Physical Medicine and Rehabilitation*, **81**, 1370–1379.

Vint, P. F., & Hinrichs, R. N. (1996). Differences between one-foot and two-foot vertical jump performances. *Journal of Applied Biomechanics*, **12**, 338–358.

Walshe, A. D., Wilson, G. J., & Murphy, A. J. (1996). The validity and reliability of a test of lower body musculotendinous stiffness. *European Journal of Applied Physiology*, **73**, 332–339.

Ward, T., & Groppel, J. (1980). Sport implement selection: Can it be based upon anthropometric indicators? *Motor Skills: Theory into Practice*, **4**, 103–110.

Watts, R. G., & Bahill, A. T. (2000). *Keep your eye on the ball: The science and folklore of baseball* (2nd ed.). New York: W.H. Freeman.

Wells, K. F., & Dillon, K. E. (1952). The sit and reach: A test of back and leg flexibility. *Research Quarterly*, **23**, 115–118.

Wells, R. P. (1988). Mechanical energy costs of human movement: An approach to evaluating the transfer possibilities of two–joint muscles. *Journal of Biomechanics*, **21**, 955-964.

Whiting, W. C., & Zernicke, R. F. (1998). *Biomechanics of musculoskeletal injury*. Champaign, IL: Human Kinetics.

Whittle, M. (1996). *Gait analysis: An introduction* (2nd ed.). Oxford: Butterworth-Heinemann.

Whittle, M. (1999). Generation and attenuation of transient impulsive forces beneath the foot: A review. *Gait and Posture*, **10**, 264–275.

Wickham, J. B., & Brown, J. M. M. (1998). Muscles within muscles: The neuromotor control of intra–muscular segments. *European Journal of Applied Physiology*, **78**, 219-225.

Wilkerson, J. D. (1997). Biomechanics. In J. D. Massengale & R. A. Swanson (Eds.), *The history of exercise and sport science* (pp. 321–366). Champaign, IL: Human Kinetics.

Wilkinson, S. (1996). Visual analysis of the overarm throw and related sport skills: Training and transfer effects. *Journal of Teaching in Physical Education*, **16**, 66–78.

Williams, E. U., & Tannehill, D. (1999). Effects of a multimedia performance principle training program on correct analysis and diagnosis of throwlike movements. *The Physical Educator*, **56**, 143–154.

Williams, J. G., & McCririe, N. (1988). Control of arms and fingers during ball catching. *Journal of Human Movement Studies*, **14**, 241–247.

Williams, K. R., & Sih, B. L. (2002). Changes in golf clubface orientation following impact with the ball. *Sports Engineering*, **5**, 65–80.

Williams, K. R., Cavanagh, P. R., & Ziff, J. L. (1987). Biomechanical studies of elite female distance runners. *International Journal of Sports Medicine*, **8**, 107–118.

Williams, P. E., & Goldspink, G. (1978). Changes in sarcomere length and physiological properties in immobilized muscle. *Journal of Anatomy*, **127**, 459–468.

Wilson, G. J. (1994). Strength and power in sport. In J. Bloomfield, T. R. Ackland, & B. C. Elliott (Eds.) *Applied anatomy and biomechanics in sport* (pp. 110–208). Melbourne: Blackwell Scientific Publications.

Wilson, G. J., Elliott, B. C., & Wood, G. A. (1991). The effect on performance of imposing a delay during a stretch-shortening cycle movement. *Medicine and Science in Sports and Exercise*, **23**, 364–370.

Wilson, G. J., Newton, R. U., Murphy, A. J., & Humphries, B. J. (1993). The optimal training load for the development of dynamic athletic performance. *Medicine and Science in Sports and Exercise*, **25**, 1279–1286.

Winter, D. A. (1984). Kinematic and kinetic patterns of human gait: Variability and compensating effects. *Human Movement Science*, **3**, 51–76.

Winter, D. A. (1990). *Biomechanics and motor control of human movement* (2nd. ed.). New York: Wiley Interscience.

Winter, D. A. (1995). Human balance and posture control during standing and walking. *Gait and Posture*, **3**, 193–214.

Winter, D. A., Wells, R. P., & Orr, G. W. (1981). Errors in the use of isokinetic dynamometers. *European Journal of Applied Physiology*, **46**, 397–408.

Winters, J. M., & Woo, S. L.-Y. (Eds.) (1990). *Multiple muscle systems*. New York: Springer.

Woo, L.-Y., Debski, R. E., Withrow, J. D., & Janaushek, M. A. (1999). Biomechanics of knee ligaments. *American Jouranal of Sports Medicine, 27,* 533–543.

Wu, G., & Cavanagh, P. R. (1995). ISB recommendations for standardization in the reporting of kinematic data. *Journal of Biomechanics, 28,* 1257–1261.

Yeadon, M. R. (1991, January–March). Coaching twisting dives. *Sports Coach,* pp. 6–9.

Yeadon, M. R. (1997). The biomechanics of human flight. *American Journal of Sports Medicine, 25,* 575–580.

Yeadon, M. R. (1998). Computer simulation in sports biomechanics. In H. J. Riehle & M. M. Vieten (Eds.), *Proceedings of the XVIth international society of biomechanics in sports symposium* (pp. 309–318). Konstanz: Universitatsverlag Konstanz.

Yeadon, M. R., & Challis, J. H. (1994). The future of performance–related sports biomechanics research. *Journal of Sports Sciences, 12,* 3-32.

Zajac, F. E. (1991). Muscle coordination of movement: A perspective. *Journal of Biomechanics, 26*(S1), 109–124.

Zajac, F. E. (2002). Understanding muscle coordination of the human leg with dynamical simulations. *Journal of Biomechanics, 35,* 1011–1018.

Zajac, F. E., & Gordon, M. E. (1989). Determining muscle's force and action in multi-articular movement. *Exercise and Sport Sciences Reviews, 17,* 187–230.

Zajac, F. E., Neptune, R. R., & Kautz, S. A. (2002). Biomechanics and muscle coordination of human walking, I: Introduction to concepts, power transfer, dynamics and simulations. *Gait and Posture, 16,* 215–232.

Zatsiorsky, V. (1995). *Science and practice of strength training.* Champaign, IL: Human Kinetics.

Zatsiorsky, V. M. (1998). *Kinematics of human motion.* Champaign, IL: Human Kinetics.

Zatsiorsky, V. (Ed.) (2000). *Biomechanics in sport: Performance enhancement and injury prevention.* London: Blackwell Science.

Zatsiorsky, V. M. (2002). *Kinetics of human motion.* Champaign, IL: Human Kinetics.

Zernicke, R. F., & Roberts, E. M. (1976). Human lower extremity kinetic relationships during systematic variations in resultant limb velocity. In P. V. Komi (Ed.), *Biomechanics V–B* (pp. 41-50). Baltimore: University Park Press.

Zernicke, R. F., Garhammer, J., & Jobe, F. W. (1977). Human patellar-tendon rupture: A kinetic analysis. *Journal of Bone and Joint Surgery, 59A,* 179–183.

Zhang, S., Bates, B. T., & Dufek, J. S. (2000). Contributions of lower extremity joints to energy dissipation during landings. *Medicine and Science in Sports and Exercise, 32,* 812–819.

Glossary

absolute angle: an angle measured to a non-moving (inertial) frame of reference

acceleration: the rate of change of velocity (vector)

accelerometer: a device that measures acceleration

actin: the thin filaments in a myofibril that interact with myosin to create muscle tension

accommodation: a decrease in biological response to an unchanging stimulus

action potential: the electrical potential change during depolarization of nerves and activated muscle fibers

active tension: the tension created by the contractile component (actin–myosin interaction) of activated muscle

affine scaling: image scaling technique used to measure in a plane not perpendicular to the optical axis of the camera in 2D cinematography/videography

agonist: an anatomical term referring to the concentric action of a muscle or muscle group for presumed to create a specific movement

aliasing: distortion of a signal by an inadequate sampling rate

analog-to-digital (A/D) conversion: the process of taking a continuous signal and sampling it over time (see "sampling rate") to create a digital (discrete numbers) representation

anatomy: the study of the structure of the body

angle–angle diagram: a kinematic graph of one variable plotted against another (not time) that is useful in the study of coordination of movements

angular acceleration: the rate of change of angular velocity (vector)

angular displacement: the change in angular position (vector)

angular momentum: the quantity of angular motion, calculated as the product of the moment of inertia times the angular velocity (vector)

angular velocity: the rate of change of angular displacement (vector)

angular impulse: the angular effect of a torque acting over time: the product of the torque and the time it acts (vector)

anisotropic: having different mechanical properties for loading in different directions

antagonist: an anatomical term referring to a muscle or muscle group that is presumed to oppose (eccentric action) a specific movement

anthropometry: the study of the physical properties of the human body

aponeurosis: connective tissue within muscle and tendon in the form of a flat sheet

Archimedes' principle: the magnitude of the buoyant force is equal to the weight of the fluid displaced

balance: a person's ability to control their body position relative to some base of support

balance principle: a biomechanical application principle which states that the stability and mobility of a body position are inversely related

ballistic: explosive, momentum-assisted movement

bandpass filter: a filter designed to pass a range (bandpass) of frequencies, removing frequencies above or below this desirable range

bending: a combination of forces on a long body that tends to bend or curve the body creating tensile loads on one side and compression loads on the other side

Bernoulli's principle: the pressure a fluid can exert decreases as the velocity of the fluid increases

Bernstein's problem: a theory of motor control in which skill learning involves the reduction of redundant degrees of freedom

bilateral deficit: simultaneous activation of two limbs that causes less force generation than the sum of the two individually activated limbs

biomechanics: study of the motion and causes of motion of living things

boundary layer: the layers of a fluid in close proximity to an object suspended in the fluid

buoyancy: the supporting or floating force of a fluid

center of buoyancy: the point at which the buoyant force acts

center of mass/gravity: the point that represents the total weight/mass distribution of a body; the mass centroid is the point where the mass of an object is balanced in all directions

center of percussion: a point on a striking object where impact with another object results in no reaction force at an associated point on the grip (see "sweet spot")

center of pressure: the location of the vertical ground reaction force vector; the center of pressure measured by a force platform represents the net forces in support and the COP may reside in regions of low local pressure

coactivation: simultaneous activation of agonist and antagonist muscles (co-contraction)

coefficient of drag: a measure of the relative fluid resistance between an object and a fluid

coefficient of friction: a measure of the resistance to sliding between the surfaces of two materials

coefficient of lift: a measure of the lift force that can be created between an object and a fluid

coefficient of restitution: a measure of the relative elasticity of the collision between two objects

common mode rejection: a measure of the quality of a differential amplifier in rejecting common signals (noise)

compression: a squeezing mechanical loading created by forces in opposite directions acting along a longitudinal axis

compliance: the ratio of change in length to change in applied force, or the inverse of stiffness (see "stiffness"); a material that is easily deformed has high compliance

components: the breaking up of a vector into parts, usually at right angles

concentric muscle action: the condition where activated muscles create a torque greater than the resistance torque (miometric)

conservation of energy: the Law of Conservation of Energy states that energy cannot be created or destroyed; instead, energy is transformed from one form to another

contourgram: exact tracings of the body positions of a movement from film/video images

contractile component: a part of the Hill muscle model that represents the active tension and shortening of actin and myosin

coordination continuum: a biomechanical application principle which states that movements requiring generation of high forces tend to utilize simultaneous segmental movements, while lower-force and high-speed movements tend to use sequential movements

couple: (1) two forces of equal size, parallel lines of actions, and opposite sense; (2) a mechanical calculation tool that is employed to represent torques without affecting linear kinetics

creep: the increase in length (strain) over time as a material is constantly loaded

cross-talk: the pick-up of EMG signals from other active muscles aside from the muscle of interest

cut-off frequency: the cutting point of a filtering technique, where frequencies above or below are removed; the lower the cut-off frequency for a lowpass filter, the greater the smoothing of the signal

deformable body: biomechanical model that documents the forces and deformations in an object as it is loaded

density: the mass of an object divided by its volume

degrees of freedom: the number of independent movements an object may make, and consequently the number of measurements necessary to document the kinematics of the object

differential amplification: EMG technique for amplifying the difference between the signals seen at two electrodes relative to a reference electrode

digital filter: a complex frequency-sensitive averaging technique used to smooth or process data

digitize (video): the A/D conversion of an analog video signal to create the discrete picture elements (pixels) used to make a video image

digitize (biomechanics): the process of measuring 2D locations of points on an image

direct dynamics: biomechanical simulation technique where the kinematics of a biomechanical model are iteratively calculated from muscle activation or kinetic inputs

direct linear transformation (DLT): a short-range photogrammetric technique to create 3D coordinates (x,y,z) from the 2D coordinates (x,y) of two or more synchronized camera views of an event

displacement: linear change in position in a particular direction (vector)

distance: liner change in position without regard to direction (scalar)

double differential amplification: EMG technique to eliminate cross-talk

drag: the fluid force that acts parallel to the relative flow of fluid past an object

dynamic flexibility: the increase in passive tension per increase in joint range of motion

dynamical systems: motor learning theory which argues that movement coordination emerges or self-organizes based on the dynamic properties of the body and environment rather than on a central motor program from the brain

dynamics: the branch of mechanics studying the motion of bodies under acceleration

dynamometer: a device that measures force or torque for muscular performance testing

eccentric muscle action: the condition where an activated muscle(s) creates a torque less than the resistance (plyometric) torque

economy: the amount of energy needed to do a specific amount of work

efficiency: in a system, the ratio of work done to work input

elastic: the resistance of a body to deformation (see "stiffness")

elastic (strain) energy: the potential mechanical work that can be recovered from restitution of a body that has been deformed by a force (see "hysteresis")

electrogoniometer: a device that makes continuous measurements of joint angle(s)

electromechanical delay: the delay between motor action potential (electric signal of muscle depolarization or EMG) and production of muscular force

electromyography (EMG): the amplification and recording of the electrical signal of active muscle

energy (mechanical): the ability to do mechanical work (potential, strain, and kinetic energy are all scalar mechanical energies)

Euler angles: a way to represent the 3D motion of an object using a combination of three rotations (angles)

excursion: the change in the length of a muscle as the joints are moved through their full range of motion

external force: a force acting on an object from its external environment

external work: work done on a body by an external force

fascicle: a bundle of muscle fibers (cells)

fast Fourier transformation (FFT): mathematical technique to determine the frequencies present in a signal

field (video): half of an interlaced video image (frame), composed of the even or odd horizontal lines of pixels

finite difference: calculating time derivative by discrete differences in kinematics divided by the time between datapoints

finite-element model: advanced biomechanical model to study how forces act within a deformable body

firing rate: the number of times a motor unit is activated per second

First Law of Thermodynamics: application of the Law of Conservation of Energy to heat systems

fluid: a substance, like water or gasses, that flows when acted upon by shear forces

force: a push, pull, or tendency to distort between two bodies

force–length relationship: skeletal muscle mechanical property that demonstrates how muscle force varies with changes in muscle length (also called the length–tension relationship)

force–motion principle: a biomechanical application principle which states that unbalanced forces are acting whenever one creates or modifies the movement of objects

force platform: a complex force transducer that measures all three orthogonal forces and moments applied to a surface

force–time principle: a biomechanical application principle which states that the time over which force is applied to an object affects the motion of that object

force–time relationship: (see "electromechanical delay")

force–velocity relationship: skeletal muscle mechanical property that shows how muscle force potential depends on muscle velocity

Fourier series: a mathematical technique for summing weighted sine and cosine terms that can be used to determine frequency content or represent a time domain signal

frame (video): a complete video image

free-body diagram: a technique for studying mechanics by creating a diagram that isolates the forces acting on a body

frequency: the inverse of time or the number of cycles of an event per second

frequency content: time-varying signals can be modeled as sums of weighted frequencies (see "Fourier series")

frequency response: the range of frequencies that are faithfully reproduced by an instrument

friction: the force in parallel between two surfaces that resists sliding of surfaces past each other

global reference frame: measuring kinematics relative to an unmoving point on the earth

Golgi tendon organ: a muscle receptor that senses muscle tension

goniometer: a device used to measure angular position

gravity: the force of attraction between objects; usually referring to the vertical force of attraction between objects and the earth

ground reaction force: the reaction (opposite) forces created by pushing against the ground (e.g., feet in running or hands in a handstand)

harmonic: a multiple of a fundamental frequency (see "frequency content")

helical (screw) axis motion: a way to represent the 3D motion of an object using an imaginary axis in space and rotations relative to that axis

highpass filter: a signal-processing technique that removes the low-frequency components of a signal

Hill muscle model: a three-component model of muscle force consisting of a contractile component, a series elastic component, and a parallel elastic component

hypertrophy: the increase in size of muscle fibers

hysteresis: the energy loss within a deformed material as it returns to its normal shape

impulse: the mechanical effect of a force acting over time (vector); $J = F \cdot t$

impulse–momentum relationship: principle which states that the change in momentum of an object is equal to the net impulse applied; the original language of Newton's second law, and equivalent to the instantaneous version: $F = ma$

inertia: the property of all matter to resist a change in its state of motion

inertial force: the mass acceleration (m**a**) term in Newton's Second Law (dynamics); the effect of inertia and acceleration on dynamic movement, but it is important to remember that its effect is not a real force acting on an object from another object

inertia principle: A biomechanical application principle which states that inertial resistance to changes in state of motion can be used to advantage in resisting motion or transferring energy

information: observations or data with unknown accuracy

in situ: Latin for "in place", or structures isolated by dissection

integrated EMG (IEMG): the area under a rectified EMG signal; correctly, the time integral reported in units of amplitude $\times$ time (mV·s); unfortunately, some studies employ outdated equipment and incorrect terminology, so that reported IEMGs are not really integrated but filtered or smoothed EMG values (mV), which is essentially a linear envelope detector

interdisciplinary: the simultaneous integrated application of several disciplines to solution of a problem

internal force: a force within an object or between the molecules of an object

internal work: work done on body segments by internal forces (muscles, ligaments, bones)

inverse dynamics: biomechanics research technique for estimating net forces and moments in a linked-segment model from measured kinematics and anthropometric data

in vitro: Latin for "in glass," or tissues removed from the body but preserved

in vivo: Latin for "in the living," or during natural movement

isokinetic ("same, or constant, motion"): the condition where activated muscles create constant joint angular velocity

isometric ("same, or constant, length"): the condition where activated muscles create a torque equal to the resistance torque, so there is no joint motion

isotonic ("same, or constant, tension"): the condition where activated muscles work against a constant gravitational resistance; muscle tension is *not* constant in these conditions

jerk: the third derivative of displacement with respect to time

joint center: an approximation of the instantaneous center of rotation of a joint

joint reaction forces: the net forces acting at joints calculated from inverse dynamics; these forces do not represent the actual bone-on-bone forces acting at joints, but a combination of bone, muscle, and ligament forces

Joule: the unit of mechanical energy and work

kinematic chain: a linkage of rigid bodies; an engineering term used to simplify the degrees of freedom needed to document the mechanical behavior of a system; Steindler (1955) proposed the terminology of a kinetic chain, and classifying chains as either open or closed; unfortunately, this has resulted in a great deal of confusion and an unclear manner of classifying movements/exercises: **open**: one end link is free to move; **closed**: constraints (forces) on both ends of the kinematic chain

kinematics: the branch of mechanics that describes the motion of objects relative to some frame of reference

kinetic energy: the capacity to do work due to the motion of an object

kinetics: the branch of mechanics that explains the causes of motion

knowledge: the application of our best logic, scientific research, and professional experience to data

laminar flow: movement of fluid in smooth, parallel layers

Law of Acceleration: Newton's Second Law of Motion, which states that the acceleration an object experiences is proportional to the resultant force, is in the same direction, and is inversely proportional to the object's mass ($\Sigma \mathbf{F} = m\mathbf{a}$)

Law of Inertia: Newton's First Law of Motion, which states that objects tend to resist changes in their state of motion; formally, we say an object will remain in a state of uniform motion (stillness or constant velocity) unless acted upon by an external force

Law of Momentum: Newton's second law written as the impulse–momentum relationship

Law of Reaction: Newton's Third Law of Motion, which states that for every force there is an equal and opposite reaction force

lever: a simple machine used to magnify motion or force; a lever consists of a rigid object rotated about an axis

lift: the fluid force that acts at right angles to the relative flow of fluid

linear envelope: EMG processing technique where a rectified signal is smoothed with a lowpass filter

linearity: a measure of the accuracy of an instrument, usually expressed as a percentage of full-scale output (FSO)

linear voltage differential transducer (LVDT): a force-measuring device

linked-segment model: a rigid body model linked together by joints

load: a force or moment applied to a material

load cell: a force-measuring device

load-deformation curve: the mechanical behavior of a material can be documented by instantaneous measurement of the deformation and load applied it

local reference frame: measuring kinematics relative to a moving point, or nearby rigid body (joint, segment, or center of mass)

lowpass filter: a signal-processing technique that removes the high-frequency components of a signal

Magnus effect: the creation of lift force on a spinning sphere

markers: high-contrast reflective materials attached to subjects to facilitate the location of segments, landmarks, or joint centers for digitizing

mass: the resistance of an object to linear acceleration

maximal voluntary contraction (MVC): the maximum force/torque a person can create with a muscle group, usually under in isometric conditions

mechanical advantage: a ratio describing the effectiveness of a lever calculated by the moment arm for the force divided by the moment arm for the resistance

mechanics: the branch of physics that deals with forces and the motion they create

mechanomyography (phonomyography, vibromyograph): the amplification and recording of the vibrations created by muscle activation

modeling: mathematical representations of the biomechanical systems used for calculations or simulations

moment (moment of force, torque): the rotating effect of a force

moment arm: the leverage of a force for creating a moment; the perpendicular distance from the axis of rotation to the line of action of the force

moment of inertia: the resistance to rotation (angular acceleration) of a body

momentum: the quantity of motion of an object calculated by the product of mass and velocity (vector)

motor action potential: the change in electrical charge about a muscle fiber as it is activated

motor unit: a motor neuron and the muscle fibers it innervates

muscle action: the activation of muscle to create tension that contributes to joint movement or stabilization

muscle inhibition: the inability to fully activate or achieve maximum muscle force during maximum voluntary contraction

muscle spindle: an intramuscular receptor that senses changes in muscle length

myofibril: the small cylindrical filaments that make up a muscle fiber/cell

myosin: the large filaments in a myofibril that interact with actin to create muscle tension

myotatic reflex: a short reflex arc that activates a muscle as it is stretched

net force: the resultant force or sum of all external forces acting on an object

Newton: the SI unit of force; 1 Newton (N) is equal to 0.22 pounds

normal reaction: the force acting at right angles to the surfaces of objects that are in contact

Nyquist frequency: a signal sampling theorem which states that the minimum digital sampling rate (Nyquist frequency) needed to accurately represent an analog signal is twice the highest frequency present in the signal

optimal projection principle: A biomechanical application principle which states that there are ranges of optimal angles for projecting objects to achieve certain goals

orthogonal: perpendicular (at right angles)

orthotics: objects/braces that correct deformities or joint positioning

overuse injury: an injury created by repetitive movements below acute injury thresholds, but due to inadequate rest and/or repetitive stress, injury develops; also known as cumulative trauma disorder or repetitive motion injury

parallel elastic component: a part of the Hill muscle model that represents the passive tension from connective tissue throughout the muscletendon unit

Pascal: the SI unit of pressure or stress (force per unit area)

passive insufficiency: the limitation of joint motion because of increases in passive tension in multiarticular muscles stretched across multiple joints

passive tension: a component of muscle tension from passive stretching of muscle, especially the connective tissue components

pennation: the angle of muscle fiber bundles relative to a tendon

piezoelectric: crystals with electromechanical properties that can be used to measure force/acceleration

point mass: a simplified mechanical model that represents an object as a point in space with a given mass

potential energy: the capacity to do work of an object due to its vertical position in a gravitational field (gravitational potential energy) or its deformation (strain energy)

potentiometer: a device that is used to measure rotation

power (mechanical): the rate of doing mechanical work; peak mechanical power represents the greatest mechanical effect, the ideal combination of force and velocity; power can be calculated as W/t or $F \cdot V$

preamplification: the amplification of small signals (EMG) close to their source before they are conducted to other devices for amplification and recording

pressure: external force divided by area over which the force acts

projectile: an object projected into space without self-propulsion capability, so the only forces acting on the object are gravity and air resistance

proprioceptive neuromuscular facilitation (PNF): specialized stretching procedures that utilize sequences of muscle actions to potentiate reflexes to relax muscles being stretched

prosthetics: artificial limbs

Pythagorean Theorem: the two sides of a right triangle forming the right angle (a and b) and the hypotenuse (c) are related as follows: $a^2 + b^2 = c^2$

qualitative analysis: systematic observation and introspective judgment of the quality of human movement for the purpose of providing the most appropriate intervention to improve performance (Knudson & Morrison, 2002)

quantitative analysis: solving a biomechanical problem using numerical measurements and calculations

quasistatic: the state of a mechanical system where the accelerations are small enough to be assumed equal to zero

radian: a dimensionless unit of rotation equal to 57.3°

radius of gyration: a convenient way to summarize an object's moment of inertia, defined as the distance from the axis of rotation at which half the object's mass must be placed in both directions to equal the object's moment of inertia

range-of-motion principle: a biomechanical application principle which states that the amount of linear and angular motion used will affect the speed and accuracy of human movement

reaction change: a method to calculate the center of gravity of static body postures

reciprocal inhibition: the inhibition of an opposing muscle group (antagonist) when a muscle group (agonist) is activated

recruitment: activation of motor units of muscles by the central nervous system

rectified EMG: a processing technique that converts negative EMG voltages to positive ones

redundancy (distribution) problem: a mathematical problem with most kinetic biomechanical models, where there are more musculoskeletal unknowns than there are equations

relative angle: an angle measured between two moving objects

residuals: difference between a smoothed and raw signal; can be used to examine the quality of the fit of the new signal to the pattern of the raw signal

resolution (video): the number of pixels available to measure a given field of view; a video image of a 3-meter wide area with a horizontal resolution of 640 pixels has a resolution for measurement of about 5 mm

resonance: frequency of vibration that matches the physical properties of a body so that the amplitudes of the vibration increase rather than decay over time

resting length: the middle of muscle range of motion where passive tension begins to rise

resultant: the addition of vectors to obtain their net effect (see "net force")

right-hand rule: a convention or standard for drawing the correct direction of angular velocity vectors

rigid body: mechanical simplification (abstraction) assuming the dimensions of an object do not change during movement or loading

root mean square (RMS): signal processing calculation that approximates the mean absolute value of a time-varying signal

rotator cuff: the four deep, stabilizing muscles of the glenohumeral joint: the infraspinatus, supraspinatus, subscapularis, and teres minor

sampling rate: the number of discrete samples per second used to represent a signal; NTSC video has an effective sampling rate of 60 Hz or 60 fields per second

sarcomere: the functional unit of a myofibril; a sarcomere is the region between two Z disks

scalar: simple quantity completely defined by a single number (magnitude)

scaling: converting image measurements to actual size

science: a systematic method for testing hypotheses with experimental evidence for the purpose of improving our understanding of reality

Second Law of Thermodynamics: no machine can convert all the input energy into useful output energy

segmental interaction principle: a biomechanical application principle which states that forces acting in a system of linked rigid bodies can be transferred through the links

segmental method: a research method used to calculate the center of gravity of a body using anthropometric data, joint coordinates, and static equilibrium

series elastic component: a part of the Hill muscle model that represents the passive tension of connective tissue in series with the contractile component

shear: mechanical loading in opposite directions and at right angles to the surface of a material

shutter speed: the period of time during which a photographic or video image is captured (e.g., 1/1000 of a second); limiting this period can prevent blurring of moving objects

simulation: use of a biomechanical model to predict motion with given input conditions in order to study the factors that affect motion (see "direct dynamics")

size principle: the orderly recruitment of motor units occurs from the smallest to the largest

smoothing: a processing technique that smooths data, removing rapid fluctuations that are not part of normal biomechanical signals

smoothing parameter: an index of the amount of smoothing allowed in splines; the larger the smoothing parameter, the more smoothing (allowable deviation between the raw and fitted curve)

snap: the fourth derivative of displacement with respect to time

speed: the rate of change of distance (scalar)

spin principle: a biomechanical application principle which states that spin is put on a projectile to affect trajectory or bounce

spline: a smoothing technique that replaces the signal with several polynomials linked together; cubic (third power) and quintic splines (fifth power) are common in biomechanics

static equilibrium: when all the forces and torques acting on an object sum to zero, meaning that the object is motionless or moving at constant velocity

static flexibility: the linear or angular measurement of the limits of motion in a joint or joint complex

statics: the branch of mechanics that studies bodies at rest or in uniform motion

stiffness: the elasticity of a material, measured as the slope of the elastic (linear) region of the stress–strain curve (Young's modulus of elasticity); a material's stiffness is usually approximated using the slope of the linear region of the load-deformation curve

strain (mechanical): the amount of deformation of a material caused by an applied force, usually expressed as a percentage change in dimensions

strain (muscular): muscular injury usually caused by large eccentric stretches of muscle fibers

strain energy: the capacity to do work of an object due to its deformation by an external force

strain gauge: a small array that is bonded to materials in order to sense the small changes in size (strain) as the material is loaded; usually used to measure force or acceleration

strength (mechanical): the toughness of a material to resist loading, usually measured as the total work or peak force required to permanently deform (yield strength) or break a material (ultimate strength)

strength (muscular): the maximum force or torque produced by a muscle group in an isometric action at a specific joint angle; research has found several domains of strength expression depending on the time, velocity, and resistance involved

stress (mechanical): the force per unit area in a material

stress fracture: a very small fracture in cortical bone caused by repetitive loading and inadequate rest

stress relaxation: the decrease in stress in a material over time when subjected to a constant force

stress–strain curve: (see "load deformation")

stretch-shortening cycle (SSC): a common coordination strategy where agonists for a movement are eccentrically loaded in a countermovement, immediately before the concentric action and motion in the intended direction; an SSC results in larger initial forces and greater concentric work than purely concentric actions

synergy: the combination of several muscle actions that serve to optimally achieve a motor task

sweet spot: striking implements (bats, rackets, etc.) have zones where impact with other objects is most effective; the term *sweet spot* tends to refer to the zone with the highest coefficient of restitution, although there are zones that minimize reaction forces (center of percussion), or minimize vibration (node)

technology: the tools and methods for applying scientific knowledge to solve problems or perform tasks

telemetry: a technique to send biomechanical signals to recording devices without wires, using an FM radio transmitter and receiver

tension: a pulling apart (making longer) of mechanical loading created by forces in opposite directions acting along the longitudinal axis of a material

tensor: a complex variable that cannot be described using only magnitude and direction

tetanus: the summation or fusion of many twitches of muscle fibers into a smooth rise in tension

thixotropy: a property of a material to change passive stiffness in response to previous loading; this history-dependent behavior is apparent in the increasing stiffness of muscle with extended inactivity

time constant: typically, an averaging/ smoothing value in EMG processing; the larger the time constant the larger the time interval averaged over, meaning more smoothing

torque (see "moment of force"): the rotating effect of a force; mechanics of materials uses torque to refer to torsion moments acting on an object

torsion: opposing loads that twist an object along its longitudinal axis

trajectory: the path in space that an object follows as it moves through the air

twitch: the force response of a muscle fiber to a single stimulation

twitch interpolation (superimposition) technique: a method used to determine the maximality of a maximum voluntary action (MVC) where stimulation is provided during an MVC

vector: a complex quantity requiring description of size and direction

viscoelastic: the property of a material where force in the material is dependent on time and deformation

weight: the downward (vertical) force action on an object due to gravity

Wolff's Law: bones remodel according to the stress in the tissue

work (mechanical): work is done when a force moves an object in the direction of the force and is calculated as the product of force and displacement

work–energy relationship: principle in physics which states that the work done on a body is equal to the net change in energy in the body

yield point: point on the load-deformation curve where a material continues to deform without increasing load

Young's modulus (see "stiffness")

Conversion Factors

Biomechanical variables are reported in traditional English units and the metric system (SI, International System). The conversion factors below appendix are useful for converting between various measurement units. It is likely you will find one unit of measurement easier to relate to, and you may need to transform some values from the literature to more convenient units of measurement.

For example, if you wanted to get a feel for how fast a person is running at 9 m/s, you could take 9 m/s times 2.23 to get 20.1 mph. If you wanted to know how fast you were running on a treadmill that reported your pace as 8.5 minutes per mile, you would first convert the pace to an average speed in miles per hour. Sixty minutes divided by 8.5 minutes would equal 7.1 mph. Next you would take 7.1 mph divided by the conversion factor (2.23) to obtain 3.2 m/s.

Variable	SI unit	×	Factor	=	Other unit
distance	m		3.28		ft
	km		0.621		miles
	radian		57.3		degrees
speed	m/s		2.23		mph
	km/hr		0.62		mph
	m/s		3.28		ft/s
	rad/s		57.3		deg/s
	rad/s		9.55		rpm
acceleration	m/s/s		0.102		g's
mass	kg		0.069		slugs
moment of inertia	kg·m^2		0.738		slugs·ft^2
force	N		0.225		pounds
torque	N·m		0.738		lbs·ft
impulse	N·s		0.225		lbs·s
energy	Joules		0.738		ft·lbs
work	Joules		0.738		ft·lbs
power	Watts		1.341		horsepower
momentum	(kg·m)/s		0.225		(slug·ft)/s
	(kg·m^2)/s		0.225		(slug·ft^2)/s
stress/pressure	Pascals		0.00015		lbs/in^2

Suggested Answers to Selected Review Questions

This appendix provides initial answers to, primarily, the odd-numbered review questions from chapters 1 through 8. The purpose of review questions is to practice and rehearse key biomechanical concepts, principles, and laws. Students are encouraged to study the topics related to each question in greater depth. The discussion questions in chapters 9 through 12 are designed for students and instructors to discuss. Discussion questions are ideal for small-group brainstorming and practice in qualitative analysis of human movement.

Chapter 1

1. Biomechanics is the study of how living things move using the science of mechanics. In the first half of the twentieth century this was synonymous with kinesiology, but now kinesiology is the academic discipline of the study of human movement.

3. The advantages of qualitative biomechanical analysis is its ease of use and flexibility, but its weaknesses are related to subjectivity and reliability. Quantitative biomechanical analysis may have greater precision and accuracy, but its weaknesses are the high cost in terms of equipment and time.

5. A wide variety of journals publish biomechanics research. These journals include specialized biomechanics, engineering, biology, medicine, strength and conditioning, and sports-medicine journals.

7. Biomechanics must be integrated with other kinesiology sciences because people are not robots that move without regard to environmental factors. Psychological, physiological, and perceptual issues are all examples of factors that might be more important than biomechanical factors in some situations.

Chapter 2

1. Biomechanics has traditionally focused on rigid body and fluid mechanics. The majority of early biomechanical studies focused on the kinematics of movement, but there are still many studies on the causes (kinetics) of movement.

3. Scalars only require knowledge of size and units. Vector variables have size, units, and direction.

5. The nine principles of biomechanics can be subdivided into principles related to human movement and projectiles.

7. Many factors affect human movement along with the principles of biomechanics. Some factors might be performer characteristics (psychological, perceptual, or social), the physical environment, the goal of the movement, and the philosophical goals of the kinesiology professional.

Chapter 3

1. There are several anatomical terms employed to describe the location and mo-

tion of body structures. Some examples include directions (anterior/posterior, medial/lateral, superior/inferior, proximal/distal) and joint movements (flexion/extension, adduction/abduction, internal rotation/external rotation).

3. Muscle fiber types and their architectural arrangement affect muscle force and range of motion. The rise and decay of muscle tension is greatest in fast-twitch fibers and decreases the greater the oxidative or slow-twitch characteristics of the fiber. Muscle fibers arranged in parallel have greater range of motion but create less force. Pennate fiber arrangements produce greater force but have less range of motion.

5. Muscle tension has active and passive components. Passive tension does not appear to play a large role in the middle of the range of motion, but does tend to limit motion when the muscle is stretched near the end of the range of motion.

7. Examples of the force–motion principle can be seen anytime an object changes its state of motion. If a dumbbell reverses direction at the bottom of an arm curl exercise, we can conclude an unbalanced upward force was applied to the dumbbell.

9. Biomechanical principles and research help the kinesiology professional to understand how human movement occurs and how movement might be improved. The major areas of biomechanics research that are the most valuable in this area are EMG, studies of anatomical variation, linked segment interactions, and modeling and simulation.

Chapter 4

1. The primary loads on body tissues are compression, tension, and shear. The combined loads are bending and torsion.

3. The tensile strengths of tendon and muscle are about 14,500 and 60 lb/in^2, respectively, while the tensile strength of bone is about 18,000 lb/in^2. These data are consistent with the higher incidence of muscle injuries compared to that for tendon or bone.

5. The Force–Velocity Relationship has several implications for resistances and speed of movement in strength-training exercises. When training for muscular strength, large resistances should be moved slowly to train the muscle where it is strongest. Training for muscular power and endurance uses smaller resistances moved at faster speeds.

7. The Force–Time Relationship defines the delay between neuromuscular signaling for creation of muscle force and a rise in that force, while the force–time principle deals with duration of force application. While these two concepts are related, the force–time principle involves adapting the timing of the application of force by a person to the demands of the task while electromechanical delay is one of the factors that affects how force can be applied.

9. The brain creates muscle tension by recruitment of motor units and modifying their firing rate or rate coding. Motor units tend to have predominantly one fiber type, so that the brain generally recruits motor units based on the size principle, from slow-twitch motor units to fast-twitch motor units.

11. Muscle spindles sense stretch and golgi tendon organs sense muscle tension.

13. Large ranges of motion allow for greater production of speed and force, while smaller ranges of motion tend to allow for more accurate movement. The weight shifts in a golf swing and baseball batting are small because of the high accuracy demands of these skills. Maximizing range of motion in the countermovement in jumps is not usually effective because of timing limitations or biomechanically weak positions in deep knee flexion.

15. A person doing a seated knee extension exercise uses concentric action of the

quadriceps groups to extend the knee, and eccentric action of the quadriceps to flex the knee. The forces acting on the lower leg include muscle forces from the hamstrings, quadriceps, ankle muscles, and gravity. If the person were exercising on a machine there would be forces applied to the leg/ankle from the machine.

Chapter 5

1. The frame of reference is the point from where motion is measured.

3. An average velocity is a velocity estimate for the middle of a time interval where displacement and time information are available ($V = d/t$). The smaller the time interval used for the calculation, the more accurate the average velocity is and the closer it gets to true instantaneous velocity. An instantaneous velocity is an exact estimate of the velocity at an instant in time, and is calculated using calculus.

5. With upward displacement as positive, the average vertical velocity ($V = d/t$) of the dumbbell for the concentric phase is $1.2/1.5 = 0.8$ m/s, while the average vertical velocity of the eccentric phase is $-1.2/2.0 = -0.6$ m/s.

7. Angular kinematics are particularly suited for analysis of human movement because joint motions are primarily rotational. Markers placed on the body can by digitized to calculate the angular kinematics of the joints during human movements.

9. Since knee extension is positive (+50 deg/s), the angular acceleration of her knee ($\alpha = \omega/t$) is: $(0 - 50)/0.2 = -250$ deg/s/s.

11. The coach could use a radar gun to measure maximum and warm-up throwing speeds. If the coach did not have a radar gun, they could measure off the standard distance and time of the throws with a stopwatch to calculate average velocities in each throwing condition.

13. To use the angular-to-linear velocity conversion formula ($V = \omega \cdot r$), the angular velocity must be in radian/second: 2000 deg/s divided by 57.3 deg (1 radian), which is equal to 34.9 radian/s. The velocity of the club head relative to the golfer's hands is: $34.9 (1.5) = 52.4$ m/s.

15. The vertical acceleration of a volleyball anywhere in flight is a downward acceleration due to gravity of -9.8 m/s/s or -32.2 ft/s/s.

Chapter 6

1. A 6-kg bowling ball has the same inertia in all states of motion. The ball's inertia is a fundamental property of matter and is measured by its mass, 6 kg. This will not change unless we get the ball rolling near the speed of light!

3. Increasing inertia is useful in movement when you want to maximize stability, or if there is time to get a larger inertia moving in a desired direction. Increasing the mass of a wrestler will make it more difficult for an opponent to move the wrestler.

5. The major determining factors of dry friction are the normal reaction and the coefficient of friction. Since adding mass to a person has other effects, the best strategy is to select a shoe with a higher coefficient of friction with common flooring.

7. If we move the shearing force to the left, we create a right triangle with a 30° angle on the right and a hypotenuse of 1000 N. The longitudinal component of the joint force (F_L) is the adjacent side, so we can use the cosine relationship to calculate: cos 30° = $F_L/1000$, so F_L = 866 N. The sine of 30° is a special value (0.5), so we can quickly see that F_S = 500 N.

9. Muscular strength is the maximum force a muscle group can create in certain conditions, usually an isometric action at a specified joint angle. Muscular power is the rate of doing muscular work. Maximum

muscular power occurs at the combination of velocity and force that maximizes muscular work. This usually occurs at moderate (about a third of maximum) velocities and muscular force.

11. Given a 800-N climber has 81.6 kg (800/9.8) of inertia and upward displacement is positive, we can use Newton's second law in the vertical direction ($F = ma$) to calculate: $-800 + 1500 = 81.6(a)$, so $a = 8.6$ m/s/s.

13. Sequential coordination of high-speed movements is advantageous because initial proximal movement contributes to SSC muscle actions, and mechanical energy can be transferred through segmental interaction.

15. Given that an upward displacement is positive and a 30-kg barbell weighs -294 N ($30 \cdot 9.8$), we can use Newton's second law in the vertical direction ($F = ma$) to calculate: $-294 + 4000 = 30(a)$, so $a = 123.5$ m/s/s or 12.6 g's of vertical acceleration.

Chapter 7

1. A torque or moment of force depends on the applied force and the moment arm.

3. The joints of the human body allow us to change our resistance to rotation or moment of inertia by moving the masses of the body segment towards or away from an axis of rotation. Bringing segments close to an axis of rotation decreases moment of inertia while extending segments away from an axis of rotation increases moment of inertia.

5. Newton's first angular analogue says that an object will stay at rest or constant rotation unless acted upon by an external torque. Newton's second angular analogue says that the angular acceleration of an object is proportional to the torque causing it, is in the same direction, and is inversely proportion to the moment of inertia. Newton's third angular analogue states that

for every torque acting on an object there is an equal and opposite torque this object applies back on the other object creating the torque.

7. The center of gravity of athletes doing a lunge-and-sprint start as illustrated below are likely the positions indicated by the dot.

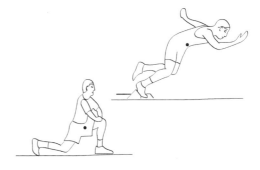

9. To maximize stability, a person can increase the size of the base of support, lower the center of gravity relative to the base of support, and position the center of gravity relative to anticipated forces. Maximizing stability tends to decrease the ability to move in all directions (mobility).

11. Given that the force applied by the student was 30 lb and we know the radius of the merry-go-round, it is easiest to find the rotary component (F_R) of the force to multiply by the radius (4 ft) to obtain the torque applied. We can calculate: $\cos 55° = F_R/30$, so $F_R = 17.2$ lb. Torque ($T = F \cdot d_\perp$) applied to the merry-go-round is: $17.2(4) = 68.8$ lb·ft. This is almost half the 120 lb·ft of torque when the force is applied at an angle that maximizes the moment arm.

13. You cannot calculate the torque because the muscle angle of pull is not known.

Chapter 8

1. The major fluid forces are buoyancy, lift, and drag. Buoyancy acts upward. Lift acts parallel to and opposing the relative

flow of fluid, while lift acts at right angles to the relative flow of fluid.

3. The center of gravity and center of buoyancy of the human body move in similar manner, following the mass shifts with moving segments. The center of gravity moves more than the center of buoyancy because the trunk volume dominates the volume of the rest of the body.

5. Optimal projection angles include the effect of fluid forces as well as the release and target locations of projection activities. For example, place-kicking has an optimal angle of projection much lower than 45° because of the fluid forces of drag.

7. The centers of buoyancy of a swimmer in three flotation positions (below) are likely the positions indicated by the dot.

9. A volleyball serve with topspin dives downward because the Magnus Effect generates a downward-and-backward-directed lift force that adds to gravity.

11. Round balls tend to curve in the direction of the spin. If the front of a ball is spinning to the right (as you observe it as it is coming toward you), the lift force will act to the right and make the ball curve to the right.

13. Swimmers and cyclists shave so as to decrease surface drag, which resists their motion, while a rougher surface of a spinning baseball will create a greater lift force. The greater Magnus Effect and lift force acting on the baseball is more important than the minor effect the roughness will have on drag.

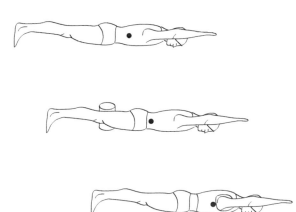

Right-Angle Trigonometry Review

Trigonometry is a branch of mathematics that is particularly useful in dealing with right-angle triangles. This is important in the study of biomechanics because vectors are usually resolved into right-angle components. This appendix provides a brief review of four trigonometric relationships for two-dimensional analysis in the first quadrant. There are many more trigonometric relationships that are fully defined for all 360° of a circle. The four relationships will be defined relative to right triangle illustrated below.

The sides of a triangle are traditionally labeled in two ways, with letters and names describing their position relative to one of the acute angles of interest (θ). The longest side of the triangle is the **hypotenuse** or **c**. The side next to the angle of interest is usually labeled **a** or the **adjacent** side. The last side is the **opposite** side or **b**.

The first relationship is the **Pythagorean Theorem**, which describes the relationship between the lengths of the sides in all right triangles. If you have knowledge of any two of the three sides of a triangle you can apply the formula $c^2 = a^2 + b^2$ to solve for the magnitude of the other side.

The sine, cosine, and tangent are the most commonly used trigonometric relationships, because they define the relationships between the acute angles and the dimensions of right triangles. The abbreviation and formula for each relationship is:

$$\sin \theta = b/c$$
$$\cos \theta = a/c$$
$$\tan \theta = b/a$$

Suppose the right triangle depicted below corresponds to the following data on the release conditions of a soccer kick: $c = 40$ m/s and $\theta = 35°$. A biomechanist wanting to determine the vertical velocity (b) in order to determine the time of flight could write:

$$\sin 35° = V_V/40,$$

and solving could yield

$$V_V = 22.9 \text{ m/s}$$

Now use the cosine, tangent, or Pythagorean Theorem to see if you can confirm if the horizontal velocity of the ball is 32.8 m/s.

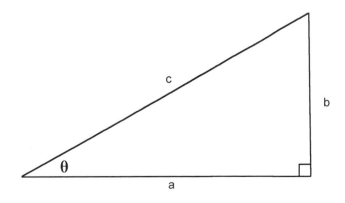

299

Qualitative Analysis of Biomechanical Principles

Principle	Body part	Rating (inadequate-normal-excessive)
Balance		
Coordination		
Force–Motion		
Force–Time		
Inertia		
Range of Motion		
Segmental Interaction		
Optimal Projection		
Spin		

Index

Lab Activities

This section of the book provides applied laboratory activities. These labs are designed to illustrate key points from the chapters of the text. The labs are also designed to be flexible enough to be used as full labs for universities with 4-credit courses or as short activities/demonstrations for 3-unit courses. The emphasis is on using actual human movements and minimal research equipment. While quantitative measurements and calculations are part of some labs, most of them focus on students' conceptual understanding of biomechanics and their ability to qualitatively analyze human movement. Most labs are structured for work in small groups of three to five students.

Citations of background information are provided for students to prepare for the labs. Space does not allow for all relevant research citations to be included on each two-page lab. If instructors assign background reading prior to labs, they should assign specific sections of the resources suggested. I am indebted to many of my peers who have shared their teaching ideas at professional meetings, especially those who have attended and contributed to the last few national conferences on teaching biomechanics.

LAB ACTIVITY 1

FINDING BIOMECHANICAL SOURCES

Biomechanics is the study of the causes of biological movement. Biomechanics is a core sub-discipline of kinesiology, the academic study of human movement. All kinesiology professions use biomechanical knowledge to inform their practice. Both scholarly and professional journals publish biomechanical research. There are many people interested in biomechanics, so biomechanical literature is spread out across many traditional scholarly areas. This lab will help you appreciate the breadth of biomechanics in your chosen career, and provide you with experience in finding biomechanical sources.

BACKGROUND READING

Chapter 1 herein: "Introduction to Biomechanics of Human Movement"

Ciccone, C. D. (2002). Evidence in practice. *Physical Therapy*, **82**, 84–88.

Minozzi, S., Pistotti, V., & Forni, M. (2000). Searching for rehabilitation articles on Medline and Embase: An example with cross-over design. *Archives of Physical Medicine and Rehabilitation*, **81**, 720–722.

TASKS

1. Identify one professional area of interest.

2. Review one year of a journal from this area of interest for biomechanical articles.

3. Identify a potential biomechanical topic of interest from your professional interests.

4. Search a computer database (Medline or SportDiscus) for biomechanical papers on your topic.

5. Answer the questions.

LAB ACTIVITY 1   NAME _____

FINDING BIOMECHANICAL SOURCES

1. What is your professional area of interest, and give a human movement topic you have a biomechanical interest in?

2. Report the name of the journal, number of articles published in a particular year, and the percentage of articles related to biomechanics.

3. Summarize the results of two searches on a literature database like Medline or SportDiscus. Be sure to specify the exact search you used, and the number and quality of citations you obtained.

4. Based on all your searches, list the two citations you believe to be most relevant to your professional interests.

5. Comment on the diversity of sources you observed in your search.

6. Rate the quality of the sources you found based on the hierarchy of evidence presented in chapter.

LAB ACTIVITY 2

QUALITATIVE AND QUANTITATIVE ANALYSIS OF RANGE OF MOTION

This text summarizes many biomechanical variables and concepts into nine principles of biomechanics. The analysis of human movements using these biomechanical principles can be qualitative (subjective) or quantitative (based on numerical measurements). All kinesiology professions have used both qualitative and quantitative analyses of human movement, but qualitative analysis is used most often. This lab will explore the Range-of-Motion Principle of biomechanics, using a variety of static flexibility tests common in physical education and physical therapy. This lab will show you there are a variety of ways to quantify range of motion and that there are strengths and weaknesses of both qualitative and quantitative analyses of human movement.

Physical therapists used to perform a standing toe touch to screen for persons with limited hamstring flexibility. Patients either passed the test by being able to touch their toes with their fingers while keeping their legs straight, or they failed to touch their toes, indicating poor hamstring flexibility. Flexible hamstrings allows a person to tilt their pelvis forward more, making it easier to touch their toes. Recently, more accurate field tests of static flexibility have been developed. The tests that will be used are the sit-and-reach test (SRT), active knee extension (AKE), and the modified Schober test (MST). The results of these flexibility tests can be analyzed qualitatively (judging if the subject has adequate flexibility) or quantitatively. Quantitative analysis can either be norm-referenced (comparing scores to all other people) or criterion-referenced. Criterion-referenced testing compares test scores to some standard of what should be. Criteria or standards are usually based on evidence on what correlates with health (health-related fitness) or with physical abilities to perform jobs safely (occupational screening). For example, physical therapists studying the sit-and-reach test suggested that subjective observation of the forward tilt of the rear of the pelvis is as effective an assessment of hamstring flexibility as the SRT score (Cornbleet & Woolsey, 1996).

BACKGROUND READING

Chapter 2 herein: "Fundamentals of Biomechanics and Qualitative Analysis"

Cornbleet, S. & Woolsey, N. (1996). Assessment of hamstring muscle length in school-aged children using the sit-and-reach test and the inclinometer measure of hip joint angle. *Physical Therapy*, **76**, 850–855.

Gajdosik, R. & Lusin, G. (1983). Hamstring muscle tightness: Reliability of an active–knee-extension test. *Physical Therapy*, **63**, 1085-1088.

Gleim, G. W., & McHugh, M. P. (1997). Flexibility and its effects on sports injury and performance. *Sports Medicine*, **24**, 289–299.

Knudson, D., Magnusson, P., & McHugh, M. (2000, June). Current issues in flexibility fitness. *The President's Council on Physical Fitness and Sports Research Digest*, pp. 1-8.

TASKS

1. Select three volunteers for flexibility testing

2. Learn how to use a sit-and-reach box, inclinometer, goniometer, and tape measure for SRT, AKE, and MST.

3. Collect the following quantitative assessments of lumbar and hamstring range of motion for one side of the body: SRT, AKE, and MST. While these measurements are being taken, have people in your lab group do a qualitative/categorical assessment (hypoflexible, normal, hyperflexible) of the subject being tested.

4. Answer the questions.

LAB ACTIVITY 2   NAME _____

QUALITATIVE AND QUANTITATIVE ANALYSIS OF RANGE OF MOTION

Ratings of Hamstring Flexibility

	Qualitative	SRT	AKE
Subject 1	_____	_____	_____
Subject 2	_____	_____	_____
Subject 3	_____	_____	_____

Ratings of Lumbar Flexibility

	Qualitative	Schober
Subject 1	_____	_____
Subject 2	_____	_____
Subject 3	_____	_____

1. Given that the healthy standard for adult (>17 years) males and females in the SRT are 17.5 and 20 cm, respectively, and a passing AKE is $\theta_K = 160°$, how well did your qualitative and quantitative ratings of hamstring flexibility agree?

2. Given that the passing score for the MST is 7 cm, how well did your qualitative and quantitative ratings of lumbar flexibility agree?

3. List the characteristics of the range of motion you evaluated in your qualitative ratings of hamstring flexibility.

4. Range of motion is a kinematic (descriptive) variable and does not provide kinetic (muscletendon resistance) information about the passive tension in stretching. Static flexibility measurements like these have been criticized for their subjectivity related to a person's tolerance for stretch discomfort (Gleim & McHugh, 1997). Are there kinetic aspects of stretching performance that can be qualitatively judged by your observations of these flexibility tests?

5. Compare and contrast the strengths and weaknesses of a qualitative versus quantitative assessment of static flexibility.

LAB ACTIVITY 3

FUNCTIONAL ANATOMY?

Anatomy is the study of the structure of the human body. The joint motions created by muscles in humans have been studied by anatomists several ways: cadaver dissection, and manipulation, observation, and palpation. Historically, anatomical analyses in kinesiology used the mechanical method of muscle action analysis to establish the agonists for specific movements. This requires a detailed knowledge of the planes of movement, joint axes, attachments, courses of the muscles, and the classification of joints. Anatomy provides only part of the prerequisite information necessary to determine how muscles create movement. A century of EMG research has clearly shown the inadequacy of functional anatomy to explain how muscles act to create human movement (Hellebrandt, 1963). Chapter 3 summarized several areas of research that show the integration of biomechanical research electromyography (EMG, kinetics, simulation) is necessary to understand the actions of muscles in human movement. This lab will review the mechanical method of muscle action analysis in functional anatomy and show why biomechanical analysis is needed to determine the actions of muscles.

BACKGROUND READING

Chapter 3 herein: "Anatomical Description and Its Limitations"

Hellebrandt, F. A. (1963). Living anatomy. *Quest*, **1**, 43–58.

Herbert, R., Moore, S., Moseley, A., Schurr, K., & Wales, A. (1993). Making inferences about muscles forces from clinical observations. *Australian Journal of Physiotherapy*, **39**, 195–202.

Knudson, D. (2001b). A lab to introduce biomechanics using muscle actions. In J. R. Blackwell and D. V. Knudson (Eds.), *Proceedings: Fifth national symposium on teaching biomechanics in sports* (pp. 36–41). San Francisco: University of San Francisco.

TASKS

1. For the anatomical plane and joint(s) specified, use functional anatomy to hypothesize a muscle involved and the muscle action responsible for the following demos and record them on the lab report.

Demo 1 — Sagittal plane	elbow joint	arm curl
Demo 2 — Sagittal plane	lumbar vertebrae	trunk flexion
Demo 3 — Sagittal plane	metacarpophalangeal	passive wrist flexion
Demo 4 — Frontal plane	hip joint	left hip adduction

2. Perform the demos:

Demo 1: Lie supine with a small dumbbell in your right hand and slowly perform arm curls. Have your lab partner palpate your upper arm, being sure to note differences in muscle activation in the first 80 and last 80° of the range of motion. Analyze only the lifting phase.

Demo 2: Lie supine with your hips flexed to 90° and your quadriceps relaxed. Cross your arms over your chest and tighten your abdominal muscles. Make a note of which end of your body is elevated. See if you can make either or both sides of your body rise.

Demo 3: In the anatomical position, pronate your right forearm and flex your elbow completely. Totally relax your right hand and wrist. In this position (hand roughly horizontal), use your left hand to extend your relaxed right wrist and let gravity passively flex the wrist. Note the motion of the fingers during wrist extension and flexion.

Demo 4: From the anatomical position, stand on your left foot and abduct your shoulders so that your arms are horizontal. Smoothly lower and raise your right hip (left hip adduction and then abduction) as many times as you can in one minute. Note the muscles that feel fatigued.

3. Answer the questions.

LAB ACTIVITY 3   NAME _____

FUNCTIONAL ANATOMY?

For the anatomical plane and joint(s) specified, use functional anatomy to hypothesize a muscle and the muscle action responsible for the following activities:

	Plane	Joint	Movement	Muscle	Action
Demo 1 —	Sagittal	elbow joint	arm curl	_____	_____
Demo 2 —	Sagittal	lumbar vertebrae	trunk flexion	_____	_____
Demo 3 —	Sagittal	metacarpophalangeal	wrist flexion	_____	_____
Demo 4 —	Frontal	hip joint	hip adduction	_____	_____

1. Functional anatomy does not consider the action of other forces (other muscles or external forces) in hypothesizing muscle actions. Describe the muscle actions throughout the range of motion in the horizontal plane arm curl, and note why an external force changes the muscle activation strategy.

2. Classifying muscle attachments as an "origin" or "insertion" is not always clear. What muscle(s) are active in the abdominal exercise, and what attachments are being pulled?

3. What muscle(s) created metacarpophalangeal extension when the wrist was passively flexed in Demo 3? What muscle(s) created metacarpophalangeal flexion when the wrist was passively extended? How does the muscle create this motion without activation?

4. Was there discomfort in the left hip adductors in Demo 4? What muscle and action was responsible for controlling left hip adduction?

5. Give a movement example (be specific) where functional anatomy may be incorrect because of:

   External forces

   Muscle synergy

   Passive tension

   Attachment stability changes

LAB ACTIVITY 4

MUSCLE ACTIONS AND THE STRETCH-SHORTENING CYCLE (SSC)

The forces muscles exert to create movement vary dramatically in terms of length, velocity of shortening or lengthening, and timing of activation. The classic *in vitro* muscle mechanical characteristics interact with other factors (activation, leverage, connective tissue stiffness, etc.) to determine the amount of torque a muscle group can create. The torque a muscle group creates naturally affects muscular strength, endurance, and other performance variables. The purpose of this lab is to demonstrate the performance consequence of muscle actions and the stretch-shortening cycle (SSC). The endurance of the elbow flexors will be examined in concentric and eccentric actions to review the Force–Velocity Relationship. Two kinds of vertical jumps will be examined to determine the functional consequences of the SSC.

BACKGROUND READING

Chapter 4 herein: "Mechanics of the Musculoskeletal System"

Komi, P. V. (Ed.) (1992). *Strength and power in sport*. New York: Blackwell Science.

Kubo, K., Kawakami, Y., & Fukunaga, T. (1999). Influence of elastic properties of tendon structures on jump performance in humans. *Journal of Applied Physiology*, **87**, 2090–2096.

Lieber, R., L., & Bodine-Fowler, S. (1993) Skeletal muscle mechanics: Implications for rehabilitation. *Physical Therapy*, **73**, 844–856.

TASKS

1. Select five volunteers for elbow flexor endurance testing. For each subject select a dumbbell with submaximal resistance (between 50 and 80% 1RM). Record the number or concentric-only repetitions (partners lower the dumbbell) for the person's stronger limb and the number or eccentric-only (partners lift the dumbbell) for their weaker limb. Attempt to keep a similar cadence for each test.

2. Perform and measure the maximum height for the countermovement jump (CMJ) and an equivalent static jump (SJ) for everyone in the lab. The SJ begins using isometric muscle actions to hold a squat position that matches the lowest point of the CMJ for that person. Observe jumps carefully since it is difficult to match starting positions, and it is difficult (unnatural) for subjects to begin the concentric phase of the SJ with virtually no countermovement.

3. Perform the calculations and answer the questions.

NAME _____

MUSCLE ACTIONS AND THE STRETCH-SHORTENING CYCLE (SSC)

Maximal Repetitions with a Submaximal Resistance

	Concentric—Stronger Side	Eccentric—Weaker Side
Subject 1	_____	_____
Subject 2	_____	_____
Subject 3	_____	_____
Subject 4	_____	_____
Subject 5	_____	_____

Pre-Stretch Augmentation in SSC

CMJ _____   SJ _____

$$PA\ (\%) = ((CMJ - SJ)/SJ) \cdot 100 \qquad \text{(Kubo } et\ al.,\ 1999)$$

My PA _____   Class Mean PA _____

QUESTIONS

1. Did the stronger side of the body have the most endurance? Explain the results of this comparison of concentric and eccentric muscles actions based on the Force–Velocity Relationship of muscle.

2. Hypothesize the likely lower extremity muscle actions in the SJ and the CMJ.

3. How much improvement in vertical jump could be attributed to using a SSC?

4. What aspects of coaching jumps and other explosive movements must be emphasized to maximize performance? Explain why your technique points may improve performance based on muscle mechanics or principles of biomechanics.

LAB ACTIVITY 5

VELOCITY IN SPRINTING

Linear kinematics in biomechanics is used to create precise descriptions of human motion. It is important for teachers and coaches to be familiar with many kinematic variables (like speeds, pace, or times) that are representative of various levels of performance. Most importantly, professionals need to understand that velocity varies over time, as well as have an intuitive understanding of where peak velocities and accelerations occur in movement. This lab will focus on your own sprinting data in a 40-meter dash and a world-class 100-meter sprint performance to examine the relationship between displacement, velocity, and acceleration. These activities provide the simplest examples of linear kinematics since the body is modeled as a point mass and motion of the body is measured in one direction that does not change.

BACKGROUND READING

Chapter 5 herein: "Linear and Angular Kinematics"

Berthoin, S., *et al.* (2001). Predicting sprint kinematic parameters from anaerobic field tests in physical education students. *Journal of Strength and Conditioning Research*, **15**, 75–80.

Mero, A., Komi, P. V., & Gregor, R. J. (1992). Biomechanics of sprint running: A review. *Sports Medicine*, **13**, 376–392.

Murase, Y., *et al.* (1976). Analysis of the changes in progressive speed during the 100-meter dash. In P.V. Komi (Ed.), *Biomechanics V-B* (pp 200–207). Baltimore: University Park Press.

TASKS

1. Estimate how fast you can run in mph _____
2. Following a warm-up, perform a maximal 40-meter sprint. Obtain times with four stopwatches for times at the 10-, 20-, 30-, and 40-meter marks.
3. Perform the calculations and answer the questions.

Kinesiology Major Normative Data

Time (s)

	Females					Males			
	10	20	30	40		10	20	30	40
Mean	2.3	3.9	5.4	7.0		2.0	3.3	4.6	5.9
sd	0.2	0.4	0.5	0.7		0.2	0.2	0.3	0.5

Maurice Greene: 1999 World Championships Seville, Spain

Meters	Seconds
0–10	1.86
10–20	1.03
20–30	0.92
30–40	0.88
40–50	0.86
50–60	0.84
60–70	0.85
70–80	0.85
80–90	0.85
90–100	0.86
	9.67

LAB ACTIVITY 5   NAME _____

VELOCITY IN SPRINTING

Record *your* times in the spaces below.

10 m	20 m	30 m	40 m
$t_1 = $ _____	$t_2 = $ _____	$t_3 = $ _____	$t_4 = $ _____

QUESTIONS

1. Calculate the average horizontal velocity in each of the 10-m intervals of your 40-m sprint ($V = \Delta d / \Delta t$). Report your answers in m/s and mph (m/s · 2.237 = mph).

2. Calculate the average velocities for the intervals of Maurice Greene's 100-m sprint. Note that the times in the table represent the change in time (time to run the interval: Δt), not the cumulative time, as in your 40-m sprint data. Average velocities are usually assigned to the midpoints of the interval used for the calculation.

Velocity (m/s) at the

5____ 15____ 25____ 35____ 45____ 55____ 65____ 75____ 85____ 95____

meter points.

3. Plot Greene's and your velocities on the following velocity-displacement graph:

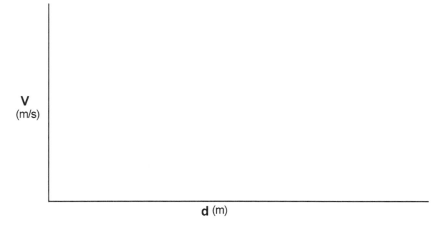

4. Give a qualitative description of the general slopes of the Greene velocity graph in question 3 (the general pattern would be same if this were a true velocity–time graph) that determine the acceleration phases of maximal sprinting. Where is acceleration the largest and why?

LAB ACTIVITY 6A

TOP GUN KINETICS: FORCE–MOTION PRINCIPLE

Newton's laws of motion explain how forces create motion in objects. The application principle related to Newton's second law is the Force–Motion Principle. The purpose of this lab is improve your understanding of Newton's laws of motion. As a candidate for the prestigious "Top Gun" kinetic scooter pilot in biomechanics class, you must not only perform the missions but use kinetics to explain your scooter's flight. Biomechanics Top Gun is like a Naval Top Gun in that skill and knowledge are required to earn the honor. It is important that you follow the instructions for each mission explicitly. Care should be taken by pilots and their ground crew to perform the task correctly and safely. Note that your multimillion-dollar scooters provide *low* (not quite zero) friction conditions, so you need to move/push briskly so you can ignore the initial effects of friction. Kinetics explains all motion: from scooters, braces, rackets, jump shots, to muscle actions. Think about the forces, what directions they act, and the motion observed in each mission. This lab is roughly based on a lab developed by Larry Abraham (Abraham, 1991).

BACKGROUND READING

Chapter 6 herein: "Linear Kinetics"

Abraham, L. D. (1991). *Lab manual for KIN 326: Biomechanical analysis of movement.* Austin, TX.

TASKS

1. Using your multimillion-dollar scooters, ropes, and spring/bathroom scales, perform the following training missions:

— Sit on the scooter and maximally push off from a wall (afterburner check). Experiment with various body positions and techniques.

— Sit on your scooter and push off from a partner on another scooter.

— Loop a rope over a bathroom scale held by a partner on a scooter. Sit on your scooter and pull your partner, who passively holds the scale, and note the largest force exerted.

— Repeat the last mission, but have your partner also vigorously pull on the scale.

2. Answer the questions.

LAB ACTIVITY 6A   NAME _____

TOP GUN KINETICS: FORCE–MOTION PRINCIPLE

1. How far were you able to glide by pushing off from the wall? What is the relationship between the direction of your push and the direction of motion?

2. How far were you able to glide by pushing off from another scooter pilot? Explain any differences from task 1 using Newton's Laws of Motion.

3. How much force was applied to pull a passive partner? Which scooter pilot moved the most and why?

4. How much force was applied when both partners vigorously pulled on the rope? Explain any differences in the observed motion from task 3 using Newton's Laws.

5. Assume the mass of your scooter cannot be modified, but you are charged with recommending technique that maximizes scooter speed and agility. Use the Force–Motion Principle to suggest why a certain body position and propulsion technique is best.

LAB ACTIVITY 6B

IMPULSE–MOMENTUM: FORCE–TIME PRINCIPLE

The timing of force application to objects affects the stress and motion created. Newton's second law applied to forces acting over time is the impulse–momentum relationship. The change in momentum of an object is equal to the impulse of the resultant force. This activity will allow you to experience some interesting real-life examples of the impulse–momentum relationship. The purpose of this lab is to improve your understanding of changing the motion of an object (specifically, it's momentum) by applying force over a period of time. In some ways body tissues are similar to water balloons in that too much force can create stresses and strains that lead to injury. It is important for teachers/coaches to understand how movement technique affects the impulse and peak force that can be applied to an object. This lab is modified from a lab proposed by McGinnis and Abendroth-Smith (1991).

BACKGROUND READING

Chapter 6 herein: "Linear Kinetics"

McGinnis, P., & Abendroth-Smith, J. (1991). Impulse, momentum, and water balloons. In J. Wilkerson, E. Kreighbaum, & C. Tant, (Eds.), *Teaching kinesiology and biomechanics in sports* (pp. 135–138). Ames: Iowa State University.

Knudson, D. (2001c). Accuracy of predicted peak forces during the power drop exercise. In J. R. Blackwell (Ed.) *Proceedings of oral sessions: XIX international symposium on biomechanics in sports* (pp. 135–138). San Francisco: University of San Francisco.

TASKS

1. Estimate how far you can throw a softball-sized water balloon. _____

2. Estimate the maximum distance you could catch a similar water balloon. ____

3. Fill several water balloons to approximately softball size ($\approx$7–10 cm in diameter).

4. Measure the maximal distance you can throw the water balloon. _____

5. Measure the maximal distance you and a partner can throw and catch a water balloon. _____

6. Answer the questions.

LAB ACTIVITY 6B   NAME _____

IMPULSE–MOMENTUM: FORCE–TIME PRINCIPLE

Distance of Throw _____   Distance of Toss & Catch _____

1. What technique factors were important in the best water balloon throws?

2. What technique factors were most important in successfully catching a water balloon?

3. Theoretically, if you could throw a water balloon 25 m, could you catch it? Why?

4. How are the mechanical behaviors of water balloons similar to muscles and tendons?

5. Below is a graph of the vertical force (**N**) measured when a medicine ball was dropped
   from the same height and bounced (●) or was caught and thrown back up in a power
   drop exercise (◆). Use the Force–Time Principle to explain the differences in the forces applied to the medicine ball. Data from Knudson (2001c).

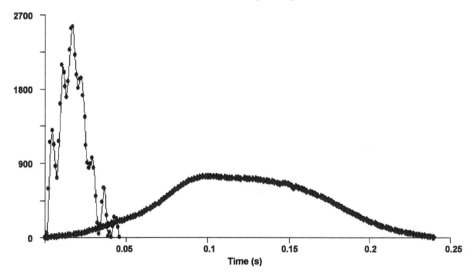

LAB ACTIVITY 7

ANGULAR KINETICS OF EXERCISE

The positions of body segments relative to gravity determine the gravitational torques that must be balanced by the muscles of the body. The purpose of this lab is to improve your understanding of torque, summation of torques, lifting, and center of gravity. These biomechanical parameters are extremely powerful in explaining the causes of human movement because of the angular motions of joints. Several classic lifting and exercise body positions are analyzed because the slow motion (very small or zero acceleration) in these movements comprise a quasi-static condition. In static conditions, Newton's second law can be simplified to static equilibrium: $\Sigma F = 0$ and $\Sigma T = 0$. Remember that a torque (T) or moment of force is the product of the force and the perpendicular distance between the line of action of the force and the axis of rotation ($T = F \cdot d_{\perp}$).

BACKGROUND READING

Chapter 7: "Angular Kinetics"

Chaffin, B. D., Andersson, G. B. J., & Martin, B. J. (1999). *Occupational biomechanics* (3rd ed.). New York: Wiley.

Hay, J. G., Andrews, J. G., Vaughan, C. L., & Ueya, K. (1983). Load, speed and equipment effects in strength-training exercises. In H. Matsui & K. Kobayashi (Eds.), *Biomechanics III-B* (pp. 939–950). Champaign, IL: Human Kinetics.

van Dieen, J. H., Hoozemans, M. J. M., & Toussaint, H. M. (1999). Stoop or squat: A review of biomechanical studies on lifting technique. *Clinical Biomechanics*, **14**, 685–696.

TASKS

1. If an athlete doubled his/her trunk lean in a squat exercise, how much more resistance would their back feel? Estimate the extra load on the lower back if a person performed a squat with a 40° trunk lean compared to a 20° trunk lean. _____ %

2. Obtain height, weight, and trunk length (greater trochanter to shoulder joint) data for a person in the lab.

3. The amount of trunk lean primarily determines the stress placed on the back and hip extensors (Hay *et al.*, 1983). Perform two short endurance tests to see how trunk lean affects muscle fatigue. Use a standard bodyweight squat technique. Hold the squats with hands on hips in an isometric position for 30 seconds and subjectively determine which muscle groups were stressed the most. Test 1 is a squat with a nearly vertical trunk and a knee angle of approximately 120°. Test 2 is a squat with a trunk lean of about 45° and a knee angle of approximately 120°. Wait at least 5 minutes between tests.

4. Perform calculations on the following simple free-body diagrams of exercise and body positions to examine how gravitational torques vary across body configurations and answer the questions.

LAB ACTIVITY 7   NAME _____

ANGULAR KINETICS OF EXERCISE

1. Where were the sites of most fatigue in the two squat tests? What muscle group feels more fatigue in a nearly vertical trunk orientation? Why?

2. Kinematic measurements from film/video and anthropometric data are often combined to make angular kinetic calculations. A static analysis can be done when the inertial forces and torques (dynamic loading from high-speed movement) are small. Assume the figure below is an image of you captured from video while performing bodyweight squats. Calculate a gravitational torque of *your* upper body about the hip (M/L) axis. Assume your head, arms, and trunk (HAT) have mass equal to 0.679 of body mass. The center of gravity of your HAT acts at 62.6% up from the hip to the shoulder.

20°

3. Calculate the gravitational torque about the hip if the bottom of your squat exercise has a trunk lean of 40°. (Show free-body diagram and work.)

4. If the weight of the head, arms, and trunk do not change during the squat exercise, what does change that increases gravitational torque as the person leans forward?

5. How different is the load on the back/hip extensors when you double your trunk lean? Is the size of this difference what you expected? Why is it different?

LAB ACTIVITY 8

MAGNUS EFFECT IN BASEBALL PITCHING

Fluid forces have dramatic effects in many human movements. Fluid dynamics is of vital interest to coaches of swimming, cycling, running, and sports where wind or ball velocities are great. The fluid forces of lift and drag increase with the square of velocity. The purpose of this lab is to improve your understanding of how fluid forces (specifically lift) can be used to affect a thrown balls trajectory. The example is in baseball pitching, although the Spin Principle applies to other ball sports. Pitching technique and the Magnus Effect are explored in the "rise" of a fastball, the "break" of a slider, and the "drop" of a curveball. Skilled performance in many sports involves appropriate application of rotation to a ball to create fluid forces for an advantageous trajectory.

BACKGROUND READING

Chapter 8 herein: "Fluid Mechanics"

Allman, W. F. (1984). Pitching rainbows: The untold physics of the curve ball. In E. W. Schrier & W. F. Allman (Eds.), *Newton at the bat: The science in sports* (pp. 3–14). New York: Charles Scribner & Sons.

Knudson, D. (1997). The Magnus Effect in baseball pitching. In J. Wilkerson, K. Ludwig, & M. Butcher (Eds.), *Proceedings of the 4th national symposium on teaching biomechanics* (pp. 121–125). Denton: Texas Woman's University Press.

TASKS

1. Set up a mock baseball pitching situation indoors with a pitching rubber and home plate about 7 m apart. Warm up the shoulder and arm muscles and gradually increase throwing intensity with whiffle balls. Exchange the whiffle ball for a styrofoam ball.

2. Hitters often perceive that a well-thrown fastball "rises" (seems to jump over their bat). The fastball is usually thrown with the index and middle fingers spread and laid across the seams of a ball, with the thumb providing opposition from the front of the ball. At release, the normal wrist flexion and radioulnar pronation of the throwing motion create downward and forward finger pressure on the ball. These finger forces create backspin on the ball. Try to increase the rate of backspin to determine if the ball will rise or just drop less than a similar pitch. Be careful to control the initial direction of the pitches by using visual references in the background. Estimate the rise or drop of the pitch relative to the initial trajectory at release.

3. A pitch that is easy to learn after the basic fastball is a slider. A slider creates a lateral "break" that can be toward or away from a batter, depending on the handedness of the pitcher and batter. The grip for a slider (right-handed pitcher) has the index and middle finger together and shifted to the right side of the ball (rear view). The thumb provides opposition from the left side of the ball. Normal wrist flexion and pronation at release now create a final push to the right side of the ball, imparting a sidespin rotation. A typical right-handed pitcher (facing a right-handed hitter) would usually direct this pitch initially toward the center to the outside corner of home plate, so the ball would break out of reach.

4. A pitch that can make a batter look foolish is the curveball. The common perception of hitters watching a well-thrown curveball is that the ball seems to "drop off the table." The ball looks like it is rolling along a horizontal table toward you and suddenly drops off the edge. The grip for a curveball is similar to a fastball grip, but with a different orientation of the seams. At release the index and middle fingers are on top of the ball, making a final push forward and downward. Common teaching cues are to pull down at release like pulling down a shade or snapping your fingers. Research has shown that radioulnar pronation is delayed in the curveball, so that at release the forearm is still in a slightly supinated position. Curveballs are thrown with forearm pronation just like other pitches; it is just delayed to near the moment of release.

5. If time is available, students can do some "show and tell" with other pitch variations. These include variations in release (sidearm, windmill softball pitch, grips, screwball, knuckleball, etc.).

6. Answer the questions.

LAB ACTIVITY 8   NAME _____

MAGNUS EFFECT IN BASEBALL PITCHING

1. Were you able to make a styrofoam fastball rise? Draw a free-body diagram of your fastball, showing all relevant forces and explain how it relates to the vertical motion of the ball you observed.

2. Could you make a styrofoam slider break sideways? If so, how much?

3. Draw a rear view of the ball from the pitcher's (your) perspective and draw on the ball the axis of ball rotation and Magnus force for your slider.

4. Draw a rear view of the ball from the pitcher's (your) perspective and draw on the ball the axis of ball rotation and Magnus force for your curveball. In what direction(s) did your curveball break?

5. Did your curveball have more lateral or downward break? Why?

6. To get a ball to curve or break to the right with the Spin Principle, describe how force is applied to the ball? Would this be the same for curves to the right in other impact and release sports?

LAB ACTIVITY 9

QUALITATIVE ANALYSIS OF LEAD-UP ACTIVITIES

An effective teaching strategy for many sports skills is to provide a sequence of lead-up activities that are similar to and build up to the skill of interest. How biomechanically similar the lead-up activities are to the sport skill of interest is important to physical educators. A qualitative answer to the similarity question will be explored in a sport skill selected by the instructor. The present lab will allow you to practice qualitative analysis of human movements using the biomechanical principles.

BACKGROUND READING

Chapter 9 herein: "Applying Biomechanics in Physical Education"

Knudson, D. V., & Morrison, C. S. (2002). *Qualitative analysis of human movement* (2nd ed.). Champaign, IL: Human Kinetics.

TASKS

1. For the sport skill identified by the instructor, identify two lead-up skills, activities, or drills.

2. Select a volunteer to perform these movements.

3. Videotape several repetitions of the movements from several angles.

4. Observe and evaluate the performance of the biomechanical principles in each movement using videotape replay.

5. Answer the questions.

NAME _____

QUALITATIVE ANALYSIS OF LEAD-UP ACTIVITIES

1. What biomechanical principles are most relevant to the sport skill of interest?

2. What was the first lead-up movement? What biomechanical principles are related to performance of this lead-up movement?

3. What was the second lead-up movement? What biomechanical principles are related to performance of this lead-up movement?

4. For the volunteer in your lab, what lead-up movement was most sport-specific? What biomechanical principles were most similar to the sport skill?

LAB ACTIVITY 10

COMPARISON OF SKILLED AND NOVICE PERFORMANCE

Coaching strives to maximize the performance of an athlete or team in competition. A key ingredient of athletic success is motor skill. Most aspects of skill are related to the biomechanical principles of human movement. A good way to practice the qualitative analysis of sport skills is to compare the application of biomechanical principles of a novice and those of a skilled performer. The purpose of this lab is to compare the application of biomechanical principles in a skilled performer and a novice performer in a common sport skill.

BACKGROUND READING

Chapter 10 herein: "Applying Biomechanics in Coaching"

Hay, J. G. (1993). *The biomechanics of sports techniques* (4th. ed.). Englewood Cliffs, NJ: Prentice-Hall.

Knudson, D. V., & Morrison, C. S. (2002). *Qualitative analysis of human movement* (2nd ed.). Champaign, IL: Human Kinetics.

TASKS

1. Select a sport skill where a novice and a skilled performer can be found from students in the lab.

2. Select two volunteers (one novice and one skilled) to perform the skill.

3. Videotape several repetitions of the skill from several angles

4. Observe and evaluate performance of the biomechanical principles in each movement using videotape replay.

5. Answer the questions.

LAB ACTIVITY 10   NAME _____

COMPARISON OF SKILLED AND NOVICE PERFORMANCE

1. What are the biomechanical principles most relevant to the sport skill of interest?

2. What biomechanical principles are strengths and weaknesses for the novice performer?

3. What biomechanical principles are strengths and weaknesses for the skilled performer?

4. What intervention would you recommend for the novice performer and why?

5. What intervention would you recommend for the skilled performer and why?

LAB ACTIVITY 11

COMPARISON OF TRAINING MODES

Strength and conditioning coaches prescribe exercises to improve performance based on the Principle of Specificity. This is often called the "SAID" principle: Specific Adaptation to Imposed Demands. There are a variety of free-weight, elastic, and mechanical resistances that coaches can prescribe to train the neuromuscular system. Qualitative analysis of exercise technique based on biomechanical principles can help a strength coach make two important evaluations: is the exercise technique safe and is it sport-specific? This lab will focus on the latter. The purpose of this lab is to compare the specificity of exercise technique in training for a sport skill.

BACKGROUND READING

Chapter 11 herein: "Applying Biomechanics in Strength and Conditioning"

Knudson, D. V., & Morrison, C. S. (2002). *Qualitative analysis of human movement* (2nd ed.). Champaign, IL: Human Kinetics.

TASKS

1. Select a sport skill of interest.

2. Select three exercises that will train the main agonists for the propulsive phase of the skill. Be sure to select an elastic resistance, inertial resistance (free weight), and an exercise machine. Strive to make the resistances about equal in these exercises.

3. Select a volunteer to perform the exercises.

4. Videotape several repetitions of the exercises perpendicular to the primary plane of movement.

5. Observe and evaluate the performance of the biomechanical principles in each exercise using videotape replay.

6. Answer the questions.

LAB ACTIVITY 11   NAME _____

COMPARISON OF TRAINING MODES

1. What are the biomechanical principles most relevant to the sport skill of interest?

2. What was the first exercise? What biomechanical principles of this exercise are similar to the sport skill?

3. What was the second exercise? What biomechanical principles of this exercise are similar to the sport skill?

4. What was the third exercise? What biomechanical principles of this exercise are similar to the sport skill?

5. Which exercise was most sport-specific? Why? (Be sure to explain based on the importance of certain biomechanical principles in terms of performance in the sport.)

LAB ACTIVITY 12

QUALITATIVE ANALYSIS OF WALKING GAIT

Sports medicine professionals qualitatively analyze movement to find clues to injury and to monitor recovery from injury. Walking is a well-learned movement that athletic trainers, physical therapists, and physicians all qualitatively analyze to evaluate lower-extremity function. There is a variety of qualitative and quantitative systems of gait analysis. This lab will focus on the qualitative analysis of two walking gaits based on biomechanical principles. Professionals qualitatively analyzing gait must remember that quantitative biomechanical analyses are needed in order to correctly estimate the loads in musculoskeletal structures, so assumptions about muscle actions in gait from body positioning alone are unwise (Herbert *et al.*, 1993).

BACKGROUND READING

Chapter 12 herein: "Applying Biomechanics in Sports Medicine and Rehabilitation"

Herbert, R., Moore, S., Moseley, A., Schurr, K., & Wales, A. (1993). Making inferences about muscles forces from clinical observations. *Australian Journal of Physiotherapy*, **39**, 195–202.

Knudson, D. V., & Morrison, C. S. (2002). *Qualitative analysis of human movement* (2nd ed.). Champaign, IL: Human Kinetics.

Whittle, M. (1996). *Gait analysis: An introduction* (2nd ed.). Oxford: Butterworth-Heinemann.

TASKS

1. Select a volunteer to perform the walking trials.

2. Have the volunteer walk in three conditions: their natural gait, as fast as they comfortably can, and simulating an injury. Injury can be easily simulated by restricting joint motion with athletic tape or a brace. Antalgic (painful) gait can be simulated by placing a small stone in a shoe.

3. Videotape several cycles of each waking gait.

4. Observe and evaluate performance related to the biomechanical principles in each gait using videotape replay.

5. Answer the questions.

NAME _____

QUALITATIVE ANALYSIS OF WALKING GAIT

1. What biomechanical principles are most evident in natural walking gait?

2. What biomechanical principles increased or decreased in importance relative to normal gait, during fast gait?

3. What injury did you simulate? What biomechanical principles increased or decreased in importance relative to normal gait, during injured gait?

4. What musculoskeletal structures are affected in your simulated injury? Hypothesize the likely changes in muscular actions and kinematics because of this injury and note where you might find biomechanical literature to confirm your diagnosis.